Advances in
ORGANOMETALLIC CHEMISTRY

VOLUME 37

Advances in Organometallic Chemistry

EDITED BY

F. GORDON A. STONE

DEPARTMENT OF CHEMISTRY
BAYLOR UNIVERSITY
WACO, TEXAS

ROBERT WEST

DEPARTMENT OF CHEMISTRY
UNIVERSITY OF WISCONSIN
MADISON, WISCONSIN

VOLUME 37

ACADEMIC PRESS
San Diego   New York   Boston
London   Sydney   Tokyo   Toronto

Academic Press, Inc.
A Division of Harcourt Brace & Company
525 B Street, Suite 1900, San Diego, California 92101-4495

United Kingdom Edition published by
Academic Press Limited
24-28 Oval Road, London NW1 7DX

International Standard Serial Number: 0065-3055

International Standard Book Number: 0-12-031137-2

PRINTED IN THE UNITED STATES OF AMERICA
94   95   96   97   98   99   EB   9   8   7   6   5   4   3   2   1

Contents

Alkyl(pentacarbonyl) Compounds of the Manganese Group Revisited

JO-ANN M. ANDERSEN and JOHN R. MOSS

σ, π-Bridging Ligands in Bimetallic and Trimetallic Complexes

SIMON LOTZ, PETRUS H. VAN ROOYEN, and RITA MEYER

Contributors

Numbers in parentheses indicate the pages on which the authors' contributions begin.

JO-ANN M. ANDERSEN (169), Department of Chemistry, University of Cape Town, Rondebosch 7700, Cape Town, South Africa

ARTHUR J. CARTY (39), Guelph-Waterloo Centre for Graduate Work in Chemistry, Waterloo Campus, Department of Chemistry, University of Waterloo, Waterloo, Ontario, Canada N2L 3G1

JOHN F. CORRIGAN (39), Guelph-Waterloo Centre for Graduate Work in Chemistry, Waterloo Campus, Department of Chemistry, University of Waterloo, Waterloo, Ontario, Canada N2L 3G1

SIMON DOHERTY (39), Guelph-Waterloo Centre for Graduate Work in Chemistry, Waterloo Campus, Department of Chemistry, University of Waterloo, Waterloo, Ontario, Canada N2L 3G1

SIMON LOTZ (219), Department of Chemistry, University of Pretoria, Pretoria 0002, South Africa

IAN MANNERS (131), Department of Chemistry, University of Toronto, Toronto, Ontario, Canada M5S 1A1

RITA MEYER (219), Department of Chemistry, University of Pretoria, Pretoria 0002, South Africa

JOHN R. MOSS (169), Department of Chemistry, University of Cape Town, Rondebosch 7700, Cape Town, South Africa

HIDEKI SAKURAI (1), Department of Chemistry and Organosilicon Research Laboratory, Faculty of Science, Tohoku University, Sendai 980-77, Japan

ENRICO SAPPA (39), Dipartimento di Chimica Inorganica, Chimica Fisica e Chimica dei Materiali, Universita di Torino, I-10125 Torino, Italy

AKIRA SEKIGUCHI (1), Department of Chemistry and Organosilicon Research Laboratory, Faculty of Science, Tohoku University, Sendai 980-77, Japan

PETRUS H. VAN ROOYEN (219), Department of Chemistry, University of Pretoria, Pretoria 0002, South Africa

ADVANCES IN ORGANOMETALLIC CHEMISTRY, VOL. 37

Cage and Cluster Compounds of Silicon, Germanium, and Tin

AKIRA SEKIGUCHI and HIDEKI SAKURAI

Department of Chemistry and Organosilicon Research Laboratory
Faculty of Science, Tohoku University
Sendai 980-77, Japan

I

INTRODUCTION

Highly symmetrical polyhedranes such as tetrahedrane, prismane, and cubane have long fascinated chemists (*1,2*). Synthesis of polyhedranes comprising Si, Ge, and Sn is one of the greatest synthetic challenges in the chemistry of higher row group 14 elements. Until relatively recently, highly strained polyhedranes of Si, Ge, and Sn were thought to be synthetically inaccessible. However, after the discovery of octasilacubane (*3*) and hexagermaprismane (*4*), the chemistry of the field developed rapidly. At present, tetrasilatetrahedrane, hexasilaprismane, and octasilacubane

1

along with germanium and tin analogs are available (except for the tetrahedranes of Ge and Sn and the prismane of Sn). Unlike polyhedranes of carbon, those of Si, Ge, and Sn exhibit quite unique physical and chemical properties owing to the highly rigid framework made up of σ bonds with low ionization potentials. Since reviewing the development of this field (5), we ourselves reported syntheses and X-ray crystallographic structures of octasilacubane, octagermacubane, hexasilaprismane, and hexagermaprismane (6,7). Three reports on the crystal structures of octasilacubanes including ours appeared almost simultaneously (6,8,9). More recently, the synthesis and crystal structure of tetrasilatetrahedrane has been reported (10). Thus, the chemistry of polyhedranes of Si, Ge, and Sn has advanced by rapid strides, providing many novel and unusual structures which are the subject of this article.

II

THEORETICAL STUDIES

For the Si_4H_4 potential energy surface, a local minimum has been predicted for tetrasilatetrahedrane (11,12). In contrast, Nagase and Nakano reported that tetrasilatetrahedrane is unlikely to be a minimum on the potential energy surface (13). Breaking of the two Si–Si bonds is predicted to occur without a significant energy barrier to form an isomer with one four-membered ring. However, it is also indicated that silyl substituents can stabilize the tetrasilatetrahedrane framework (14a,b). Therefore, the derivative $(H_3SiSi)_4$ is an interesting synthetic target. For Si_6H_6 (15a–d) and Si_8H_8 (16a,b), hexasilaprismane and octasilacubane are highly feasible molecules according to the theoretical calculations.

Geometries are compared by theoretical calculations of tetrahedrane, prismane, and cubane and silicon analogs. Figure 1 shows calculated geometries of the both the carbon and silicon compounds. The Si–Si bond lengths increase in the order Si_4H_4 (2.314 Å) < Si_6H_6 (2.359 and 2.375 Å) < Si_8H_8 (2.396 Å) (16a). The Si–Si bond lengths in the three-membered rings are shorter than the typical Si–Si single bond, whereas those in the four-membered rings are longer.

Table I summarizes the calculated strain energies of compound of the type E_nH_n, where n is 4, 6, and 8 and E is C, Si, Ge, and Pb, derived from homodesmic reactions (16b). The strain energies of silicon and carbon tetrahedranes are very similar (140 kcal/mol). However, replacement of the three-membered rings by four-membered rings results in a significant decrease in strain energy: 140.3 (Si_4H_4) > 118.2 (Si_6H_6) > 99.1 (Si_8H_8)

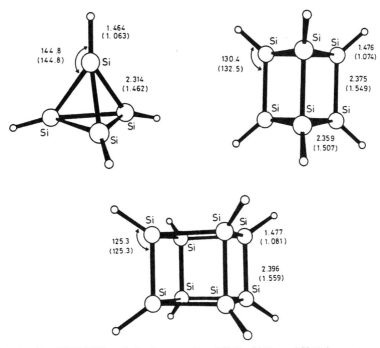

FIG. 1.   The HF/6-31G* optimized geometries of Si_4H_4, Si_6H_6, and Si_8H_8 in angstroms and degrees. Values in parentheses are for the carbon compounds. Reprinted with permission from Nagase et al. (16a), J. Chem. Soc., Chem. Commun., **1987**, 60. Copyright by The Chemical Society.

kcal/mol, whereas those in carbon compounds remain roughly unchanged. Nagase proposed that this was caused by an increasing tendency to maintain the $(ns)^2(np)^2$ valence electron configuration in compounds with the heavier atoms (17). As a result, a four-membered ring with 90° bond angles is made favorable. The substituent effect is very

TABLE I

STRAIN ENERGIES (kcal/mol) OF POLYHEDRANES E_nH_n (HF/DZ + d)^a

E_nH_n	C^b	Si	Ge	Sn	Pb
Tetrahedrane (E_4H_4)	141.4	140.3	140.3	128.2	119.3
Prismane (E_6H_6)	145.3	118.2	109.4	93.8	65.2
Cubane (E_8H_8)	158.6	99.1	86.0	70.1	59.6

^{a} From Nagase (16b).
^{b} HF/6-31G* values.

important for the stabilization of polyhedranes. It is calculated that the SiH_3 substituent leads to a remarkable relief of the strain: 114.5 kcal/mol for $(SiSiH_3)_4$ (tetrasilatetrahedrane), 95.7 kcal/mol for $(SiSiH_3)_6$ (hexasilaprismane), and 77.9 kcal/mol for $(SiSiH_3)_8$ (octasilacubane) (14).

Among polyhedranes, prismane is of special interest since hexasilabenzene, hexagermabenzene, and other heavier atom analogs of benzene are still elusive. Benzene is situated at the minimum of the C_6H_6 potential energy surface. Strained isomers such as prismane, Dewar benzene, and benzvalene lie at higher energy (1,2). However, the relative energies for the heavier elements are sharply different from those for carbon. Table II gives the calculated relative energies of E_6H_6 valence isomers, indicating the saturated prismanes to be the most stable valence isomers among E_6H_6 compounds (E = Si, Ge, Sn, Pb) (15a). This is caused by the fact that the higher row group 14 elements are rather reluctant to form double bonds (18a–e). Thus, it is quite reasonable to assume that the tetrahedrane, prismane, and cubane analogs of the heavier atoms are reasonably accessible and can be synthetic targets, provided that appropriate bulky groups can stabilize the polyhedranes.

TABLE II

RELATIVE ENERGIES (kcal/mol) OF E_6H_6 VALENCE ISOMERS

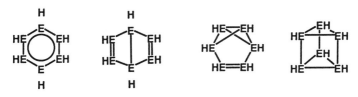

E	Benzene	Dewar benzene	Benzvalene	Prismane
C^a	0.0	88.1	84.5	127.6
Si^a	0.0	3.7	0.5	−9.5
Ge^b	0.0	−1.8	−1.2	−13.5
Sn^b	0.0	−6.5	−11.0	−31.3
Pb^b	0.0	−10.6	—	−67.0

[a] From Nagase et al. (15a).
[b] From Nagase et al. (15d).

III

SUBSTITUENT, PRECURSOR, AND REDUCING REAGENT

A. *Substituent*

The selection of substituent is critical for the successful isolation of polyhedranes composed of higher row group 14 elements. Figure 2 shows the first ionization potentials of Me_3E-EMe_3 (Ip, E = Si, Ge, and Sn), which are appreciably lower than those of the carbon analogs (e.g., 8.69 eV for $Me_3Si-SiMe_3$, 12.1 eV for H_3C-CH_3) (*19*). In fact, the E-E bonds of the small ring compounds comprising Si, Ge, and Sn are readily oxidized because of the existence of high-lying orbitals and the inherent strain. Therefore, full protection of the framework by the bulky substituent is required to suppress the attack by external reagents.

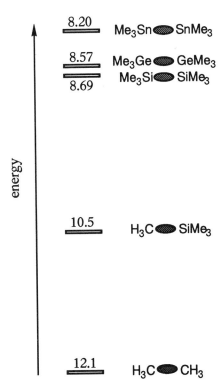

FIG. 2.   First ionization potentials (eV) of σ bonds of group 14 elements.

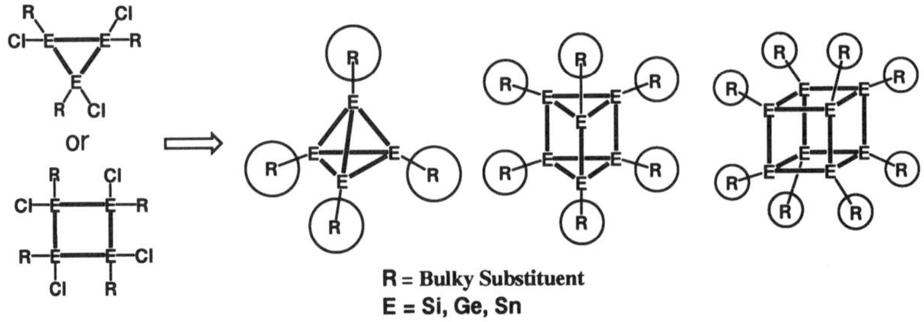

R = Bulky Substituent
E = Si, Ge, Sn
SCHEME 1

B. Precursor

The proper choice of starting compounds with suitable bulky substituents is of crucial importance. The most logical precursors of polyhedranes are three-membered and four-membered compounds bearing halogens (Scheme 1). These ring compounds may give cubanes and prismanes made up of the heavier atoms through the dimerization reaction of the reactive E=E double bond. Tetrahalogenocyclobutanes consisting of the heavier elements will be candidate precursors of tetrahedranes since the higher row group 14 elements prefer saturated structures to unsaturated ones. Compounds of the type REX_3 and REX_2-EX_2R, with judicious selection of R, also can serve as precursors of polyhedranes through multistep reactions. The steric bulkiness of the substituent R determines the ring size and the shape of polyhedranes.

C. Reducing Reagent

The role of metals as reducing reagents is also crucial. In general, alkali metals such as Li, Na, Na/K, K, and lithium naphthalenide (LiNp) are employed as coupling reagents for chlorosilanes and chlorogermanes. However, these reducing reagents are sometimes powerful enough to cause cleavage of the resulting Si–Si or Ge–Ge bonds. In contrast with the alkali metals, magenesium metal does not cleave the strained Si–Si and Ge–Ge bonds. In particular, $Mg/MgBr_2$ is an excellent reagent. For instance, cyclotrigermanes and cyclotetragermanes are readily available by reductive coupling of dichlorogermanes with $Mg/MgBr_2$ in tetrahydrofuran (THF) (20). Reactions with dihalogenogermanes bearing bulky substituents afforded cyclotrigermanes, whereas the less hindered dichlorogermanes gave cyclotetragermanes (Table III).

The reaction of dichlorodimesitylsilane with the $Mg/MgBr_2$ reagent did

TABLE III

REDUCTIVE COUPLING OF DIHALOGENOGERMANES BY Mg AND MgBr$_2$[a]

$$R_2GeX_2 + Mg + MgBr_2 \xrightarrow{\text{THF}} (R_2Ge)_n \; n = 3,4; \; X = Cl, Br$$

Halogermane	Product[b]	Yield (%)
Ph$_2$GeCl$_2$	(Ph$_2$Ge)$_4$	46
Mes$_2$GeCl$_2$	(Mes$_2$Ge)$_3$	54
Xy$_2$GeCl$_2$	(Xy$_2$Ge)$_3$	62
Ar$_2$GeCl$_2$	(Ar$_2$Ge)$_3$	32
Mes(^{t}Bu)GeCl$_2$	[Mes(^{t}Bu)Ge]$_3$	10
^{i}Pr$_2$GeCl$_2$	(^{i}Pr$_2$Ge)$_4$	9
(Me$_3$SiCH$_2$)$_2$GeCl$_2$	[(Me$_3$SiCH$_2$)$_2$Ge]$_4$	15
Ar$_2$GeBr$_2$	(Ar$_2$Ge)$_3$	15

[a] From Ando and Tsumuraya (20).
[b] Mes, 2,4,6-Trimethylphenyl; Xy, 2,6-dimethylphenyl; Ar, 2,6-diethylphenyl.

not take place even under vigorous conditions. However, the reaction of dichlorodiphenylsilane with Mg/MgBr$_2$ smoothly proceeded to give octaphenylcyclotetrasilane (**1**) (Scheme 2) (*21a*). The presence of anthracene led to the formation of 2:3,5:6-dibenzo-7,7,8,8-tetraphenyl-7,8-disilabicyclo[2.2.2]octa-2,5-diene (**2**) (*21b*). Likewise, 2:3,5:6-dibenzo-7,8-dimethyl-7,8-diphenyl-7,8-disilabicyclo[2.2.2]octa-2,5-dienes (**3a** and **3b**) were produced as a steroisomeric mixture by the reaction of 1,2-diphenyl-1,2-dimethyl-1,2-dichlorodisilane with the Mg/MgBr$_2$ reagent (Scheme 3) (*22*). When the reaction was applied to 1,2-ditolyl-1,1,2,2-tetrachlorodisilane, a formal disilyne adduct to anthracene (**4**) was produced along with significant amounts of polymeric substances (*23*).

The reactive species of the Mg/MgBr$_2$ system is presumed to be a univalent MgBr(I) (*24*). The first step of the reaction involves one-electron

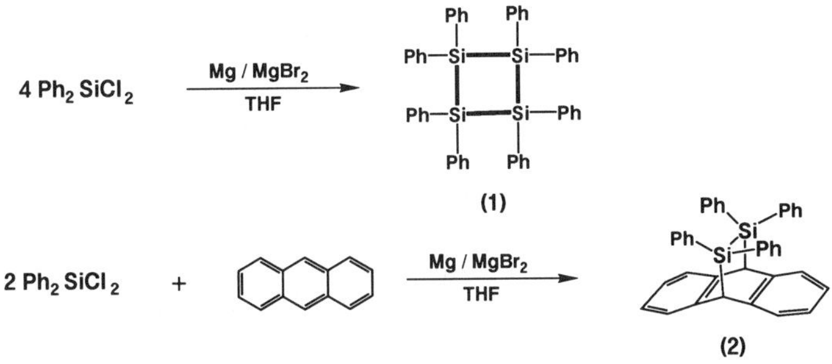

SCHEME 2

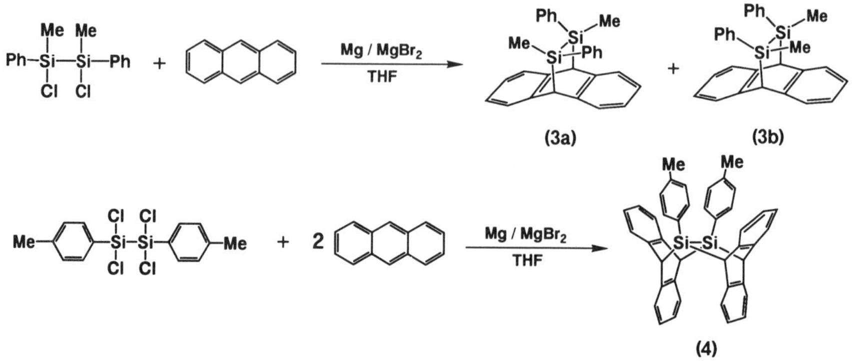

SCHEME 3

transfer from Mg Br(I) to 1,2-dichlorodisilane to give a reactive disilene, which undergoes Diels–Alder reaction with anthracene to produce the disilene adduct. Therefore, polysilapolyhedranes are expected from the reductive oligomerization of $RSiCl_2-SiCl_2R$ or $RSiCl_3$ if the choice of the bulky substituent R is reasonable. For the preparation of tetrasilatetrahedrane, $tert$-Bu_3SiNa was used as the electron transfer reagent (10).

IV

SYNTHESIS OF PRECURSOR

A. Tetrachlorocyclotetragermane

Although the logical synthetic intermediates for the cage compounds should be compounds of the types $[R(X)E]_3$ and $[R(X)E]_4$, where X represents halogens or other negative groups, no example of such compounds had been reported prior to 1992. Most of the extensive work on three- and four-membered rings was limited to $(R_2E)_3$ and $(R_2E)_4$ (E = Si, Ge). Here we describe the strategies used for synthesis of cyclotetrasilane and cyclotetragermane containing four chlorines at each Si or Ge atom.

In the germanium series, phenylated cyclotetragermane (6) was first prepared (Scheme 4) (25). The coupling of $tert$-butyldichloro(phenyl)germane (5a) with a mixture of $Mg/MgBr_2$ in THF produced 1,2,3,4-tetra-$tert$-butyl-1,2,3,4-tetraphenylcyclotetragermane (6a) as a mixture of geometrical isomers in 71% yield. Although four geometrical isomers are possible for 6a, as depicted in Fig. 3, only three stereoisomers are formed in an approximate ratio of 70:20:10. The ^1H and ^{13}C NMR spectra of the

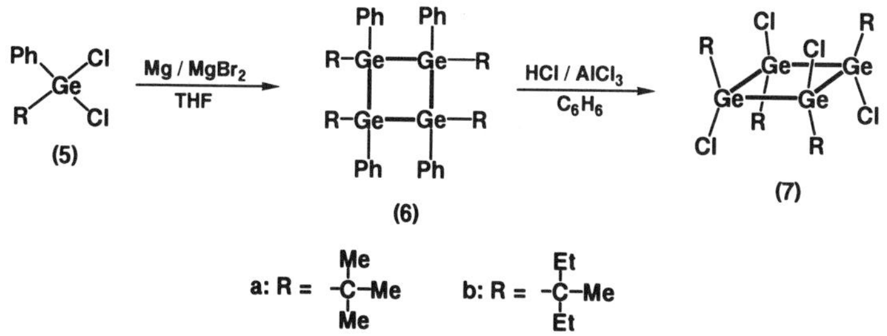

SCHEME 4

major isomer of **6a** showed the existence of the *tert*-butyl groups in a ratio of 1:1:2 in accordance with the cis–cis–trans configuration. The other two isomers were assigned as all-trans and cis–trans–cis. The predominant formation of the cis–cis–trans isomer is caused by statistical factors (*26,27*). Formation of the all-cis isomer is disfavored for steric reasons.

The four phenyl groups were readily replaced by chlorine, without cleavage of the Ge–Ge linkage, by the action of hydrogen chloride in the presence of aluminum chloride (Scheme 4). Among the four possible geometrical isomers of 1,2,3,4-tetra-*tert*-butyl-1,2,3,4-tetrachlorocyclotetragermane (**7a**), only one isomer was obtained in 86% yield (*25*).

Figure 4 shows the all-trans configuration of **7a** established by X-ray diffraction. The molecule has crystallographic mirror symmetry and the four heavy atoms (two chlorines and two germaniums) lie in the same plane. The cyclotetragermane ring is slightly puckered with a dihedral angle of 21° for Ge(1)–Ge(3)–Ge(2)/Ge(1)–Ge(3′)–Ge(2). The chlorine atoms and the *tert*-butyl groups occupy the pseudoaxial and the less hindered pseudoequatorial positions, respectively. The Ge–Ge–Ge bond angles constructing the four-membered ring are in the range of 88.8–89.5° (average 89.1°), and the Ge–Ge bond lengths are 2.455–2.471 Å (average

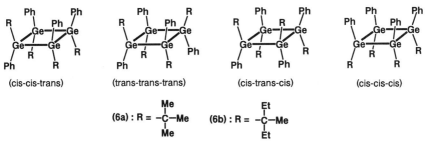

FIG. 3.   Possible steroisomers of cyclotetragermane (**6**).

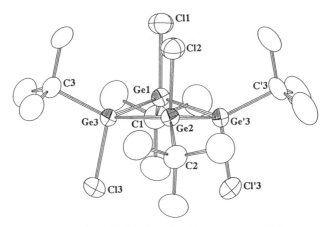

FIG. 4.   ORTEP drawing of cyclotetragermane (**7a**).

2.463 Å). The Ge–Ge–C bond angle (125°) are largely expanded owing to the steric repulsion, whereas Ge–Ge–Cl bond angles (105°) are contracted.

The formation of the all-trans isomer **7a** indicates geometrical isomerization during the reaction. To clarify this point, cyclotetragermane (**6a**) was reacted with 4 equiv of trifluoromethanesulfonic acid (Scheme 5). The ¹H NMR spectrum of the reaction mixture showed three sets of *tert*-butyl signals together with a resonance of the liberated benzene. This indicates the complete replacement of the four phenyl groups by the triflate substituents with the cis–cis–trans configuration. The resulting germyl triflate (**8**) was converted to all-trans isomer **7a** with acetyl chloride.

The cyclotetragermane 1,2,3,4-tetrachloro-1,2,3,4-tetra(1-ethyl-1-methylpropyl)cyclotetragermane (**7b**) with the all-trans configuration was pre-

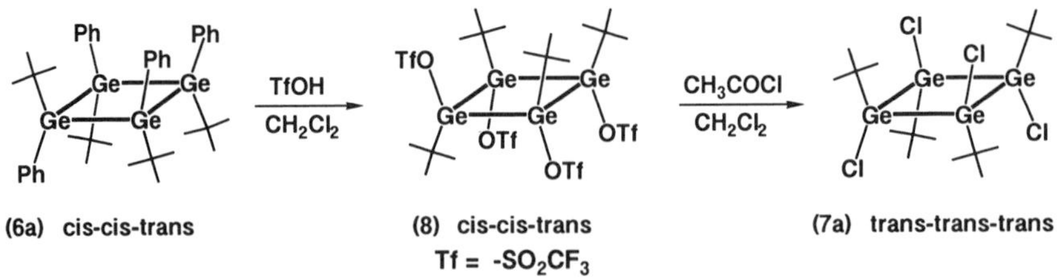

(6a) cis-cis-trans	(8) cis-cis-trans	(7a) trans-trans-trans
	Tf = -SO₂CF₃	

$$Tf = -SO_2CF_3$$

SCHEME 5

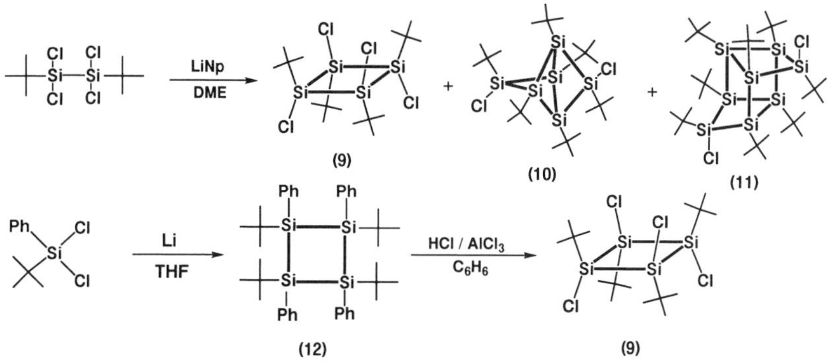

SCHEME 6

pared in a similar manner (Scheme 4). Thus, 1,2,3,4-tetra(1-ethyl-1-methylpropyl)-1,2,3,4-tetraphenylcyclotetragermane (**6b**) (main configuration, cis–cis–trans) was allowed to react with hydrogen chloride in the presence of aluminum chloride in benzene to afford all-trans **7b** (45%). Treatment of **6b** with trifluoromethanesulfonic acid followed by acetyl chloride also gave **7b** in 41% yield.

B. *Tetrachlorocyclotetrasilane*

The functionalized cyclotetrasilane 1,2,3,4-tetra-*tert*-butyl-1,2,3,4-tetrachlorocyclotetrasilane (**9**) was also available (Scheme 6). Treatment of 1,2-di-*tert*-butyl-1,1,2,2-tetrachlorodisilane with 2.5 equiv of LiNp gave rise to **9** with the all-trans configuration together with tricyclo-[2.2.0.0^{2,5}]hexasilane (**10**) and tetracyclo[3.3.0.0^{2,7}.0^{3,6}]octasilane (**11**) (28). 1,2,3,4-Tetra-*tert*-butyl-1,2,3,4-tetra(phenyl)cyclotetrasilane (**12**) was prepared in 50% yield by treatment of *tert*-butyldichloro(phenyl)silane with an excess amount of lithium metal at −10°C in THF (Scheme 6) (29). Three isomers of **12** were isolated as pure compounds: cis–cis–trans (16%), cis–trans–cis (2%), and all-trans (33%). Each isomer was subjected to chlorodephenylation with HCl/AlCl$_3$ in benzene to give all-trans **9** exclusively. Like cyclotetragermane (**7a**), cyclotetrasilane **9** has a folded structure with a dihedral angle of 26.6°. The chlorine atoms and the *tert*-butyl groups occupy the pseudoaxial and the less hindered pseudoequatorial positions, respectively. The average bond distances are 2.375 Å for Si–Si, 1.905 Å for Si–C (*tert*-butyl), and 2.086 Å for Si–Cl (29b). The cyclotetrasilane **9** was nicely utilized as a precursor of a ladder polysilane. Thus, the cross-coupling of **9** with Cl(*i*-Pr)$_2$SiSi(*i*-Pr$_2$)Cl with lithium produced *anti*-1,2,5,6-tetra-*tert*-butyl-3,3,4,4,7,7,8,8-octaisopropyltricyclo[4.2.0.0^{2,5}]octasilane in 40% yield (*30*).

C. Cage and Ladder Polygermane

The cyclotetragermane **7a** was reacted with 2 equiv of sodium in refluxing THF for 24 hr to give 4,8-dichloroocta-*tert*-butyltetracyclo [3.3.0.0^{2,7}.0^{3,6}]octagermane (**13**) (8%) and 3,4,7,8-tetrachloroocta-*tert*-butyltricyclo[4.2.0.0^{2,5}]octagermane (**14**) (21%) (Scheme 7) (*31*). The reaction of **7a** with lithium metal gave similar results, but only a trace of **14** was obtained. Polyhedral germane **13** was also obtained by reductive reactions of either 1,2-di-*tert*-butyl-1,1,2,2-tetrachlorodigermane or *tert*-butyltrichlorogermane with LiNp (*32*).

The ladder polygermane **14** is pale yellow, and the UV spectrum shows absorption bands at 226 nm (ε 111,000) and 341 nm (ε 4700). The cyclotetragermane **7a** and the crushed cubane **13** show absorptions at 216 nm (ε 61,000) and 224 nm (ε 101,000), respectively. An appreciable bathochromic shift is observed in the case of the antiladder polygermane.

Figures 5 and 6 show ORTEP drawings of **13** and **14**. The skeleton of **13** is constructed from three squares and two five-membered rings. An interesting feature is the position of the two chlorine atoms and two *tert*-butyl groups. The latter are directed toward the inside of the framework, the former toward the outside. In the antiladder polygermane **14**, the central

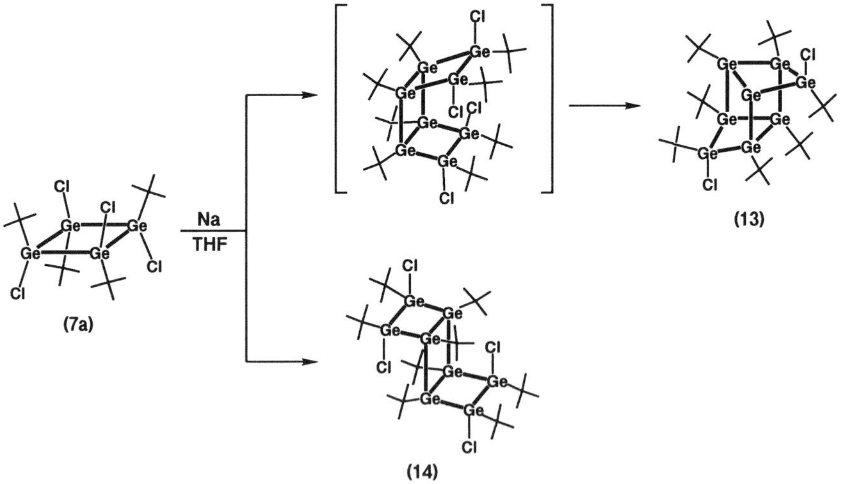

SCHEME 7

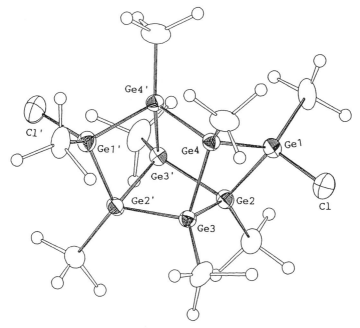

FIG. 5.   ORTEP drawing of crushed cubane (**13**).

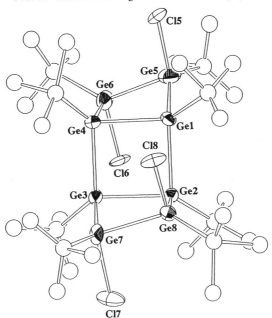

FIG. 6.   ORTEP drawing of ladder polygermane (**14**).

four-membered ring is almost planar, while the other two four-membered rings are puckered with a dihedral angle of 27°. The reaction mechanism for the formation of polygermanes **13** and **14** is not clear. Dimerization of cyclotetragermane **7a** by alkali metals is likely to provide syn and anti dimers (Scheme 7). The syn dimer could be a synthetic intermediate to octa-*tert*-butyloctagermacubane. However, the coupling reaction of the syn dimer occurred in the wrong direction to give the crushed cubane **13**, instead of octagermacubane. In contrast, the coupling of cyclotetrasilane **9** led to the formation of octa-*tert*-butyloctasilacubane.

V

CUBANES COMPRISING SILICON, GERMANIUM, AND TIN

A. *Octasilacubane*

Trihalogenosilanes ($RSiX_3$) or tetrahalogenodisilanes ($RSiX_2$—SiX_2R) bearing appropriate substituents R can serve as precursors to octasilacubane by reductive coupling reactions. For electronic reasons, a silyl-substituted octasilacubane, $Si_8(SiH_3)_8$, is calculated to be 10 kcal/mol less strained than an alkyl-substituted derivative, Si_8Me_8 (*14*). In fact, in 1988 Matsumoto *et al.* reported the successful synthesis of the first octasilacubane, octakis(*tert*-butyldimethylsilyl)pentacyclo[4.2.0.0^{2,5}.0^{3,8}.0^{4,7}]octasilane (**15**) (Scheme 8) (*3*). The octasilacubane **15** was synthesized in one step by condensation of 2,2,3,3-tetrabromo-1,4-di-*tert*-butyl-1,1,4,4-tetramethyltetrasilane or 1,1,1-tribromo-2-*tert*-butyl-2,2-dimethyldisilane with sodium in toluene in 55 and 72% yields, respectively. The octasilacubane forms bright yellow prisms and is stable in an inert atmosphere. In air, the strained framework is oxidized to give colorless solids. The ^{29}Si NMR chemical shift of the cubane framework appeared at −35.03 ppm. The

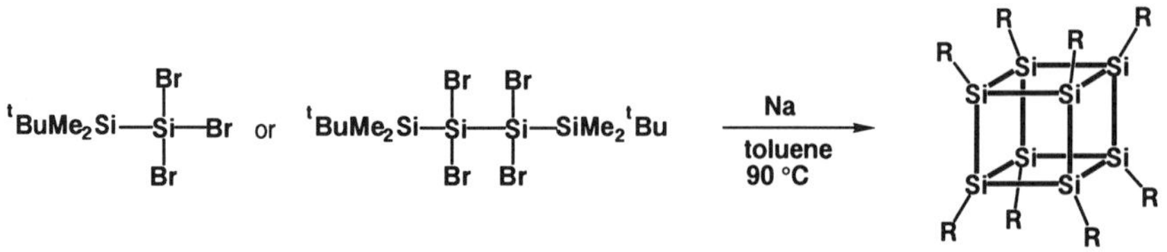

(15) R = SiMe$_2$^{t}Bu

SCHEME 8

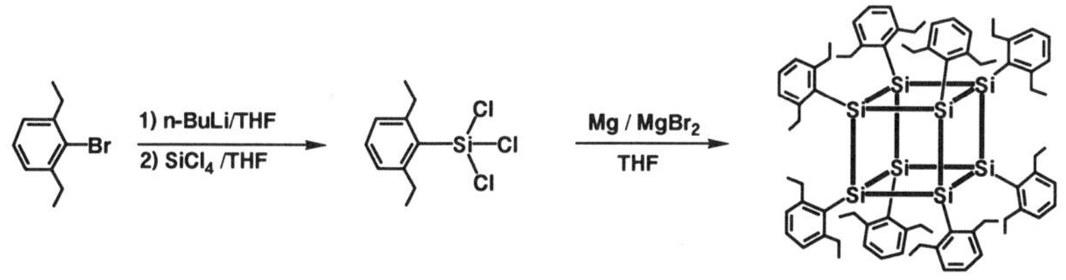

(16)

SCHEME 9

UV–Vis spectrum tailed off into the visible region. However, no structural information based on X-ray crystallographic analysis has been reported for **15**.

For the first precise X-ray crystal structure of an octasilacubane, an aryl group was an appropriate substituent. Thus the 2,6-diethylphenyl group was chosen as the protecting substituent for an octasilacubane. Octakis (2,6-diethylphenyl)octasilacubane (**16**) was synthesized by a dechlorinative coupling reaction with the $Mg/MgBr_2$ reagent in THF (Scheme 9) (6). An exothermic reaction occurred when a THF solution of (2,6-diethylphenyl)trichlorosilane was added to a mixture of $Mg/MgBr_2$. Chromatography on silica gel with a hexane/toluene eluent under argon gave the cubane **16** in 1% yield as orange crystals. The mass spectrum showed the M^+ cluster ion in the range of 1288–1294 in agreement with the calculated formula of $C_{80}H_{104}Si_8$. The ^1H, ^{13}C, and ^{29}Si NMR spectra are fully consistent with the highly symmetrical structure: ^1H NMR (C_6D_6, δ) 0.81 (t, 48 H, $J = 7.4$ Hz), 2.82 (q, 32 H, $J = 7.4$ Hz), 6.96 (d, 16 H, $J = 7.6$ Hz), 7.16 (t, 8 H, $J = 7.6$ Hz); ^{13}C NMR (C_6D_6, δ) 13.8, 31.9, 124.2, 129.2, 139.5 (ipso C), 150.0 (ortho C). The ^{29}Si NMR resonance of the cubic framework appeared at 0.36 ppm. The cubane **16** was gradually oxidized by air.

Figure 7 shows ORTEP drawings of **16** together with selected bond lengths and angles. The X-ray structure indicates the almost perfect cubic structure of the skeleton. The Si–Si–Si bond angles are 88.9–91.1° (average 90.0°). The Si–Si bond lengths are in the range 2.384–2.411 Å (average 2.399 Å) and longer than the normal bond length (2.34 Å). The Si–Si bond lengths of **16** are in close agreement with those calculated for Si_8H_8 (2.396 Å for HF/6-31G*, 2.382 Å for HF/DZ + d) (*16a,b*). The lengths of exocyclic Si–C_{ar} bonds of 1.911 Å (average) are somewhat elongated relative to the normal Si–C bond length (1.88 Å). The Si–Si–C_{ar} bond angles of 124.4° are significantly expanded owing to the endocyclic angular constraint. The 2,6-diethylphenyl groups lie in alternative planes with dihe-

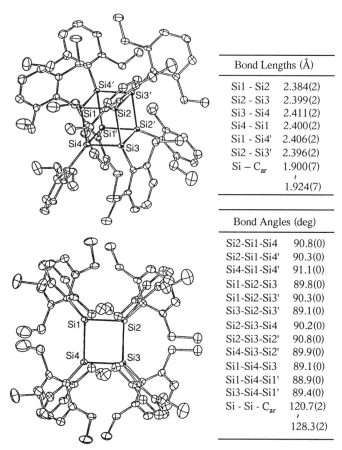

Bond Lengths (Å)	
Si1 - Si2	2.384(2)
Si2 - Si3	2.399(2)
Si3 - Si4	2.411(2)
Si4 - Si1	2.400(2)
Si1 - Si4'	2.406(2)
Si2 - Si3'	2.396(2)
Si – C_{ar}	1.900(7)
	₹
	1.924(7)

Bond Angles (deg)	
Si2-Si1-Si4	90.8(0)
Si2-Si1-Si4'	90.3(0)
Si4-Si1-Si4'	91.1(0)
Si1-Si2-Si3	89.8(0)
Si1-Si2-Si3'	90.3(0)
Si3-Si2-Si3'	89.1(0)
Si2-Si3-Si4	90.2(0)
Si2-Si3-Si2'	90.8(0)
Si4-Si3-Si2'	89.9(0)
Si1-Si4-Si3	89.1(0)
Si1-Si4-Si1'	88.9(0)
Si3-Si4-Si1'	89.4(0)
Si - Si - C_{ar}	120.7(2)
	₹
	128.3(2)

FIG. 7.   ORTEP drawings of octasilacubane (**16**).

dral angles of around 90°. As a result, the cubic skeleton of **16** is efficiently protected by the eight 2,6-diethylphenyl groups.

As stated before, the cyclotetrasilane bearing four chlorines at each silicon is a logical precursor of octasilacubane. Reductive coupling of 1,2,3,4-tetra-*tert*-butyl-1,2,3,4-tetrachlorocyclotetrasilane (**9**) with sodium in the presence of 12-crown-4 in refluxing toluene led to the formation of the octakis(*tert*-butyl)octasilacubane (**17**) (Scheme 10) (*33*). The cubane **17** was also synthesized in 20% yield by one-step condensation of *tert*-butyltrichlorosilane with sodium/12-crown-4 (*9*). Cubane **17** is a purple crystalline compound and is sparingly soluble in organic solvents. The ^{29}Si NMR signals were observed at 6.6 and 10.6 ppm with a relative intensity 1:2 by CPMAS NMR spectroscopy. A diffuse reflection absorp-

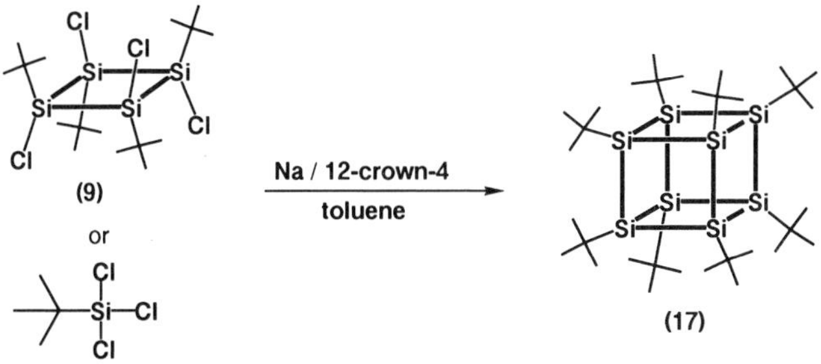

SCHEME 10

tion spectrum of **17** showed a broad absorption between 450 and 650 nm.
An alkyl-substituted octasilacubane, octakis(1,1,2-trimethylpropyl)oc-
tasilacubane (**18**), was also synthesized in 2.6% yield by the condensation
of (1,1,2-trimethylpropyl)trichlorosilane with sodium in toluene (Scheme
11) (*8*). The cubane forms red-orange prisms and is moderately soluble in
common organic solvents. The ^{29}Si NMR signal was observed at 22.24
ppm. The UV–Vis spectrum of **18** exhibits absorption maxima at 252
(ε 30,800), 350 (ε 850), and around 500 nm (ε 70). It is very stable in air and
survives for 2 weeks in the solid state.

The structures of the octasilacubanes **17** and **18** determined by X-ray
diffraction revealed a slightly distorted cubic geometry: Si–Si bond
lengths of 2.35–2.43 Å and Si–Si–Si bond angles of 89.5–91.3° (average
90.0°) for **17** (*9*); Si–Si bond lengths of 2.398–2.447 Å (average 2.424 Å)
and Si–Si–Si bond angles of 87.2–92.6° (average 90.0°) for **18** (*8*). The
distortion from an ideal cube is caused by the steric congestion of eight
very bulky substituents.

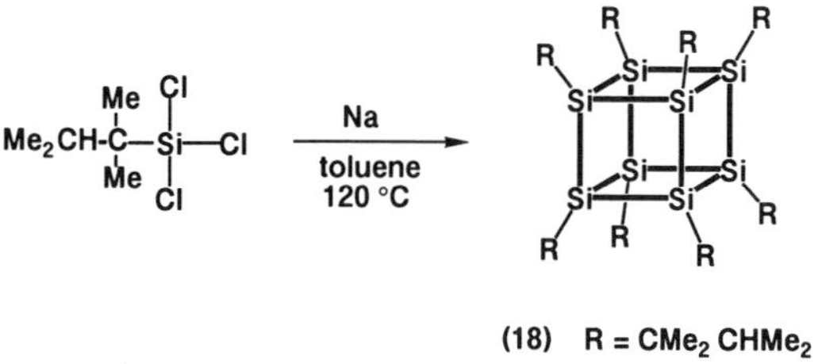

(18)   R = $CMe_2 CHMe_2$

SCHEME 11

B. Octagermacubane

The choice of the substituent R is critical for the cubane synthesis. For
instance, the cyclotetragermane **7a** bearing four *tert*-butyl groups and four

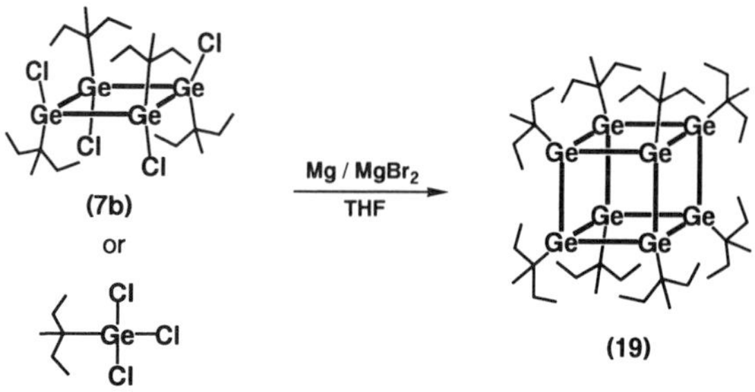

SCHEME 12

chlorines gave unsatisfactory results (*31*). A more bulky substituent, the 1-ethyl-1-methylpropyl group, was then chosen as a protecting group. Octakis(1-ethyl-1-methylpropyl)pentacyclo[4.2.0.0^{2,5}.0^{3,8}.0^{4,7}]octagermane (**19**) was synthesized by dechlorinative coupling of the cyclotetragermane **7b** with the Mg/MgBr$_2$ reagent in 16% yield as yellow crystals (Scheme 12) (*6*). After isolation and purification, cubane **19** in the solid state is fairly stable to atmospheric oxygen and moisture. The octagermacubane **19** exhibits thermochromism; crystals of **19** reversibly change color from pale yellow at $-100°C$ to orange at 100°C.

The octagermacubane **19** was also obtained by the one-step coupling condensation of (1-ethyl-1-methylpropyl)trichlorogermane with Mg/MgBr$_2$ in 3% yield. Although the yield is low, the "one-step" synthesis is quite useful since the trichlorogermane is readily available from tetrachlorogermane. Similar to octasilacubane **16**, octakis(2,6-diethylphenyl)-octagermacubane (**20**) was synthesized by one-step coupling of (2,6-diethylphenyl)trichlorogermane with the Mg/MgBr$_2$ reagent in 1% yield (Scheme 13). Crystals of **20** were less stable to atmospheric oxygen than those of **19**. Therefore, the steric bulkiness of the 2,6-diethylphenyl ligand

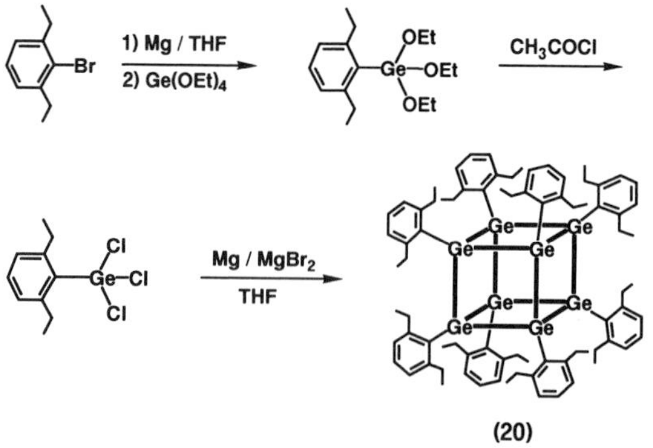

SCHEME 13

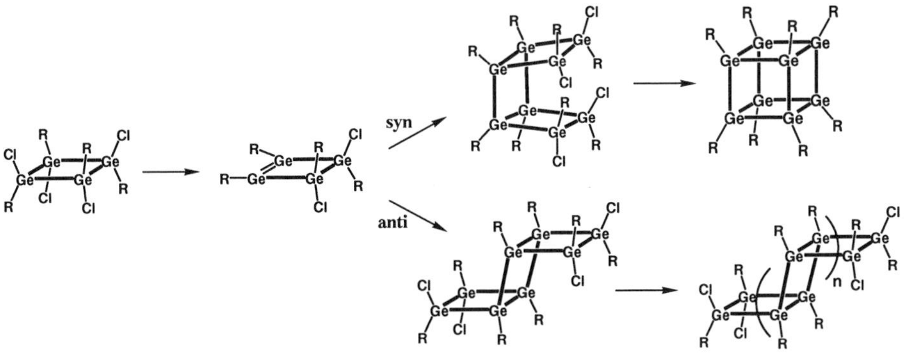

SCHEME 14

is much less than that of the 1-ethyl-1-methylpropyl ligand for the cubane framework.

The stereochemical inversion of 1-ethyl-1-methylpropyl groups is involved in the formation of the cubane **19**. The reactive cyclotetragermene may be formed as a likely intermediate, and subsequent [2 + 2] dimerization can produce syn and anti dimers (Scheme 14). The former gives rise to the octagermacubane, whereas the latter may produce ladder polygermanes.

A satisfactory refinement of the X-ray analysis was accomplished by using octakis(2,6-diethylphenyl)octagermacubane (**20**). The crystal of **20** is crystallographically isomorphous with that of the octasilacubane **16**. Figure 8 shows the ORTEP drawing with structural parameters. The Ge–Ge–Ge bond angles are 88.9–91.1° (average 90.0°). The Ge–Ge bond lengths are 2.478–2.503 Å (average 2.490 Å), which are close to but slightly shorter than that of 2.527 Å calculated for Ge_8H_8 (*16a,b*). The exocyclic Ge–C_{ar} bonds (1.982 Å) are slightly elongated compared to the normal Ge–C_{ar} length (1.95 Å). Owing to geometric reasons, the Ge–Ge–C_{ar} bond angles (average 124.6°) are expanded.

A single crystal of **19** has a 3-fold axis and belongs to the rhombohedral space group *R3c*. However, structural refinement was unsuccessful because of the disorder of the substituents. Preliminary results show the evident cubic structure of **19**: Ge–Ge bond lengths of 2.534 Å (average) and Ge–Ge–Ge bond angles of 90.0° (*34*).

C. Octastannacubane

Thermolysis of hexakis (2,6-diethylphenyl)cyclotristannane (**21**) in naphthalene at 200–220°C gave red octakis(2,6-diethylphenyl)octastanacubane (**22**) in 0.76% yield together with dark blue violet pentastanna-

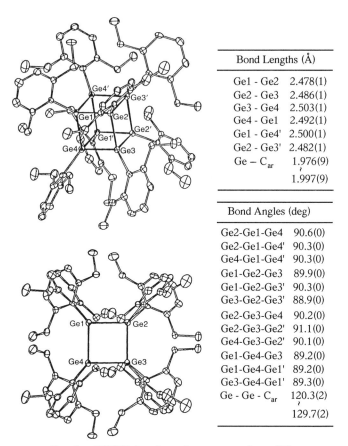

Bond Lengths (Å)	
Ge1 - Ge2	2.478(1)
Ge2 - Ge3	2.486(1)
Ge3 - Ge4	2.503(1)
Ge4 - Ge1	2.492(1)
Ge1 - Ge4'	2.500(1)
Ge2 - Ge3'	2.482(1)
Ge – C_{ar}	1.976(9)
	1.997(9)

Bond Angles (deg)	
Ge2-Ge1-Ge4	90.6(0)
Ge2-Ge1-Ge4'	90.3(0)
Ge4-Ge1-Ge4'	90.3(0)
Ge1-Ge2-Ge3	89.9(0)
Ge1-Ge2-Ge3'	90.3(0)
Ge3-Ge2-Ge3'	88.9(0)
Ge2-Ge3-Ge4	90.2(0)
Ge2-Ge3-Ge2'	91.1(0)
Ge4-Ge3-Ge2'	90.1(0)
Ge1-Ge4-Ge3	89.2(0)
Ge1-Ge4-Ge1'	89.2(0)
Ge3-Ge4-Ge1'	89.3(0)
Ge - Ge - C_{ar}	120.3(2)
	129.7(2)

FIG. 8.   ORTEP drawings of octagermacubane (20).

[1.1.1]propellane (23) (Scheme 15) (35,36). Both 22 and 23 are highly air-sensitive. The ^{119}Sn NMR spectrum of 22 appeared at 44.3 ppm with satellite signals due to $^1J(^{119}Sn-^{117}Sn)$ and $^2J(^{119}Sn-^{117}Sn)$ couplings of 1576 and 1345 Hz, respectively. As found for sila- and germacubanes, the stannacubane 22 also has thermochromic properties.

Although the mechanism to give the stannacubane 22 from the cyclotristannane 21 remains unclear, the reactive "RSn" species is postulated to form via disproportionation: $2 R_2Sn: \rightarrow RSn + R_3Sn\cdot$. Thermal decomposition of 21 leads to the formation of $R_2Sn:$ and $R_2Sn=SnR_2$. The latter also can generate the divalent $R_2Sn:$. Oligomerization of RSn may produce 22. Thermolysis of the cyclotristannane in the presence of benzophenone increased the yield of 22. In addition, decakis(2,6-diethyl-

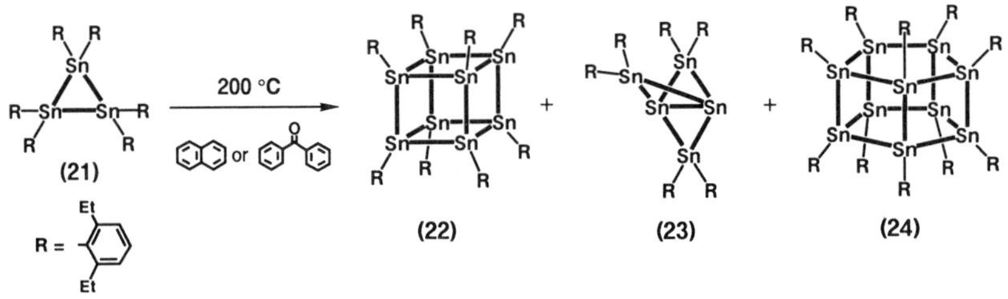

SCHEME 15

phenyl)decastanna[5]prismane (24) was also formed (Scheme 15) (37). Complexation of RSn by the benzophenone stabilizes the divalent species leading to $(RSn)_n$ ($n = 8$, 10). The cubic arrangement of the eight tin atoms was determined by X-ray diffraction of 22. The Sn–Sn bond lengths are in the range of 2.839–2.864 Å (average 2.854 Å) and the Sn–Sn–Sn bond angles are 89.1–91.1° (average 90.0°) (35). The structure of [5]prismane 24 was also verified by X-ray diffraction. The Sn–Sn bond lengths of 24 (average 2.856 Å) are the same as those of 22 (37).

D. Comparison of Structures

A geometric comparison of cubanes C_8H_8 and E_8R_8 [E = Si (16), Ge (20, and Sn (22), R = 2,6-$Et_2C_6H_3$] has been made. Table IV gives the structural parameters of the cubanes determined by X-ray diffraction together with the calculated values for C_8H_8 (38) and E_8H_8 (16b). The E–E–E bond angles for all cubanes range from 89 to 91°; thus, the skeletons are made up of an almost perfect cubic arrangement of group 14 elements. The E–E bond lengths of R_8E_8 (2.399 Å for Si, 2.490 Å for Ge, and 2.854 Å for Sn) are apparently longer than the normal values owing to the inherent bond strain. These values are in close agreement with those calculated for E_8H_8 (2.382 Å for Si, 2.527 Å for Ge, and 2.887 Å for Sn) (16b). The range of E–E–C_{ar} bond angles increases in the following order: 121–128° for Si < 120–130° for Ge < 117–133° for Sn. This implies a decrease of the steric congestion among the neighboring ligands caused by elongation of the E–C_{ar} bond: 1.911 Å for Si < 1.982 Å for Ge < 2.193 Å for Sn.

E. Electronic Absorption Spectra

All cubanes composed of the higher row group 14 elements are colored from yellow to purple. The aryl-substituted silacubane 16 (6) is orange and

TABLE IV

STRUCTURAL PARAMETERS OF CUBANES COMPRISING GROUP 14 ELEMENTS

E = C, R = H ;

E = Si, Ge, Sn, R = (Et / Et aryl)

	$C_8H_8{}^a$	Si_8R_8	Ge_8R_8	$Sn_8R_8{}^b$
E–E (Å)				
X-Ray	1.551 (av.)	2.3999 (av.)	2.490 (av.)	2.854 (av.)
	(1.549–1.553)	(2.384–2.411)	(2.478–2.503)	(2.839–2.864)
Calculated^c	1.559	2.382	2.527	2.887
Normal	1.54	2.34	2.40	2.78
E–C_{ar} (Å)	1.06 (av.)	1.911 (av.)	1.982 (av.)	2.193 (av.)
E–E–E (°)	89.3–90.5	88.9–91.1	88.9–91.1	89.1–91.1
E–E–C_{ar} (°)	123–127	121–128	120–130	117–133

^{a} From Fleischer (38).
^{b} From Sita and Kinoshita (35).
^{c} From Nagase (16b).

the silyl-substituted one **15** (3) is yellow, whereas the *tert*-butyl-substituted **17** (9) is reported to be purple. Figure 9 shows the UV–Vis spectra of cubanes **16, 19,** and **20.**

An absorption maximum at 240 nm (ε 48,700), tailing into the visible region (500 nm), was observed for **19.** The absorption at 240 nm belongs to the σ–σ* transition. Aryl substitution caused electronic perturbation. Thus, the aryl-substituted **20** exhibits three absorption bands at 243 nm (ε 111,000), 293 nm (ε 45,500), and 404 nm (ε 3500). The absorption band at 243 nm is attributed to the σ–σ* transition, and that at 293 nm is caused by σ–π mixing between Ge–Ge bonds and the aromatic π system. The aryl-substituted silacubane **16** also showed three absorption bands at 234 nm (ε 87,000), 284 nm (ε 42,000), and 383 nm (ε 5000). These absorptions are somewhat blue-shifted from germacubane **20.** Octastannacubane **22** had absorption bands at 275 nm (ε 112,000), tailing into the visible region [320 nm (sh) (ε 32,000), 450 nm (sh) (ε 2000)]. The electronic absorptions assigned to the σ–σ* transition appear to be reflected in the order of HOMO levels in E–E σ bonds (E = Si, Ge, Sn) (Fig. 2).

The lowest energy absorption band is forbidden when the cubane has high symmetry but can be weakly allowed when it has low symmetry. Semiempirical CNDO/S calculation shows that the spectrum shape in the low energy region depends strongly on substituents (39). Among *tert*-

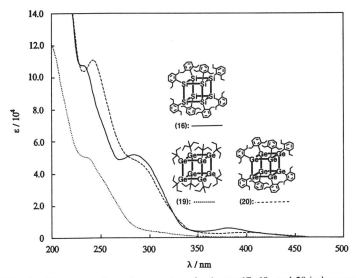

FIG. 9.   Electronic absorption spectra of cubanes **17, 19,** and **20** in hexane.

butyl; trimethylsilyl-, and phenyl-substituted octasilacubanes, *tert*-butyl-substituted cubane **17** has the lowest energy absorption band.

Cubic Si-skeleton clusters were used to model the optical properties of porous silicon (*40*). In the octasilacubane **15,** the absorption edge is observed at approxiamtely 3.2 eV, and a broad photoluminescence spectrum is also observed with a peak of 2.50 eV. Both porous silicon and octasilacubane show broad photoluminescence spectra and large Stokes shifts.

VI

PRISMANES COMPRISING SILICON AND GERMANIUM

A. *Hexasilaprismane*

For the synthesis of hexasilaprismane, tetrachlorodisilane ($RSiCl_2$—$SiCl_2R$) and trichlorosilane ($RSiCl_3$) were used. The $Mg/MgBr_2$ reagent is quite useful for the hexasilaprismane synthesis. The hexasilaprismane hexakis (2,6-diisopropylphenyl)tetracyclo[2.2.0.0^{2,6}.0^{3,5}]hexasilane **(25)** was thus prepared by the dechlorinative coupling reation of 1,2-bis(2,6-diisopropylphenyl)-1,1,2,2-tetrachlorodisilane with the $Mg/MgBr_2$ reagent in THF (Scheme 16) (*7*). Purification by chromatography on silica

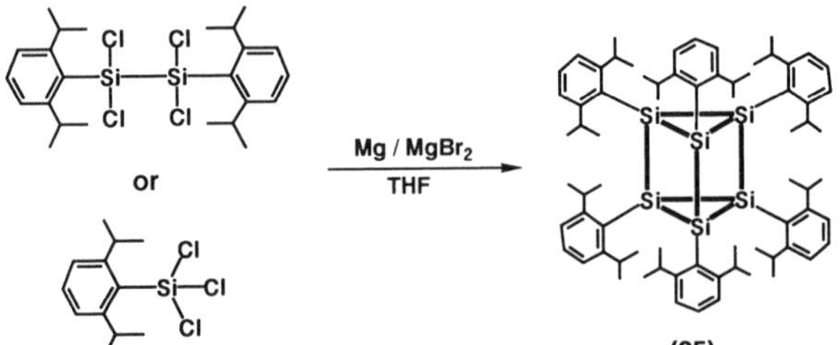

SCHEME 16

gel with hexane/toluene afforded the hexasilaprismane **25** in 7% yield as orange crystals: mp $>220°C$; MS (FAB) 1134–1140 (M^+ cluster), consistent with the calculated formula of $C_{72}H_{102}Si_6$. Previous study demonstrated that reaction of 1,2-bis(2,6-diisopropylphenyl)-1,1,2,2-tetrachlorodisilane with lithium naphthalenide did not give the hexasilaprismane **25** (*18e*). Therefore, the choice of reducing reagent is critical. The reaction of (2,6-diisopropylphenyl)trichlorosilane with $Mg/MgBr_2$ also gave the prismane **25**, but in very low yield (1%). Contrary to other polyhedranes, the hexasilaprismane **25** is thermally and oxidatively fairly stable in the solid state. No change was observed even after several months in air.

The six, 2,6-diisopropylphenyl groups attached to the each silicon in **25** are equivalent on the NMR time scale. ^{29}Si NMR spectroscopy in solution demonstrated that the six silicons of the skeleton are equivalent, with a single resonance appearing at -22.3 ppm. However, CPMAS ^{29}Si NMR of solid **25** showed signals at -22.2 and -30.8 ppm with a relative intensity of 2:1. The ^{29}Si resonance is shifted considerably upfield from that of the octasilacubane **16**.

The six 2,6-diisopropylphenyl groups attached to each silicon are equivalent on the NMR scale. However, owing to the large steric congestion, the ^1H NMR spectrum of **25** at 25°C shows that two isopropyl and aryl protons are magnetically nonequivalent, indicating significant hindered rotation of the substituents around the axis of the $Si-C_{ar}$ bonds (Fig. 10). One of the set of methine protons appears at 3.45 ppm, and the two methyl protons fall at 0.64 and 1.10 ppm. The other set of methine protons appears at 4.82 ppm along with the two methyl protons at 1.01 and 1.62 ppm. Owing to the restricted rotation of the aryl groups, the ^1H NMR spectrum is temperature-dependent. The coalescence temperatures were 328 K for the meta protons ($\Delta\nu = 33$ Hz) and 363 K for the isopropyl methine protons ($\Delta\nu = 411$ Hz). Thus, the barrier hight ($\Delta G\ddagger$) for the rotation was estimated to be 16.5 kcal/mol.

Figure 11 gives ORTEP drawings together with structural parameters for **25**. The crystal has a 2-fold axis of symmetry. The skeleton has a

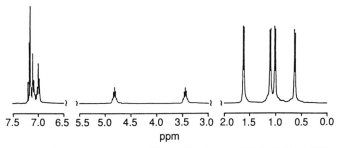

FIG. 10.   ¹H NMR spectrum of hexasilaprismane (**25**) in C₆D₆ at 25°C.

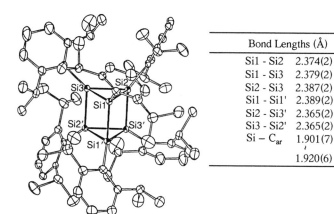

Bond Lengths (Å)	
Si1 - Si2	2.374(2)
Si1 - Si3	2.379(2)
Si2 - Si3	2.387(2)
Si1 - Si1'	2.389(2)
Si2 - Si3'	2.365(2)
Si3 - Si2'	2.365(2)
Si – C$_{ar}$	1.901(7)
	1.920(6)

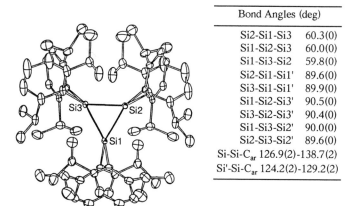

Bond Angles (deg)	
Si2-Si1-Si3	60.3(0)
Si1-Si2-Si3	60.0(0)
Si1-Si3-Si2	59.8(0)
Si2-Si1-Si1'	89.6(0)
Si3-Si1-Si1'	89.9(0)
Si1-Si2-Si3'	90.5(0)
Si3-Si2-Si3'	90.4(0)
Si1-Si3-Si2'	90.0(0)
Si2-Si3-Si2'	89.6(0)
Si-Si-C$_{ar}$ 126.9(2)-138.7(2)	
Si'-Si-C$_{ar}$ 124.2(2)-129.2(2)	

FIG. 11.   ORTEP drawings of hexasilaprismane (**25**).

slightly distorted prismane structure made from two triangular units [Si–Si length 2.374–2.387 Å (average 2.380 Å) and Si–Si–Si angle 59.8–60.3 (average 60.0°)] and three rectangular units Si–Si' length 2.365–2.389 Å (average 2.373 Å and Si–Si–Si' angle 89.6–90.5 (average 90.0°)]. The lengths of the exocyclic Si–C bonds are 1.901–1.920 Å (average 1.908 Å). Because of the geometry, the exocyclic bond angles are significantly expanded: Si–Si–C_{ar} angle 126.9–1.38.7° (average 133.5°) and Si'–Si–C_{ar} angle 124.2–129.2° (average 126.9°). The distortion of the skeleton is also suggested by the observation of two different ^{29}Si resonances at −22.2 and −30.8 ppm with an intensity ratio of 2:1.

All the Si–Si bonds in 25 are elongated from the normal Si–Si bond length (2.34 Å) but are shorter than those in octasilacubane 16 (average 2.399 Å) (6) and cyclotrisilane ($R_2Si)_3$ (R = 2,6-dimethylphenyl: average 2.407 Å) (41). The Si–Si bond lengths of 25 are somewhat longer than those of calculated for Si_6H_6 (2.359 Å for triangular units and 2.375 Å for rectangular units) (16a,b). The aryl planes are arranged in a screw-shaped manner around the three-membered ring so that the skeleton of the prismane is effectively covered with the six 2,6-diisopropylphenyl groups.

B. Hexagermaprismane

The first hexagermaprismane, hexakis[bis(trimethylsilyl)methyl]tetracyclo[2.2.0.0^{2,6}.0^{3,5}]hexagermane, was reported in 1989 (4). For the synthesis of hexagermaprismanes, trichlorogermanes [$RGeCl_3$, R = 2,6-i-Pr$_2C_6H_3$, CH(SiMe$_3)_2$] were used. Magnesium metal is quite effective for the synthesis of hexagermaprismane. However, the Mg/MgBr$_2$ reagent is too reactive, and none of the hexagermaprismane is produced.

2,6-(Diisopropylphenyl)trichlorogermane, the requisite precursor, was prepared by chlorination of (2,6-diisopropylphenyl)triethoxygermane with acetyl chloride. The aryl-substituted hexagermaprismane, hexakis(2,6-diisopropylphenyl)tetracyclo[2.2.0.0^{2,6}.0^{3,5}]hexagermane (26), was thus synthesized by dechlorinative coupling of (2,6-diisopropylphenyl)trichlorogermane with magnesium (Scheme 17) (7). Like hexasilaprismane 25, the ^{1}H and ^{13}C NMR spectra of 26 at −40°C show that the two isopropyl and aryl protons are magnetically nonequivalent (Fig. 12). The coalescence temperatures were −10°C for the meta protons (Δv = 33 Hz) and 12°C for the isopropyl methine protons (Δv = 408 Hz). Thus, the $\Delta G\ddagger$ value of 13.1 kcal/mol for the rotation is much smaller than that of hexasilaprismane 25, owing to the increased E–C_{ar} bond length. As evidenced by the dynamic NMR, the hexagermaprismane 26 is less stable than the hexasilaprismane 25 toward atmospheric oxygen.

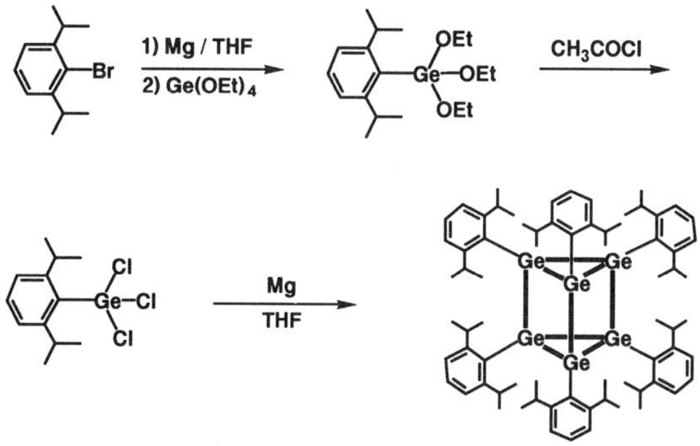

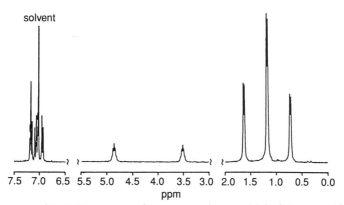

SCHEME 17

The action of the reducing reagent is quite mysterious. The reaction of (2,6-diisopropylphenyl)trichlorogermane with the Mg/MgBr$_2$ reagent resulted in the formation of tetrakis(2,6-diisopropylphenyl)digermene (27) instead of the prismane 26 (Scheme 18). The reaction probably proceeds via a distribution reaction promoted by MgBr$_2$ to produce R$_2$GeX$_2$ (X = Cl, Br) which reacts reductively, giving Ge=Ge double-bonded species.

Figure 13 shows ORTEP drawings of hexagermaprismane 26. The prismane skeleton of 26 is made up from two triangular units [Ge–Ge length 2.497–2.507 (average 2.503 Å) and Ge–Ge–Ge angle 59.8–60.1 (average 60.0°)] and three rectangular units [Ge–Ge' length 2.465–2.475 (average 2.468 Å) and Ge–Ge–Ge' angle 89.0–91.1 (average 90.0°)]. Unlike silaprismane 25, the cyclopropyl Ge–Ge bond lengths are distinctly longer

FIG. 12. ^{1}H NMR spectrum of hexagermaprismane (26) in C$_7$D$_8$ at −40°C.

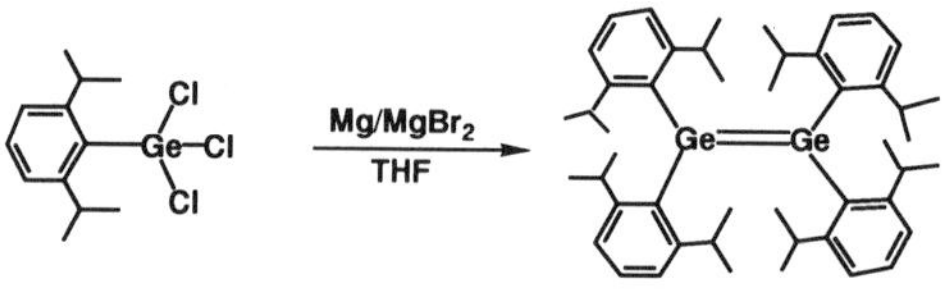

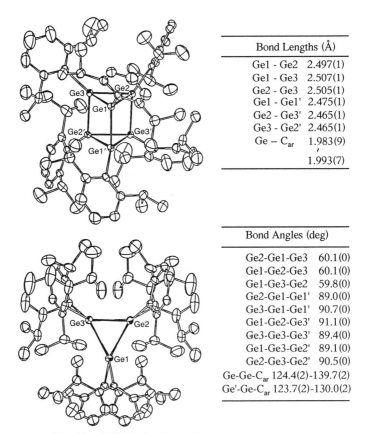

(27)

SCHEME 18

than those linking two cyclopropyl rings. The exocyclic bond lengths are 1.983–1.993 Å (average 1.988 Å), and the exocyclic bond angles are largely expanded (Ge–Ge–C_{ar}, average 133.0°; Ge′–Ge–C_{ar}, average (126.9°). All the Ge–Ge bonds in **26** are elongated from the normal Ge–Ge bond length (2.40 Å), but they are shorter than those in cyclotrigermane ($R_2Ge)_3$ (R = 2,6-dimethylphenyl: average 2.541 Å) (*42*). The Ge–Ge bond

Bond Lengths (Å)	
Ge1 - Ge2	2.497(1)
Ge1 - Ge3	2.507(1)
Ge2 - Ge3	2.505(1)
Ge1 - Ge1'	2.475(1)
Ge2 - Ge3'	2.465(1)
Ge3 - Ge2'	2.465(1)
Ge – C_{ar}	1.983(9)
	1.993(7)

Bond Angles (deg)	
Ge2-Ge1-Ge3	60.1(0)
Ge1-Ge2-Ge3	60.1(0)
Ge1-Ge3-Ge2	59.8(0)
Ge2-Ge1-Ge1'	89.0(0)
Ge3-Ge1-Ge1'	90.7(0)
Ge1-Ge2-Ge3'	91.1(0)
Ge3-Ge3-Ge3'	89.4(0)
Ge1-Ge3-Ge2'	89.1(0)
Ge2-Ge3-Ge2'	90.5(0)
Ge-Ge-C_{ar}	124.4(2)-139.7(2)
Ge'-Ge-C_{ar}	123.7(2)-130.0(2)

FIG. 13.   ORTEP drawings of hexagermaprismane (**26**).

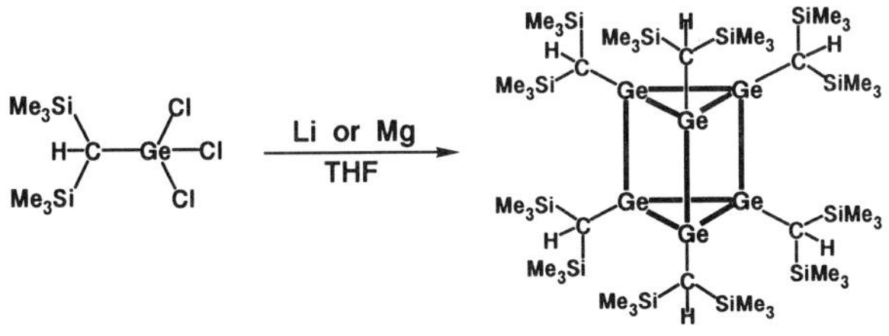

(28)

SCHEME 19

lengths of **26** are somewhat shorter than those calculated for Ge_6H_6 (2.502 Å for triangular units and 2.507 Å for rectangular units (*16b*).

The hexagermaprismane **28** bearing bis(trimethylsilyl)methyl groups was synthesized by condensation of bis(trimethylsilyl)methyltrichlorogermane. Thus, treatment of bis(trimethylsilyl)methyltrichlorogermane with excess magnesium metal in THF at 0°C provided yellow-orange crystals of the hexagermaprismane **28** in 24% yield (Scheme 19). Lithium metal works similarly in THF at −78°C to give the prismane **28** in 12% yield. However, reaction of the trichlorogermane with lithium metal was very delicate and unsteady owing to the cleavage of the highly strained Ge–Ge bond. Reproducible results were obtained by the use of mgnesium metal. However, the $Mg/MgBr_2$ reagent gave none of the prismane **28**.

The structural features of hexagermaprismane **28** are similar to those of **26**. However, the Ge–Ge bond lengths of **28** are longer than those of **26** because of the increased steric bulkiness of the bis(trimethylsilyl)methyl groups. The skeleton of the prismane **28** is constructed essentially by two equilateral triangles [Ge–Ge length 2.578–2.584 Å (average 2.580 Å) and Ge–Ge–Ge angle 60.0–60.1° (average 60.0°)] and three rectangles [Ge–Ge length 2.516–2.526 Å (average 2.522 Å) and Ge–Ge–Ge angle 88.5–91.4° (average 89.9°)].

C. Comparison of Structures

Table V gives structural parameters of prismanes (C, Si, Ge) together with calculated values. The E–E bond length of the three-membered unit is denoted by the symbol *a* and that linking the triangular unit is indicated as *b*. The calculations of C and Si prismanes predict shortening of the bond length of the three-membered unit (*a*) relative to that of the four-membered unit (*b*) (*16a,b*). Indeed, unsubstituted prismane (*43*) and derivatives of the type C_6R_6 [R = Me (*44*) and R = $SiMe_3$ (*45*)] are in accordance with this prediction. However, owing to steric repulsions the skeleton of Si_6R_6 (**25**: R = 2,6-*i*-$Pr_2C_6H_3$) (*a* = 2.374–2.387 Å, *b* = 2.365–

TABLE V

STRUCTURAL PARAMETERS OF PRISMANES COMPRISING GROUP 14 ELEMENTS

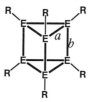

E	R	a (Å)	b (Å)	Method
C	H	1.507	1.549	Calculated[a]
		1.500	1.585	Electron diffraction[b]
	Me	1.540	1.551	Electron diffraction[c]
	SiMe$_3$	1.510	1.582	X-Ray diffraction[d]
Si	H	2.359	2.375	Calculated[e]
	i-Pr	2.380	2.373	X-Ray diffraction
		(2.374–2.387)	(2.365–2.389)	
Ge	H	2.502	2.507	Calculated[c]
	i-Pr	2.503	2.468	X-Ray diffraction
		(2.497–2.507)	(2.465–2.475)	
		2.580	2.522	X-Ray diffraction
		(2.578–2.584)	(2.516–2.526)	

[a] From Nagase *et al.* (*16a*).
[b] From Karl *et al.* (*43*).
[c] From Karl *et al.* (*44*).
[d] From Sekiguchi *et al.* (*45*).
[e] From Nagase (*16b*).

2.389 Å) is slightly distorted from an ideal triangular prism geometry. Both bond a and bond b in **25** are longer than those calculated for Si$_6$H$_6$ (a = 2.359 Å, b = 2.375 Å). In contrast, bond a is distinctly longer than b in Ge$_6$R$_6$ **26** (R = 2,6-i-Pr$_2$C$_6$H$_3$) (a = 2.503 Å, b = 2.468 Å). Owing to the

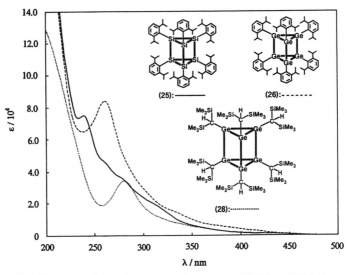

FIG. 14.   Electronic absorption spectra of prismanes **25**, **26**, and **28** in hexane.

bulky substituents, both bond a and bond b in Ge$_6$R$_6$ **28** [R = CH(SiMe$_3$)$_2$] (a = 2.580 Å, b = 2.522 Å) are longer than those in **26**, but bond a is appreciably longer than bond b as found in **26**.

D. *Electronic Absorption Spectra*

As found in the cubanes, all the prismanes composed of Si and Ge are yellow to orange in color. Figure 14 shows the UV–Vis spectra of prismanes **25**, **26**, and **28**. In general, the prismanes have absorptions tailing into the visible region. For example, the hexasilaprismane **25** has an absorption band with a maximum at 241 nm (ε 78,000) tailing to around 500 nm. Hexagermaprismane **26** exhibits an absorption band with a maximum at 261 nm (ε 84,000), which is red-shifted compared to that of **28** because of the high-lying orbitals of the Ge–Ge bonds. The absorption band with a maximum at 280 nm (ε 32,200) observed for the hexagermaprismane **28** is further red-shifted relative to that of **26**. This is caused by the destabilization of the HOMO resulting from Ge–Ge bond elongation owing to a very bulky substituent, the bis(trimethylsilyl)methyl group.

E. *Photochemical Reactions*

Hexasilaprismane **25** is photosensitive (*7*). On irradiation of **25** with light having a wavelength of 360–380 nm in solution (3-MP or 2-MeTHF) at −50°C or in a glass matrix at 77 K, new absorption bands appeared at 335, 455, and 500 nm assignable to the absorption bands of hexasila-

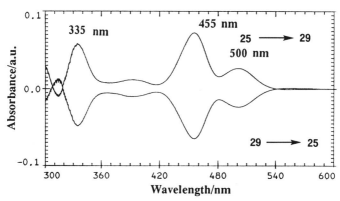

FIG. 15.   UV–Vis difference spectra showing the formation of hexasila-Dewar benzene (**29**) and subsequent photochemistry (2-MeTHF, 100 K).

Dewar benzene **29**. Excitation of the bands with wavelengths longer than 460 nm resulted in the immediate regeneration of **25**.

A single chemical species is produced during the reaction since the two bands appeared and disappeared simultaneously (Fig. 15). The folding angle of hexasila-Dewar benzene Si_6H_6 is calculated to be 120° by the ab initio calculation method. The spatial interaction of the two Si=Si double bonds should split the π-MO into bonding (π_S) and antibonding (π_A) sets, which, depending on symmetry, should further be perturbed by the bridged Si–Si (σ) orbital (*46*). The allowed transitions from π_S to π^*_S and from π_A to π^*_A correspond to the experimental absorption bands of 455 and 335 nm, respectively. However, the transition from π_A to π^*_S is forbidden, and this transition is weakly allowed when the hexasila-Dewar benzene has less than C_{2v} symmetry. The relatively weak absorption at 500 nm is assigned to this transition (Table VI).

For reasons of symmetry, the bridged Si–Si bond interacts with π_S and π_S^* MOs, and thus the former is destabilized and the latter is stabilized (Fig. 16). The splittings of π_A/π_S and π^*_A/π^*_S can be estimated to be 0.25 and 1.22 eV, respectively. Due to the reason of the symmetry, the photochemical excitation of the hexasila-Dewar benzene **29** readily occurs to give the hexasilaprismane **25** (Scheme 20).

The hexasila-Dewar benzene **29** is thermally stable at $-150°C$, however, it gradually reverted to the hexasilaprismane **25**. The half-life is $t_{\frac{1}{2}} = 0.52$ min at 0°C in 3-MP (first-order rate constant, $k = 2.21 \times 10^{-2}$ s^{-1}). The activation parameters for hexasila-Dewar benzene **29** to the hexasilaprismane **25** are Ea = 13.7 kcal/mol, $\Delta H^{\pm} = 13.2$ kcal/mol, $\Delta S^{\pm} = -17.8$ eu. The small Ea value is consistent with the high reactivity of Si=Si

TABLE VI
TRANSITION ENERGIES OF HEXASILA-DEWAR BENZENES (Si₆H₆ AND 29)

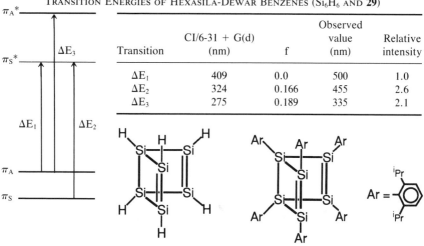

Transition	CI/6-31 + G(d) (nm)	f	Observed value (nm)	Relative intensity
ΔE_1	409	0.0	500	1.0
ΔE_2	324	0.166	455	2.6
ΔE_3	275	0.189	335	2.1

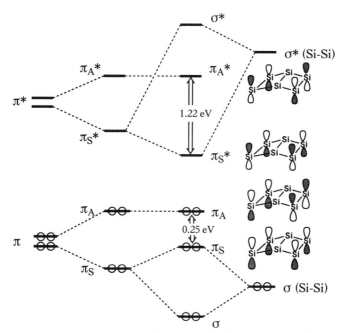

FIG. 16.   Qualitative molecular orbital diagram of hexasila-Dewar benzene.

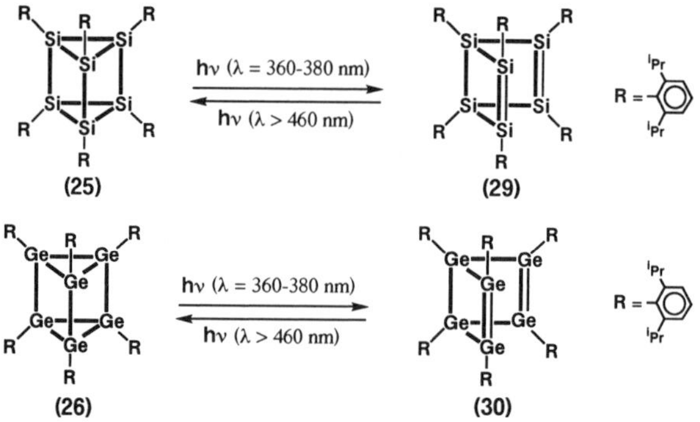

SCHEME 20

double bonds. The large negative activation entropy suggests the signifi-cant decrease of the freedom in the transition state to form **25**. All at-tempts to trap the hexasila-Dewar benzene **29** with methanol or sulfur failed, probably due to the rapid intramolecular reaction to regenerate **25**.

Irradiation of hexagermaprismane **26** (R = 2,6-i-Pr$_2$C$_6$H$_3$) also produced new absorption bands at 342, 446, and 560 nm due to hexagerma-Dewar benzene **30** at low temperature. Excitation of these bnads with λ > 460 nm light led to the regeneration of the original absorption due to the prismane **26**. The photochemical generation of hexagerma-Dewar benzene **30** and the subsequent [2 + 2] reaction to regenerate hexagermaprismane may be involved (Scheme 20). Hexagerma-Dewar benzene **30** was also gradually reverted to the prismane **26** above $-160°$C.

VII

TETRASILATETRAHEDRANE

Tetrasilatetrahedrane is accessible if the appropriate substituent is se-lected. As found in the case of the octasilacubane (*tert*-BuMe$_2$SiSi)$_8$ (**15**), silyl substituents decrease the strain in the polyhedrane (*14*). Thus, the synthesis of tetrasilatetrahedrane has been accomplished by use of the "supersilyl" group (*tert*-Bu$_3$Si) (*10*).

The dehalogenation of *tert*-Bu$_3$Si—SiCl$_3$ with sodium at 80°C led to the formation of various products: 1,2-bis(supersilyl)disilane (**31**) and tris(supersilyl)cyclotrisilane (**32**). The disilane **31** was brominated to give tetrabromodisilane (**33**), reaction of which with the supersilyl anion (*tert*-Bu$_3$SiNa) in THF at $-20°$C gave tetrakis(tri-*tert*-butylsilyl)-

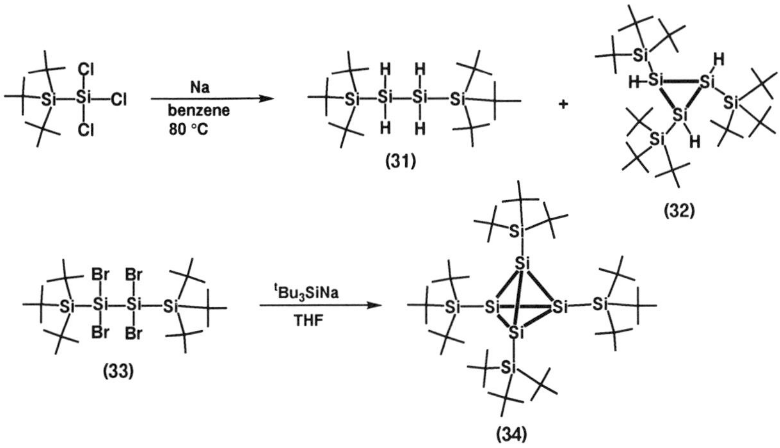

(31)

(32)

(33)

(34)

SCHEME 21

tricyclo[1.1.0.0^{2,4}]tetrasilane (**34**) as yellow-orange crystals in 57% yield (Scheme 21). The reaction mechanism for **34** is not clear. However, the reactive disilyne (*tert*-Bu$_3$)Si—Si≡Si—Si(*tert*-Bu$_3$) is a possible intermediate which undergoes dimerization to give tetrasilacyclobutadiene, thereby leading to **34**.

Tetrasilatetrahedrane **34** is unexpectedly stable to water, air, and light. It cannot be reduced by sodium but reacts with tetracyanoethylene (TCNE) and Br$_2$. Unlike the *tert*-butyl-substituted tetrahedrane of carbon (*47*), tetrahedrane **34** is thermally stable, and the crystals do not melt below 350°C. Steric bulkiness evidently stabilizes the skeleton of the tetrahedrane. As found in octasilacubanes and hexasilaprismane, the color of the crystals changes reversibly. The ^{29}Si resonance of the framework appeared at 38.89 ppm, shifted downfield relative to octasilacubanes and hexasilaprismane.

Recrystallization of a mixture of **34** and hexa-*tert*-butyldisilane from C$_6$D$_6$ formed two (*tert*-Bu$_3$Si)$_4$Si$_4$·(*tert*-Bu$_3$Si)$_2$·C$_6$D$_6$ compounds (**34A** and **34B**) suitable for X-ray crystallography (Fig. 17). The skeleton has different Si–Si distances: Si–Si bond lengths of 2.320 and 2.315 Å and Si–Si–Si bond angles of 59.9° for **34A**; Si–Si bond lengths of 2.326 and 2.341 Å and Si–Si–Si bond angles of 60.4° for **34B**. These Si–Si bond lengths are somewhat longer than the calculated value of 2.314 Å for Si$_4$H$_4$ owing to steric reasons; however, they are apparently shorter than the typical Si–Si single bond. The exocyclic Si–Si bonds (2.355 and 2.365 Å for **34A**, 2.371 and 2.356 Å for **34B**) are stretched relative to the normal Si–Si bond.

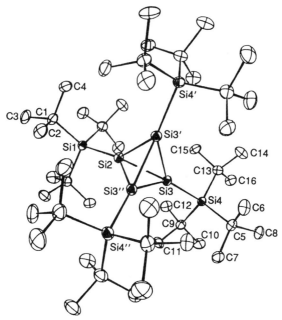

FIG. 17.   ORTEP drawing of tetrasilatetrahedrane (**34**). Reprinted with permission from Wiberg *et al.* (*10*), *Angew, Chem., Int. Ed. Engl.*, **1993**, *32*, 1054. Copyright by VCH.

VIII

OUTLOOK

Since the discovery of octasilacubane in 1988 and hexagermaprismane in 1989, the chemistry of polyhedranes composed of higher row group 14 elements has been growing rapidly. The isolation of hexasilaprismane and tetrasilatetrahedrane may open routes to the still elusive hexasilabenzene, tetrasilacyclobutadiene, and disilyne with a Si≡Si triple bond. Also, it is quite interesting whether clusters such as Si_{60} and Ge_{60} corresponding to C_{60} (fullerene) can exist (*48*). Laser vaporization of elemental silicon was reported to form various silicon clusters including Si_{60}; however, structural characterization has not yet been established (*49*). The synthesis of such giant clusters composed of higher row group 14 elements seems to be the next experimental target (*14b,50*).

ACKNOWLEDGMENTS

We are grateful to Dr. T. Yatabe, Dr. C. Kabuto, H. Yamazaki, and S. Doi for collaborations. We are particularly grateful to Professors Shigeru Nagase and Robert West for helpful discussions and useful advice. This work was financially supported by the Ministry of

Education, Science and Culture of Japan (Specially Promoted Research No. 02102004). We also thank the ASAI Germanium Research Institute for the gift of tetrachlorogermane.

REFERENCES

(1) Greenberg, A.; Liebman, J. F. In "Strained Organic Molecules"; Academic Press: New York, 1978.
(2) Balaban, A. T.; Banciu, M.; Ciorba, V. In "Annulenes, Benzo-, Hetero-, Homo-Derivatives, and Their Valence Isomers"; CRC Press: Boca Raton, Florida, 1987.
(3) Matsumoto, H.; Higuchi, K.; Hoshino, Y.; Koike, H.; Naoi, Y.; Nagai, Y. *J. Chem. Soc., Chem. Commun.* **1988**, 1083.
(4) Sekiguchi, A.; Kabuto, C.; Sakurai, H. *Angew. Chem., Int. Ed. Engl.* **1989**, *28*, 55.
(5) Sekiguchi, A.; Sakurai, H. In "The Chemistry of Inorganic Ring Systems"; Steudel, R., Ed.; Elsevier: New York, 1992; Chapter 7; Sakurai, H.; Sekiguchi, A. In "Frontiers of Organogermanium, -Tin and -Lead Chemistry"; Lukevics, E.; Ignatovich, L., Eds.; Latvian Institute of Organic Synthesis: Riga, 1993.
(6) Sekiguchi, A.; Yatabe, T.; Kamatani, H.; Kabuto, C.; Sakurai, H. *J. Am. Chem. Soc.* **1992**, *114*, 6260.
(7) Sekiguchi, A.; Yatabe, T.; Kabuto, C.; Sakurai, H. *J. Am. Chem. Soc.* **1993**, *115*, 5853.
(8) Matsumoto, H.; Higuchi, K.; Kyushin, S.; Goto, M. *Angew. Chem., Int. Ed. Engl.* **1992**, *31*, 1354.
(9) Furukawa, K.; Fujino, M.; Matsumoto, N. *Appl. Phys. Lett.* **1992**, *60*, 2744.
(10) Wiberg, N.; Finger, C. M. M.; Polborn, K. *Angew. Chem., Int. Ed. Eng.* **1993**, *32*, 1054.
(11) Clabo, D. A.; Jr.; Schaefer III, H. F. *J. Am. Chem. Soc.* **1986**, *108*, 4344.
(12) Sax, A.F.; Kalcher, J. *J. Chem. Soc., Chem. Commun.* **1987**, 809; Sax, A. F.; Kalcher, J. *J. Comput. Chem.* **1989**, *10*, 309.
(13) Nagase, S.; Nakano, M. *Angew. Chem., Int. Ed. Engl.* **1988**, *27*, 1081.
(14a) Nagase, S.; Kobayashi, K.; Nagashima, M. *J. Chem. Soc., Chem. Commun.* **1992**, 1302.
(14b) Nagase, S. *Pure Appl. Chem.* **1993**, *65*, 675.
(15a) Nagase, S.; Kudo, T.; Aoki, M. *J. Chem. Soc., Chem. Commun.* **1985**, 1121.
(15b) Sax, A.; Janoschek, R. *Angew, Chem., Int. Ed. Engl.* **1986**, *25*, 651.
(15c) Sax, A.; Janoschek, R. *Phosphorus Sulfur* **1986**, *28*, 151.
(15d) Nagase, S.; Kobayashi, K.; Kudo, T. *Main Group Met. Chem.* **1994**, *17*, 171.
(16a) Nagase, S.; Nakano, M.; Kudo, T. *J. Chem. Soc., Chem. Commun.* **1987**, 60.
(16b) Nagase, S. *Angew. Chem., Int. Ed. Engl.* **1989**, *28*, 329.
(17) Nagase, S. *Polyhedron,* **1991**, *10*, 1299.
(18a) West, R. *Pure Appl. Chem.,* **1984**, *56*, 163.
(18b) Raabe, G.; Michl, J. *Chem. Rev.* **1985**, *85*, 419.
(18c) West, R. *Angew. Chem., Int. Ed. Eng.,* **1987**, *26*, 1201.
(18d) Raabe, G.; Michl, J. In "The Chemistry of Organic Silicon Compounds"; Patai, S.; Rappoport, Z. Eds.; Wiley: New York, 1989; Part 2.
(18e) Tsumuraya, T.; Batcheller, S. A.; Masamune, S. *Angew. Chem., Int. Ed. Engl.* **1991**, *30*, 902.
(19) First ionization potentials of Me_3E—EMe_3 in PE spectra (eV): Me_3Si—$SiMe_3$ (8.69), Bock, H.; Ensslin, W. *Angew. Chem., Int. Ed. Engl.* **1971**, *10*, 404; Me_3Ge—$GeMe_3$ (8.57), Mochida, K.; Masuda, S.; Harada, Y. *Chem. Lett.* **1992**, 2281; Me_3Sn—$SnMe_3$ (8.20), Szepes, L.; Kóranyi, T.; Náray-Szabó, G.; Modelli, A.; Distefano, D. *J. Organomet. Chem.* **1981**, *217*, 35.

(20) Ando, W.; Tsumuraya, T. *J. Chem. Soc., Chem. Commun.* **1987**, 1514.
(21a) Smith, C. L.; Gooden, R. *J. Organomet. Chem.* **1974**, *81*, 33.
(21b) Smith, C. L.; Pounds, J. *J. Chem. Soc., Chem. Commun.* **1975**, 910.
(22) Sekiguchi, A.; Yamazaki, H.; Sakurai, H. unpublished results.
(23) Sekiguchi, A.; Yamazaki, H.; Sakurai, H. unpublished results.
(24) Gomberg, M. *Recl. Trav. Chim.* **1929**, *48*, 847; Gomberg, M.; Bachmann, W. E. *J. Am. Chem. Soc.* **1927**, *49*, 236.
(25) Sekiguchi, A.; Yatabe, T.; Naito, H.; Kabuto, C.; Sakurai, H. *Chem. Lett.* **1992**, 1697.
(26) Biernbaum, M.; West, R. *J. Organomet. Chem.* **1977**, *131*, 179; Helmer, B. J.; West, R. *Organometallics*, **1982**, *1*, 1458.
(27) Kawase, T.; Batcheller, S. A.; Masamune, S. *Chem. Lett.* **1987**, 227.
(28) Kabe, Y.; Kuroda, M.; Honda, Y.; Yamashita, O.; Kawase, T.; Masamune, S. *Angew. Chem., Int. Ed. Engl.* **1988**, *27*, 1725.
(29a) Matsumoto, H.; Sakamoto, A.; Minemura, M.; Sugaya, K.; Nagai, Y. *Bull. Chem. Soc. Jpn.* **1986**, *59*, 3314.
(29b) Kyushin, S.; Kawabata, M.; Sakurai, H.; Matsumoto, H.; Miyake, M.; Sato, M.; Goto, M. *Organometallics*, **1994**, *13*, 795.
(30) Kyushin, S.; Kawabata, M.; Yagihashi, Y.; Matsumoto, H.; Goto, M. *Chem. Lett.* **1994**, 997.
(31) Sekiguchi, A.; Naito, H.; Kabuto, C.; Sakurai, H. *Nippon Kagaku Kaishi (J. Chem. Soc. Jpn., Chem. Ind)* **1994**, 248.
(32) Sekiguchi, A. Naito, H.; Nameki, H.; Ebata, K.; Kabuto, C.; Sakurai, H. *J. Organomet. Chem.* **1989**, *368*, C1.
(33) Sekiguchi, A.; Kamatani, H.; Sakurai, H. unpublished results.
(34) Sekiguchi, A.; Yatabe, T.; Kabuto, C.; Sakurai, H. unpublished results.
(35) Sita, L. R.; Kinoshita, I. *Organometallics*, **1990**, *9*, 2865.
(36) Sita, L. R.; Bickerstaff, R. D. *J. Am. Chem. Soc.* **1989**, *111*, 6454.
(37) Sita, L. R.; Kinoshita, I. *J. Am. Chem. Soc.* **1991**, *113*, 1856.
(38) Fleischer, E. B. *J. Am. Chem. Soc.* **1964**, *86*, 3889.
(39) Furukawa, K.; Teramae, H.; Matumoto, N. *65th Annu. Meet. Jpn. Chem. Soc.* **1993**, 4F(Abstract 1), 342.
(40) Kanemitsu, Y.; Suzuki, K.; Uto, H.; Masumoto, Y.; Matsumoto, T.; Kyushin, S.; Higuchi, K.; Matsumoto, H. *Appl. Phys. Lett.* **1992**, *61*, 2446; Kanemitsu, Y.; Suzuki, K.; Uto, H.; Masumoto, Y.; Higuchi, K.; Kyushin, S.; Matsumoto, H. *Jpn. J. Appl. Phys.* **1993**, *32*, 408.
(41) Masamune, S.; Hanzawa, Y.; Murakami, S.; Bally, T.; Blount, J. F. *J. Am. Chem. Soc.* **1982**, *146*, 1150.
(42) Masamune, S.; Hanzawa, Y.; Williams, D. J. *J. Am. Chem. Soc.* **1982**, *104*, 6136.
(43) Karl, R. R.; Gallaher, K. L.; Wang, Y. C.; Bauer, S. H. unpublished results cited in *J. Am. Chem. Soc.* **1974**, *96*, 17.
(44) Karl, R. R.; Wang, Y. C.; Bauer, S. H. *J. Mol. Struct.* **1975**, *25*, 17.
(45) Sekiguchi, A.; Ebata, K.; Kabuto, C.; Sakurai, H. unpublished results.
(46) Gleiter, R.; Schafer, W. *Acc. Chem. Res.* **1990**, *23*, 369.
(47) Maier, G. *Angew. Chem. Int. Ed. Engl.* **1988**, *27*, 309.
(48) Jarrold, M. F. *Science* **1991**, *252*, 1085; Lange, T.; Martin, T.P. *Angew. Chem. Int. Ed. Engl.* **1992**, *31*, 172; Zybill, C. *Angew. Chem. Int. Ed. Engl.* **1992**, *31*, 173.
(49) Zhang, Q.-L.; Liu, Y.; Curl, R. F.; Tittel, F. K.; Smalley, R. E. *J. Chem. Phys.* **1988**, *88*, 1670.
(50) Nagase, S.; Kobayashi, K. *Chem. Phys. Lett.* **1991**, *187*, 291.

Homometallic and Heterometallic Transition Metal Allenyl Complexes: Synthesis, Structure, and Reactivity

SIMON DOHERTY, JOHN F. CORRIGAN, and
ARTHUR J. CARTY

Guelph-Waterloo Centre for Graduate Work in Chemistry
Waterloo Campus, Department of Chemistry
University of Waterloo, Waterloo
Ontario, Canada N2L 3G1

ENRICO SAPPA

Dipartimento di Chimica Inorganica, Chimica Fisica e Chimica dei Materiali
Universita di Torino, Via Pietro Giuria 7, I-10125 Torino Italy

I

INTRODUCTION

Over the past 30 years a diverse array of unsaturated C_1 ($=CR^1R^2$, $\equiv CR^1$) and C_2($=C=CR^1R^2$, $-CR^1=CR^2R^3$, $C\equiv CR$) hydrocarbyl-containing transition metal complexes has been described (*1a–p*). Indeed, the synthesis, chemical reactivity, stoichiometric and catalytic applications of such compounds have played a prominent role in the rapid development of organometallic chemistry during this time (2). In sharp contrast to the effort devoted to C_1 and C_2 ligands, complexes bearing C_3 fragments have received far less attention. We also note in passing that although a strong theoretical base for understanding structure and reactivity for unsaturated C_1 and C_2 ligands has been developed (3), there have to our knowledge been no in-depth theoretical studies of allenyl, propargyl, or allenylidene reactivity. Of the unsaturated C_3 hydrocarbyl ligands I–IV (Fig. 1), only mononuclear and to a lesser extent polynuclear derivatives of the allyl group I have been thoroughly investigated (4a,b). Metal complexes of the propargyl ligand IV have come under increased scrutiny because of their synthetic utility (5a,b). The remaining C_3 hydrocarbyl groups II (allenyl), III (allenylidene), V (propenylidyne), and VI (propynylidyne) have been less explored.

This article is prompted by developments in the chemistry of allenyl complexes: their systematic synthesis, structural diversity, and patterns of chemical reactivity. The increasing attention being focused on allenyl and allenylidene compounds can be attributed to several factors: an upsurge in interest in cumulene ligands in general and electronic communi-

$$-CH_2-CH=CH_2 \qquad -CH=C=CH_2$$

I **II**

$$=C=C=CH_2 \qquad -CH_2-C\equiv CH$$

III **IV**

$$\equiv C-CH=CH_2 \qquad \equiv C-C\equiv CH$$

V **VI**

$$M-CH=C=CH_2 \qquad M-CH_2-C\equiv CH$$
$$\alpha \quad \beta \quad \gamma \qquad\qquad \alpha \quad \beta \quad \gamma$$

VII **VIII**

FIG. 1. Unsaturated C_3 hydrocarbyl ligands.

cation along metal–polyunsaturated chains; the recognition that unsaturated C_3 ligands other than allyl may have utility as carbon transfer agents in synthesis; the interest of organometallic chemists in probing the reactivity of more complex hydrocarbyls at polynuclear centers; and the possibility that C_3 ligands such as **II** and **III** may play a role in carbon–carbon chain growth processes where the constituent fragments (CH_2, CH, C) are thought to be actively involved. Although the review concentrates on allenyl complex chemistry, discussion of related propargyl (**IV**) and allenylidene (**III**) systems is included where necessary. Literature up to early 1993 is cited. (See note added in proof on page 130 for references up to 1994.)

II

RELATIONSHIPS AMONG ALLENYL, PROPARGYL, AND ALLENYLIDENE COMPLEXES

Allenyl complexes of the type $[ML_n(CH{=}C{=}CH_2)]$ are tautomeric with the propargyl species $[ML_n(CH_2{-}C{\equiv}CH)]$. The σ-bound propargyl complexes have a rich and varied chemistry (*5a,b*), the early developments of which included protonation (*6a,b*) and cycloaddition reactions (*7*). The allenyl to propargyl rearrangement is well known in organic chemistry (*8*), but only recently was the equivalent organometallic tautomerization demonstrated (*9a–c*). Pathways for this transformation include 1,3-hydrogen and 1,3-metal sigmatropic rearrangements. Kinetic measurements favor the latter mechanism, with estimates for the barrier to activation in the region of 19–23 kcal/mol. Preliminary studies indicate a thermodynamic preference for the allenyl tautomer, arising from the stronger M–C(allenyl) bond relative to that of M–C(propargyl).

The allenylidene $[M{=}C{=}C{=}CR^1R^2]$ unit represents a member of the class of carbon-rich unsaturated hydrocarbyl fragments, and it is capable of binding one to four metal centers. In sharp contrast to the rapidly developing chemistry of mononuclear allenylidene complexes, the scarcity of bi- and polynuclear species of this type has hindered the development of an extensive chemistry. Selected examples from the allenylidene literature are cited in this article owing to the close structural analogy with allenyl derivatives. The reader is referred to a review article by Bruce covering the relevant literature of this metallacumulene up to 1989 (*1a*).

III

SYNTHESIS

A. Preparation of Mononuclear Propargyl, Allenyl, and Allenylidene Complexes

1. *Nucleophilic Attack by Metal Anions on Propargyl Halides*

The most direct and successful synthesis of metal propargyl complexes involves nucleophilic attack of a metal carbonylate anion at a propargyl halide (*9a–c,10*) [Eqs. (1) and (2)]. The propargylic products isolated

$$[CpRu(CO)_2]^- + XCH_2C\equiv CPh \longrightarrow CpRu(CO)_2(\eta^1\text{-}CH_2C\equiv CPh) \qquad (1)$$

1

$$[CpW(CO)_3]^- + BrCH_2C\equiv CH \longrightarrow CpW(CO)_3(\eta^1\text{-}CH_2C\equiv CH) \qquad (2)$$

2

(e.g., **1,2**) are often stable; however, in several instances rearrangement to the tautomeric allenyl derivative has been observed (*vide infra*) (*11*). Collman and co-workers reported that oxidative addition of α-halo acetylenes to iridium(I) Vaska's type complexes proceeded via an apparent S_N2' process affording the six-coordinate allenyl complexes [IrCl$_2$(CO)(P-Ph$_3$)$_2$(η^1-HC=C=CR^{1}R^2)] (R^1 = R^2 = H, Me, R^1 = H, R^2 = Me) (**3**) [Eq. (3)] (*12*). Other transition metal centers susceptible to oxidative addition

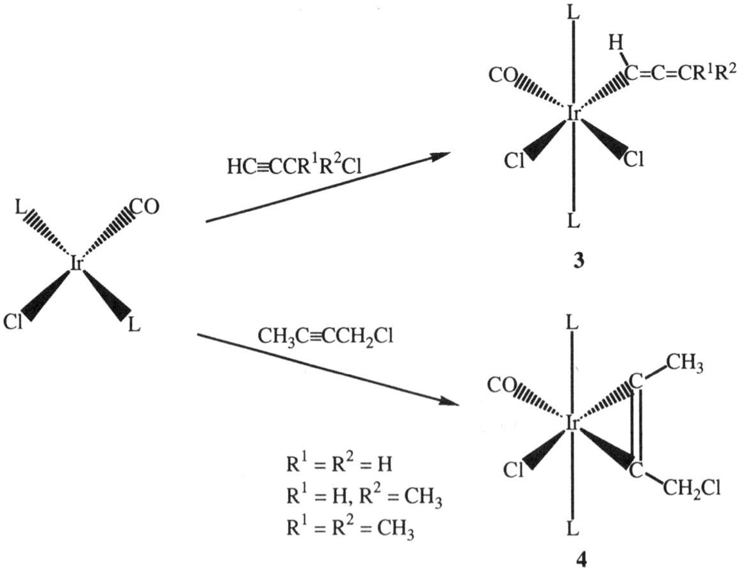

$$(3)$$

R^1 = R^2 = H
R^1 = H, R^2 = CH$_3$
R^1 = R^2 = CH$_3$

form similar allenyl products $[PtCl(PPh_3)_2(\eta^1\text{-}HC{=}C{=}CR^1R^2)]$ (5) and $[Co(dmgh)_2(py)(\eta^1\text{-}HC{=}C{=}R^1R^2)]$ (dmgh = dimethylglyoxime; py = pyridine) (6) [Eqs. (4) and (5)]. The formation of σ-bonded allenyl com-

$$Pt(PPh_3)_4 \ + \quad \begin{array}{c} HC{\equiv}CCR^1R^2CCl \\ \text{or} \\ HClC{=}C{=}CR^1R^2 \end{array} \quad \longrightarrow \quad \begin{array}{c} Cl \diagdown \quad \diagup PPh_3 \\ Pt \\ Ph_3P \diagup \quad \diagdown C{=}C{=}CR^1R^2 \\ H \end{array} \qquad (4)$$

5

$$[Co] \ + \ HC{\equiv}CCR^1R^2X \quad \longrightarrow \quad [Co] \diagdown \begin{array}{c} H \\ C{=}C{=}CR^1R^2 \\ \mathbf{6} \end{array}$$

$$\begin{array}{c} R^1 = R^2 = H \\ R^1 = R^2 = Me \end{array} \qquad (5)$$

[Co] = pyridinebis(dimethylglyoxime)cobalt(I)hydride

plexes occurs for unhindered, terminal acetylenes only. More bulky reagents give the η^2-coordinated complexes $[IrCl(CO)(PPh_3)_2(\eta^2\text{-}CH_3C{\equiv}CCH_2Cl)]$ (4), the stabilities of which depend largely on the alkyne substituent. The same allenyl-containing products 3 were isolated from the reaction of an appropriate chloroallene with the corresponding metal complex, presumably via an oxidative addition pathway.

2. From η^2-Alkyne Complexes

Other methods for accessing propargyl and allenyl complexes involve hydride abstraction from η^2-alkyne compounds. Thus the complex $[Re(\eta^5\text{-}C_5Me_5)(CO)_2(\eta^2\text{-}MeC{\equiv}CMe)]$ (7) afforded the cationic propargyl complex $[Re(\eta^5\text{-}C_5Me_5)(CO)_2(\eta^3\text{-}CH_2C{\equiv}CCH_3)][PF_6]$ [8(I)] (Scheme 1) (13). Based on ^{13}C NMR spectroscopic characteristics, Casey proposed that the propargylic structure 8(I) contains a significant contribution from the allenyl resonance form 8(II).

Watson and Bergman first provided evidence that four-electron donor π-bound alkynes of the cationic transition metal complex [Mo-$(CO)Cp(\eta^2\text{-}CH_3C{\equiv}CCH_3)_2]^+$ could be deprotonated to give η^2-allenyl-containing species (14) by effecting catalytic deuterium–hydrogen exchange between free but-2-yne and acetone-d_6 on addition of NEt_3. Later Green and co-workers demonstrated the generality of this process by preparing a series of $\eta^2(3e)$-bonded allenyl complexes (15). They showed that the addition of NEt_3 to an acetone-d_6 solution of $[MoCp\{P(O\text{-}Me)_3\}_2(\eta^2\text{-}PhC{\equiv}CCH_2Ph)]^+$ (9a) selectively deuterated the methylenic

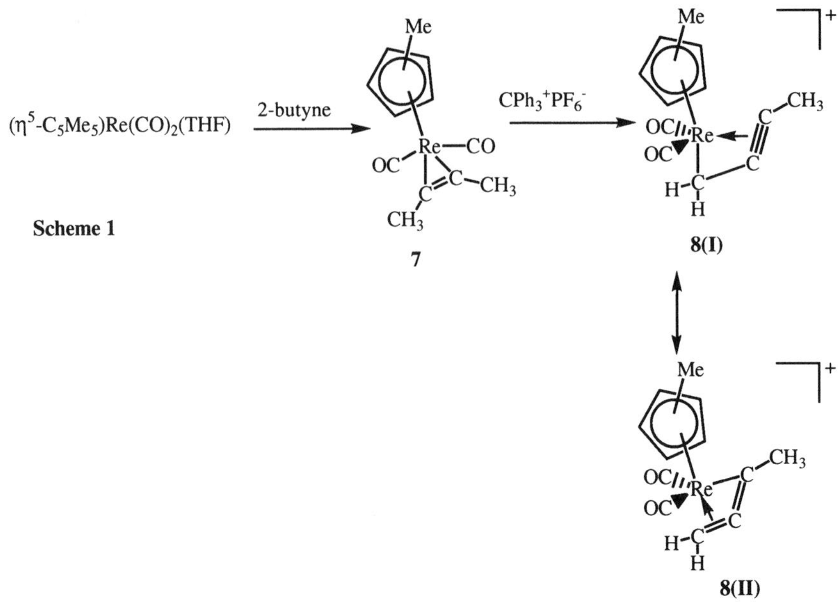

$(\eta^5\text{-}C_5Me_5)Re(CO)_2(THF)$ $\xrightarrow{\text{2-butyne}}$ **7** $\xrightarrow{CPh_3^+PF_6^-}$ **8(I)**

Scheme 1

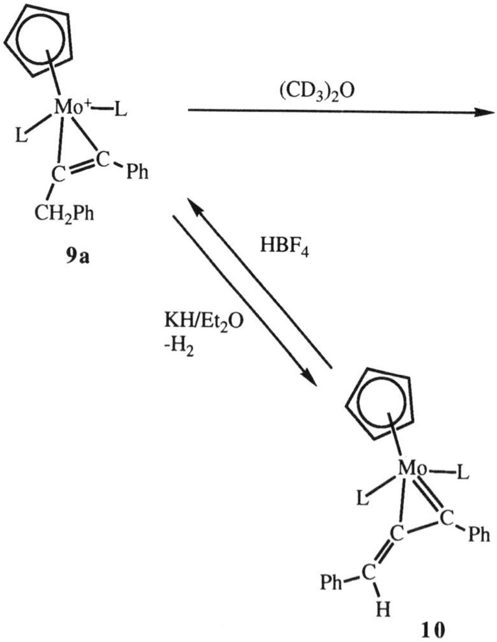

8(II)

SCHEME 1

9a $\xrightarrow{(CD_3)_2O}$ **b**

HBF$_4$

KH/Et$_2$O
-H$_2$

10

SCHEME 2

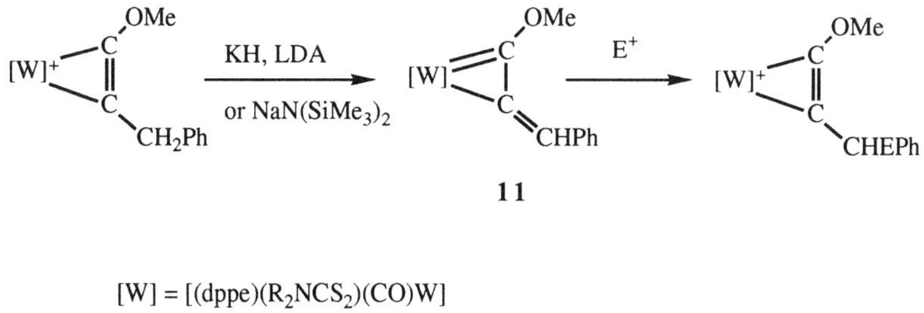

$$[W]^+ \underset{CH_2Ph}{\overset{OMe}{\underset{C}{\overset{C}{\rightthreetimes}}}} \quad \xrightarrow[\text{or } NaN(SiMe_3)_2]{KH, LDA} \quad [W] \underset{CHPh}{\overset{OMe}{\underset{C}{\overset{C}{\rightthreetimes}}}} \quad \xrightarrow{E^+} \quad [W]^+ \underset{CHEPh}{\overset{OMe}{\underset{C}{\overset{C}{\rightthreetimes}}}}$$

11

$[W] = [(dppe)(R_2NCS_2)(CO)W]$

SCHEME 3

protons, whereas treatment with $KH/^tBuOH-Et_2O$ resulted in elimination of dihydrogen and formation of the $\eta^2(3e)$-allenyl complex $[MoCp\{P(O-Me)_3\}_2\{\eta^2-C(Ph)C=CHPh\}]$ (**10**) (Scheme 2), its isolation supporting the proposal of an intermediate allenyl complex in the deuteration.

Templeton employed a similar procedure to generate allenyl complexes to compare their reactivity with the related enolate counterparts. Deprotonation of the η^2-alkyne ligand in $[W(CO)(dppe)(dtc)(\eta^2-MeOC\equiv CCH_2Ph)]^+$ [dppe $= Ph_2P(CH_2)_2PPh_2$; dtc $= R_2NCS_2^-$, $R_2 = Me_2$, $^iPr_2$, C_4H_4] gave the corresponding neutral η^2-allenyl complexes $[W(CO)(dppe)(dtc)(\eta^2-MeOC=C=CHPh)]$ (**11**), present as a mixture of two isomers corresponding to the relative orientation of the allenyl fragment with respect to the metal center (*16a*). Complex **11** was shown to readily add electrophiles at C_γ, providing a convenient route to functionalized η^2-alkyne complexes (Scheme 3). Similar chemistry was subsequently extended to $[WI(CO)Tp'(\eta^2-PhC\equiv CCH_3)]$ [Tp' $=$ hydridotris(3,5-dimethylpyrazolyl)borate], producing the highly nucleophilic $\eta^2(3e)$-allenyl complex $[WI(CO)Tp'(\eta^2-PhC=C=CH_2)]^-$ (**12**) (*16b*).

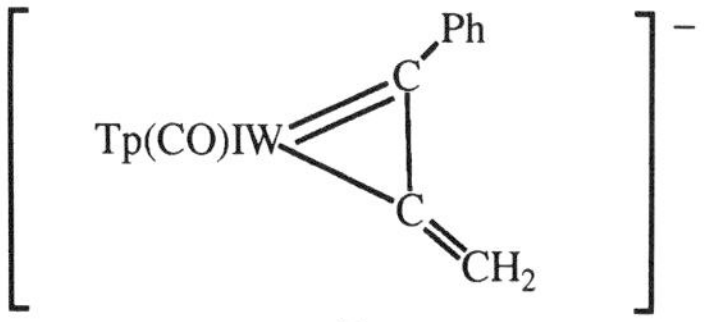

$$\left[Tp(CO)IW \underset{C}{\overset{C}{\rightthreetimes}} \begin{matrix} Ph \\ \\ CH_2 \end{matrix} \right]^-$$

12

3. Early Transition Metal Propargyl and Allenyl Complexes

Only recently have early transition metal propargylic complexes been recognized. The lanthanide alkyls $[LnCH(SiMe_3)_2(\eta^5-C_5Me_5)_2]$ (Ln $=$ La, Ce) react with the 2-alkynes $MeC\equiv CR$ (R $=$ Me, Et, ^nPr) to afford 1,2-disubstituted 3-alkylidenecyclobutenes. The first step in this catalytic cyclodimerization reaction was clearly indicated by NMR spectroscopy to be propargylic metallation of the α-methyl group to give [Ln(η^5-C$_5$Me$_5$)$_2$(CH$_2$C≡CMe)] (**13**), liberating CH$_2$(SiMe$_3$)$_2$ [Eq. (6)] (*17*). A

$$(\eta^5\text{-}C_5Me_5)_2LnCH(SiMe_3)_2 \longrightarrow (\eta^5\text{-}C_5Me_5)_2Ln(CH_2C≡CMe)$$

$$+ \qquad\qquad\qquad\qquad\qquad\qquad \textbf{13}$$

$$MeC≡CMe \qquad\qquad\qquad\qquad\qquad + \qquad\qquad\qquad (6)$$

$$CH_2(SiMe_3)_2$$

$$Ln = La, Ce$$

similar σ-bond metathesis was established as the dominant pathway for reaction of the sterically demanding alkenyl complex [Zr(η^5-C$_5$Me$_5$)$_2$(η^1-MeC=CMe$_2$)] with but-2-yne. In contrast to the double-insertion products [ZrCp$_2$(η^5-CH$_2$C(Me)=C(Me)C(Me)=CHMe)]$^+$ (**14**) observed with the sterically less crowded [ZrCp$_2$(η^1-MeC=CMe$_2$), a σ-bond metathesis reaction between [Zr(η^5-C$_5$Me$_5$)$_2$(η^1-MeC=CMe$_2$)] and but-2-yne afforded the propargylic complex [Zr(η^5-C$_5$Me$_5$)$_2$(η^3-CH$_2$C≡CMe)]$^+$ (**15**) and 2-methyl-2-butene (Scheme 4) (*18*). These σ-bond metathesis reactions have not been fully exploited as possible synthetic routes to propargylic and allenylic complexes, and early transition metal complexes of this type remain elusive. An absence of structural data for the complexes also renders the nature of their bonding somewhat speculative.

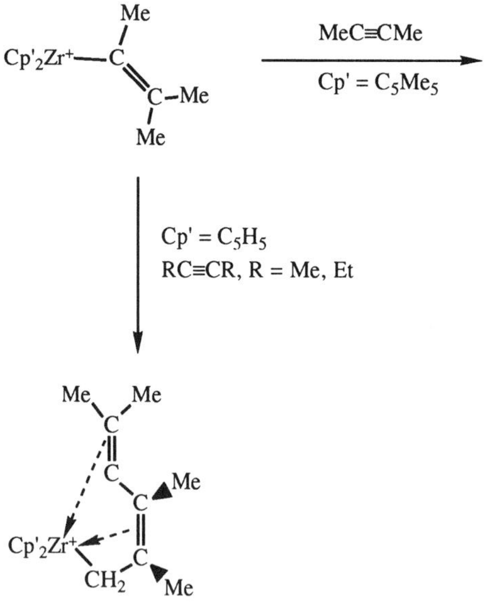

14

SCHEME 4

A rather elegant entry to these early transition metal complexes has been developed by Wojcicki and co-workers (*19*). Reacting propargylic Grignard reagents with $[Zr(CH_3)ClCp_2]$ and $[ZrCl_2Cp_2]$ gave the unusual η^3-allenyl complexes $[Zr(CH_3)Cp_2(\eta^3\text{-}PhC{=}C{=}CH_2)]$ (**16**) and $[ZrCp_2(\eta^1\text{-}CH_2C{\equiv}CPh)(\eta^3\text{-}PhC{=}C{=}CH_2)]$ (**17**), respectively [Eqs. (7) and (8)]. The spectroscopic properties of the former complex indicated

$$\begin{array}{c} Cp_2ZrClCH_3 \\ + \\ PhC{\equiv}CCH_2MgBr \end{array} \longrightarrow Cp_2Zr(CH_3)\{\eta^3\text{-}C(Ph){=}C{=}CH_2\} \qquad (7)$$

$$\mathbf{16}$$

$$\begin{array}{c} Cp_2ZrCl_2 \\ + \\ PhC{\equiv}CCH_2MgBr \end{array} \longrightarrow Cp_2Zr\{\eta\text{-}C(Ph){=}C{=}CH_2\}_2 \qquad (8)$$

$$\mathbf{17}$$

bonding characteristics consistent with an η^3-coordination mode (*vide supra*). We anticipate that such versatile preparative methods will pave the way for an explosive growth in the organometallic chemistry of early transition metal propargyl and allenyl complexes, a particularly promising area of research both for synthesis and catalysis.

A potentially powerful and widely applicable procedure for the preparation of σ-allenyl or σ-bound propargyl complexes involves the deprotonation of η^2-bound allene complexes (Scheme 5). During attempts to prepare Re σ-bound allenyl complexes from the η^2-bound allene precursor [Re-Cp(NO)(PPh$_3$)(η^2-CH$_2$=C=CH$_2$)] (**18**), Gladysz and co-workers demonstrated (*9b*) that deprotonation by ^{t}BuOK gave the methylacetylide complex [ReCp(NO)(PPh$_3$)(η^1-C≡CCH$_3$)] (**19**), via the intermediacy of a σ-bound allenyl complex. Treatment of **18** with the stronger base methyllithium gave [ReCp(NO)(PPh$_3$)(η^1-CH=C=CH$_2$)] (**20**), a σ-bound allenyl complex which was readily transformed to **19** on treatment with HBF$_4$ and ^{t}BuOK, via the intermediate η^2-bound alkyne complex [ReCp(NO)(PPh$_3$)(η^2-CH≡CCH$_3$)]$^+$. The same studies provided a further example of a propargyl to allenyl rearrangement [Eq. (9)]. [ReCp(NO)

$$\underset{\mathbf{21}}{\left[\text{NO}{-}\overset{\text{Re}}{\underset{\text{PPh}_3}{\big|}}\,,\ \underset{\text{H}^{-}C{\equiv}C}{\overset{CH_2}{}}\right]} \xrightarrow{80\ ^\circ C,\ C_6D_6} \underset{\mathbf{20}}{\left[\text{NO}{-}\overset{\text{Re}}{\underset{\text{PPh}_3}{\big|}}\,,\ \underset{\text{H}}{\overset{H_{/\!/}{}_{\cdot}C{=}C{=}C{-}H}{}}\right]} \qquad (9)$$

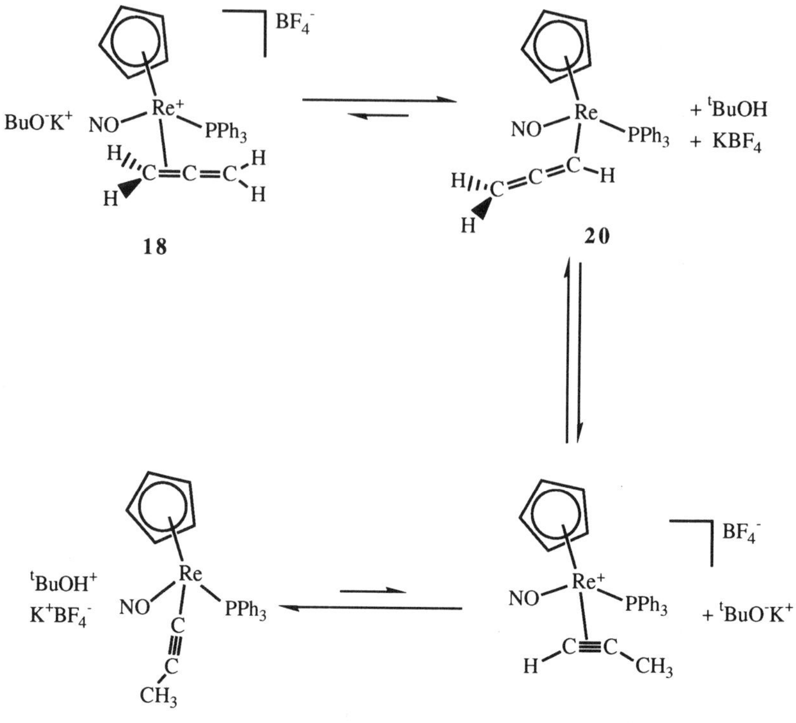

SCHEME 5

$(PPh_3)(\eta^1\text{-}CH_2C\equiv CH)]$ **(21)** was prepared by treating propargyl tosylate with the anion $[ReCp(NO)(PPh_3)]^-$ at $-15°C$. The thermal equilibration of propargylic **21** and allenylic **20** was shown to be slow on the time scale of η^2-allene complex deprotonation.

4. *Mononuclear Allenylidene Complexes*

The expectation that organometallics containing ligands with conjugated and delocalized π systems will provide useful applications for the synthesis of new polyunsaturated organic substrates or serve as precursors to organometallic polymers displaying electronic conductivity and nonlinear optical properties has stimulated intense research efforts into their synthesis and properties.

Mononuclear allenylidene complexes have been prepared by either (i) reaction of a deprotonated propargyl alcohol with a binary metal carbonyl or (ii) dehydration of the acetylenic alcohol $HC\equiv CCR_2OH$ with an appro-

$$M(CO)_6 + \left[C\equiv C-CPh_2O \right]^{2-} \longrightarrow (CO)_5M = C \underset{O^-}{\overset{}{\diagdown}} \underset{C\equiv C-CPh_2O^-}{\overset{}{}}$$

$$-CO \downarrow$$

$$\left[(CO)_5M-C\equiv C-CPh_2O \right]^{2-}$$

$$\downarrow COCl_2$$

$$(CO)_5M = C = C = CPh_2$$

22

SCHEME 6

priate metal fragment in the presence of $NaPF_6$. The latter process has been reported to occur via initial alkyne coordination, subsequent tautomerization to the η^1-hydroxyvinylidene intermediate, and dehydration (*21c*). Berke has successfully prepared various mononuclear allenylidene complexes by reacting binary metal carbonyls (or lightly stabilized analogs) with deprotonated 2-propyn-1-ols,$[C\equiv CCR_2O]^{2-}$ (*20a,b*). The intermediate acetylide dianions readily deoxygenate on treatment with $COCl_2$ (Scheme 6). For example, reaction of $W(CO)_6$ and $[C\equiv CCPh_2O]^{2-}$ followed by treatment of the reaction mixture with phosgene gave $[W(CO)_5(\eta^1\text{-}C=C=CPh_2)]$ (**22**) from which the dimer $[\{W(CO)_5\}_2(\mu\text{-}\eta^1,\eta^1\text{-}C=C=CPh_2)]$ (**23**) could be obtained by metal fragment condensation

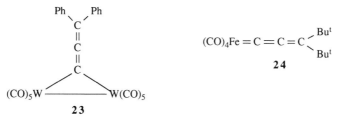

(Section III,B,4) (*20a*). A similar procedure proved successful for the synthesis of $[Fe(CO)_4(\eta^1\text{-}C=C=C^tBu_2)]$ (**24**) from the acetonitrile adduct $Fe(CO)_4(MeCN)$ (*20b*).

Selegue originally demonstrated that propargyl alcohols could be dehydrated by the metal fragments $[RuClCpL_2]$ [L = $P(OMe)_3$, PPh_3 (*21a*), resulting in the isolation of the η^1-allenylidene complexes $[RuCpL_2\{\eta^1\text{-}C=C=CR_2\}]^+$ (**25**) [Eq. (10)]. Analogous reactions with $HC\equiv$

$$CpL_2RuCl + HC\equiv CC(OH)R_2 \xrightarrow{NH_4PF_6} [CpL_2Ru=C=CR_2]^+PF_6^-$$

(10)

25a; L = P(OMe)$_3$, R = Ph
25b; L = PPh$_3$, R = Me

CCMe$_2$OH gave an apparent dimerization of the intermediate allen-ylidene [RuCp(PPh$_3$)$_2$\{η^1-C=C=CMe$_2$\}]$^+$, affording [Ru$_2$Cp$_2$(PPh$_3$)$_4$ (μ-C$_{10}$H$_{12}$)]$^{2+}$ (**26**) (*21b*).

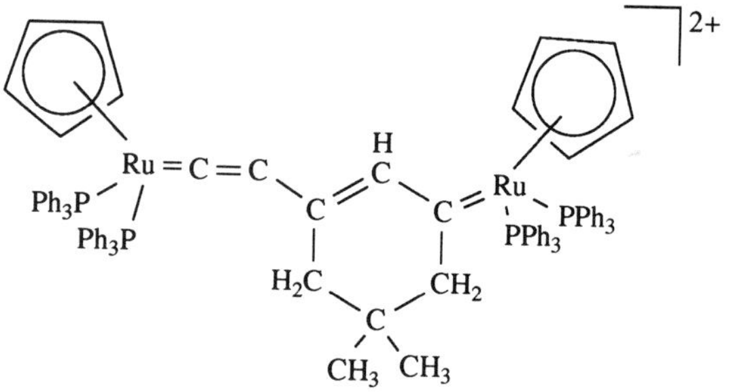

26

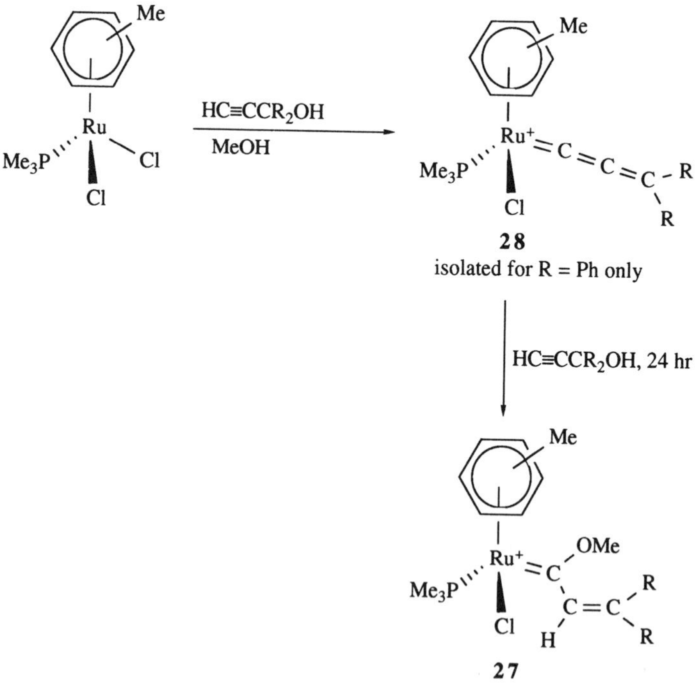

28
isolated for R = Ph only

HC≡CCR$_2$OH, 24 hr

27

SCHEME 7

Dixneuf and co-workers have performed detailed investigations into the analogous chemistry of the isoelectronic derivative $[RuCl_2(\eta^6\text{-arene})(PR_3)]$ (*21c–e*), and they revealed that by varying both the electronic and steric properties of the metal fragment markedly different reactivities from those observed by Selegue could be achieved. The less electron-rich [Ru-$Cl_2(\eta^6\text{-}C_6Me_6)(PMe_3)]$ was reported to react with $HC{\equiv}CCR_2OH$ (R $\neq$ H) and $NaPF_6$ in methanol to give the vinylcarbene complex [RuCl$(\eta^6\text{-}C_6Me_6)(PMe_3)\{\eta^1\text{-}C(OMe)CH{=}CR_2\}]^+$ (**27**) (*21c*). Shorter reaction times involving $HC{\equiv}CCPh_2OH$ enabled the isolation of the intermediate allenylidene cation $[RuCl(\eta^6\text{-}C_6Me_6)(PPh_3)\{\eta^1\text{-}C{=}C{=}CPh_2\}]^+$ (**28**) (Scheme 7). In related studies the ferrocenylallenylidene complex [Ru-Cl$(\eta^6\text{-}C_6Me_6)(PMe_3)\{\eta^1\text{-}C{=}C{=}CPh(C_5H_4)Fe(C_5H_5)\}]^+$ (**29**) was isolated from the reaction of $[RuCl_2(\eta^6\text{-}C_6Me_6)(PMe_3)]$ and $HC{\equiv}CCPhOH(C_5H_4)$-Fe(C_5H_5), with no evidence for the formation of the corresponding alkenylcarbene species [Eq. (11)] (*21e*). The enhanced stability of **29** was attributed to a contribution from an alternative acetylide resonance structure.

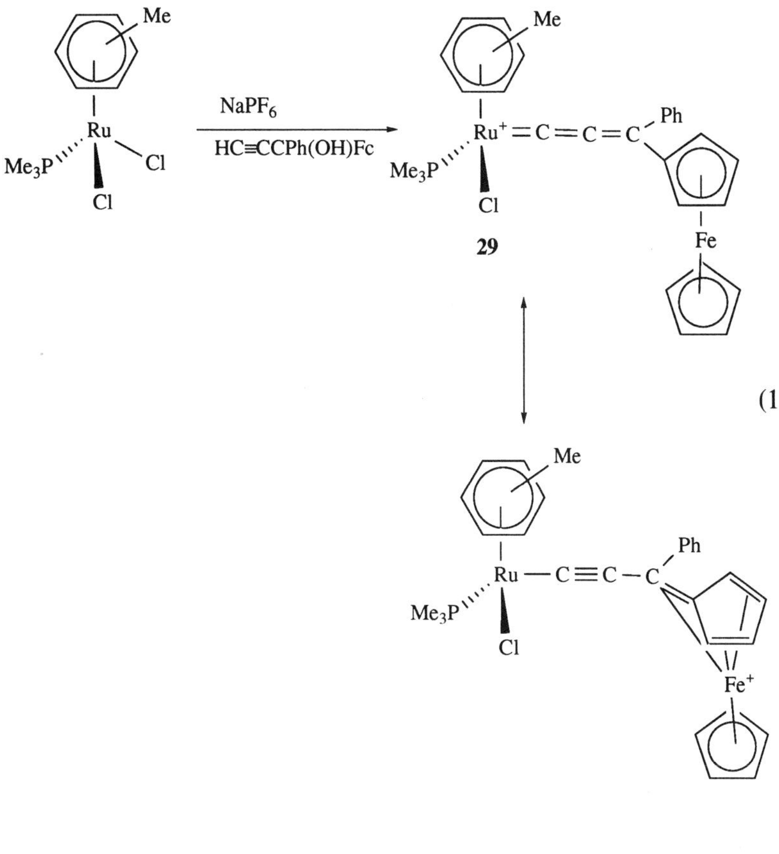

(11)

The more electron-rich fragment $[RuCl(NP_3)]^+$ $[NP_3 = N(CH_2\text{-}CH_2PPh_2)_3]$ reacts with the propargyl alcohols $HC\equiv CCPh_2OH$ and $HC\equiv CC\equiv CCPh_2OH$ to give the stable allenylidene $[RuCl(NP_3)\{\eta^1\text{-}C=C=CPh_2\}]^+$ (30) and alkenylallenylidene $[RuCl(NP_3)\{\eta^1\text{-}C=C=C(O\text{-}Me)CH=CPh_2\}]^+$ (31) monomers, respectively, reinforcing the impor-

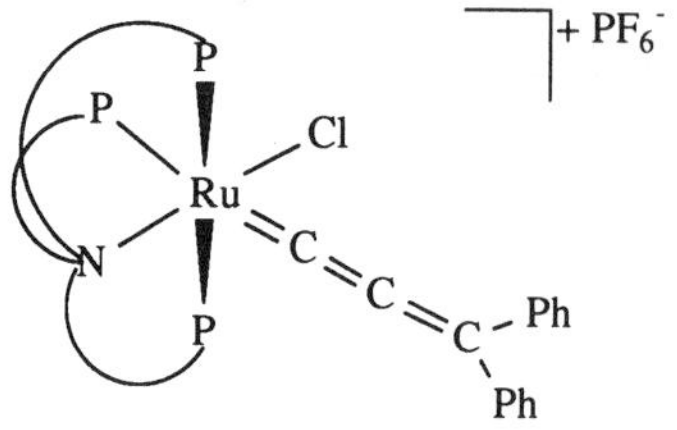

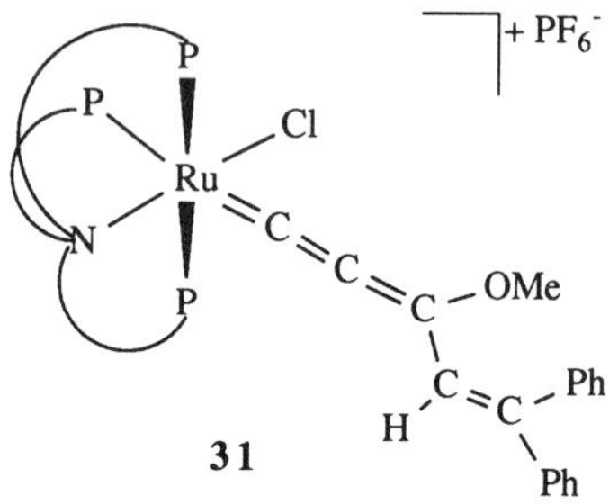

tance of the electronic and steric properties of the central metal fragment in these dehydration reactions (21d). More recently the bis-alkenyl allenylidene complex $[Ru(dppm)_2\{\eta^1\text{-}C=C=C(OMe)(CH=CPh_2)\}_2]^{2+}$ (33) was prepared from the trans-bisdiynyl ruthenium monomer $[Ru(dppm)_2\{C\equiv CC\equiv CCPh_2(OSiMe_3)\}_2]$ (32) (dppm = $Ph_2PCH_2PPh_2$) [Eq. (12)] (22).

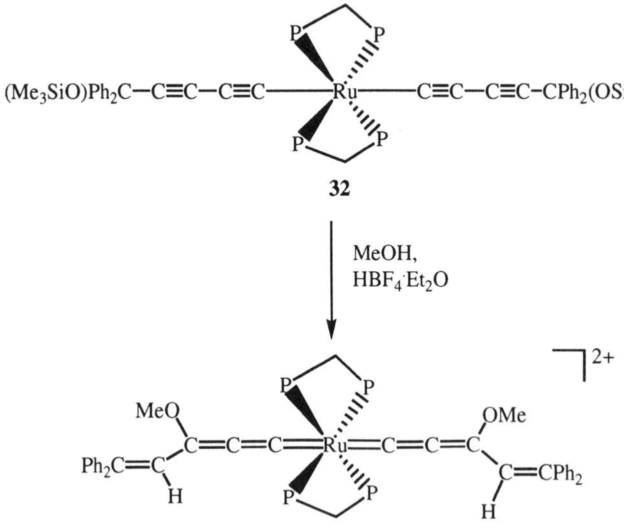

(12)

B. *Propargyl Complexes as Source of Allenyl Groups*

1. *From Propargyl Ligands*

Seyferth and co-workers were the first to apply the methodology of metal carbonylate anion attack at a propargyl halide for the synthesis of a series of dimeric thiolate-bridged allenyl complexes (*23*). Reaction of $[Fe_2(CO)_7(\mu\text{-}SR)]^-$ with a variety of propargyl halides afforded the binuclear μ-allenyl complexes $[Fe_2(CO)_6(\mu\text{-}SR)(\mu\text{-}\eta^1,\eta^2\text{-}RC{=}C{=}CR'_2)]$ (**34**) in exceptional yields (90–95%), conclusively demonstrating that nucleophilic attack proceeded via an S_N2' type mechanism (Scheme 8). Further attempts to extend this chemistry to the reaction of (allenyl) lithium reagents and $Fe_3(CO)_{12}$ followed by addition of electrophiles failed to afford allenyl-containing products, with only substituted ferroles being isolated (*24*).

The conversion of a dimeric π-bound methyl propargyl ether complex $[Mo_2(CO)_4Cp_2(\mu\text{-}\eta^1,\eta^2\text{-}CH{\equiv}CCH_2OMe)]$ (**35**) to the cationic allenyl complex $[Mo_2(CO)_4Cp_2(\mu\text{-}\eta^2, \eta^3\text{-}CH{=}C{=}CH_2]^+$ (**36**) has been described by Curtis *et al.* (*25*). Protonation of **35** with HBF_4 induced the loss of methanol and formation of the required complex. Alternatively the same complex was reported accessible via the acid-promoted elimination of methanol from $[Mo_2(CO)_4Cp_2\{\mu\text{-}\eta^2,\eta^2\text{-}MeO(H)C{=}C{=}CH_2\}]$, a rare example of an intact η^2,η^2-bound allene (Scheme 9).

The light yellow complex $[W(CO)_3Cp(\eta^1\text{-}CH{=}C{=}CH_2)]$ [**2(II)**] was isolated as the major product of the thermal transformation of its propargyl tautomer $[W(CO)_3Cp(\eta^1\text{-}CH_2C{\equiv}CH)]$ [**2(I)**] via one of two possible pathways: (i) a 1,3-hydrogen or (II) a 1,3-tungsten sigmatropic rear-

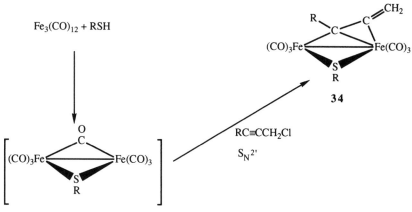

SCHEME 8

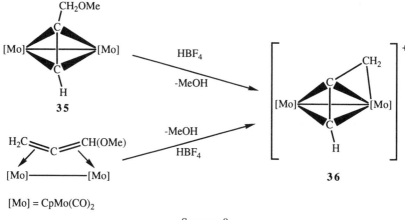

SCHEME 9

rangement [Eq. (13)], the latter being favored (*9a*). Furthermore it was demonstrated that *trans*-[W(CO)$_2$Cp{P(OMe)$_3$}(η^1-CH$_2$—C≡CH)] [**2(I)a**] was converted to a cis and trans isomeric mixture of [W(CO)$_2$Cp{P

$$ \text{2(I)} \xrightarrow[\text{MeCN}]{\Delta} \text{2(II)} \tag{13} $$

2(I)   2(II)

(OMe)$_3$}(η^1-CH=C=CH$_2$)] [**2(II)a**] providing more evidence for the 1,3 tungsten migration (*9c*). Phosphine-induced isomerization of the γ-substituted η^1-propargyl complex [PtBr(PPh$_3$)$_2$(η^1-CH$_2$C≡CPh)] (**37**) to the α-substituted η^1-allenyl complex [Pt(PMe$_3$)$_3$(η^1-CH=C=CH$_2$)]$^+$ (**38**) implicated a 1,3-platinum sigmatropic rearrangement as the likely mechanism (*26a–c*). Complex **37** was successfully converted to [Pt(PPh$_3$)$_2$(η^3-C(Ph)=C=CH$_2$)]$^+$ (**39b**), an η^3-propargyl/allenyl complex from which both the η^1-propargyl and η^1-allenyl complexes [Pt(CO)(PPh$_3$)$_2$(η^1-CH$_2$C≡CH)]$^+$ [**40(I)**] and [Pt(CO)(PPh$_3$)$_2$(η^1-CH=C=CH$_2$)]$^+$ [**40(II)**] were readily obtained via reaction with carbon monoxide (Scheme 10). Such transformations support the idea that η^3-propargyl/allenyl intermediates play a major role in propargyl/allenyl tautomerizations. Similar η^3-propargyl-bound complexes have been successfully prepared from η^1-allenyl precursors [Eq. (14)] (*26b*).

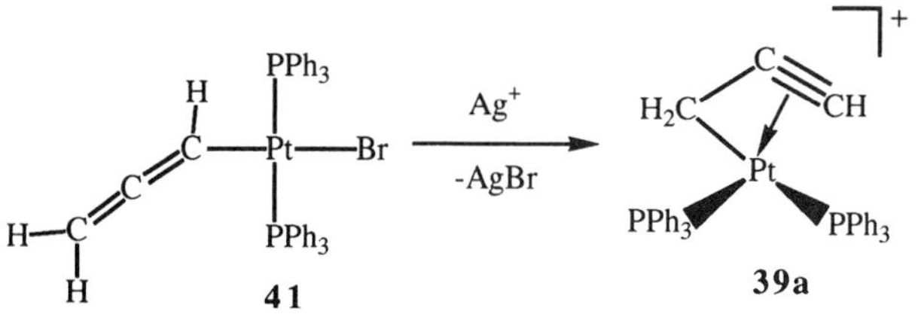

$$\begin{array}{c} \text{H}\quad\text{PPh}_3 \\ \text{C}\!-\!\text{Pt}\!-\!\text{Br} \\ \text{H}\!-\!\text{C}\quad\text{PPh}_3 \\ \text{H} \end{array} \quad\xrightarrow[\text{-AgBr}]{\text{Ag}^+}\quad \left[\begin{array}{c} \text{C} \\ \text{H}_2\text{C}\!=\!\!\!=\!\text{CH} \\ \text{Pt} \\ \text{PPh}_3\quad\text{PPh}_3 \end{array}\right]^+ \qquad (14)$$

$$\mathbf{41} \qquad\qquad\qquad \mathbf{39a}$$

Osella and co-workers have demonstrated the effective dehydroxylating ability of $Fe(CO)_5$. After refluxing $[Co_2(CO)_6(\mu\text{-}\eta^2,\eta^2\text{-}EtC\equiv C\text{-}CH(OH)CH_3)]$ with $Fe(CO)_5$ in acetone, high yields of the heterometallic dimer $[CoFe(CO)_6(\mu\text{-}\eta^2,\eta^3\text{-}MeHCC\equiv CEt)]$ (**42**) were isolated (*27a*) [Eq. (15)]. These investigators further showed that heterobimetallic propargyl complexes of the type $[FeCo(CO)_6(\mu\text{-}\eta^2,\eta^3\text{-}R_2CC\equiv CR)]$ (**42**) and [FeCo-

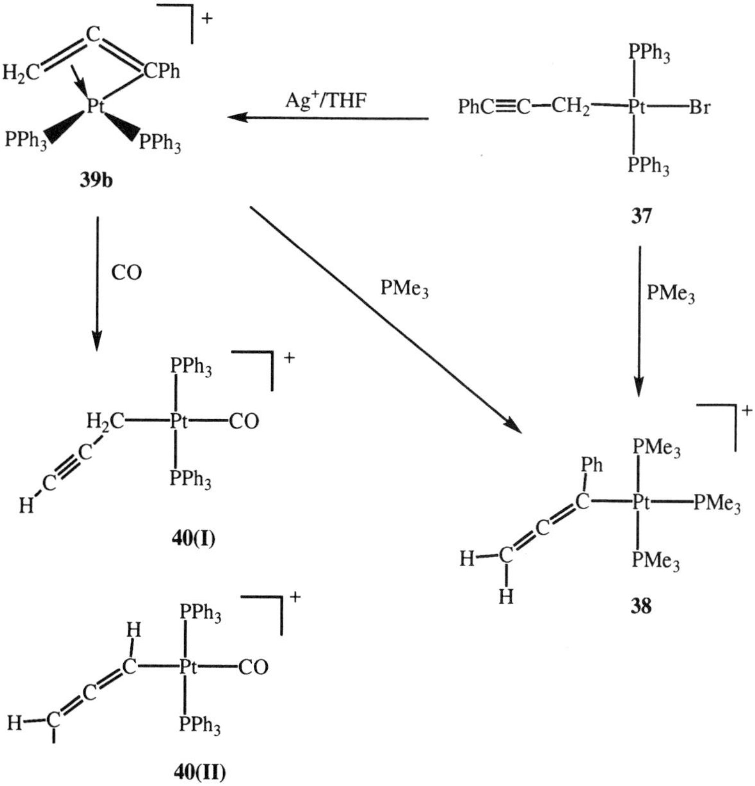

SCHEME 10

$[CO)_6(\mu-\eta^2,\eta^3-R_2CC\equiv CCR_2OH)]$ **(43)** were formed via metal atom exchange and dehydroxylation of the organic fragment in the mono- and diols of $[Co_2(CO)_6(\mu-\eta^2,\eta^2-(HO)R_2CC\equiv CR)]$ and $[Co_2(CO)_6(\mu-\eta^2,\eta^2-HOR_2CC\equiv CCR_2OH)]$ (Scheme 11). Double dehydroxylation of the latter afforded the six-electron donor butatriene ligand in $[Fe_2(CO)_6(\mu-\eta^3,\eta^3-R_2C=C=CR_2)]$ **(44)** (*27b*).

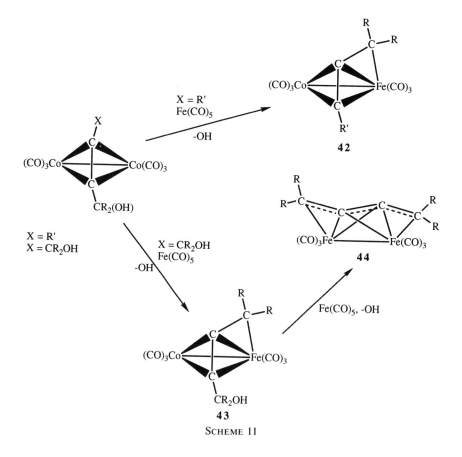

SCHEME 11

2. Metal Fragment Condensation Reactions with Propargylic Complexes

An increasingly popular route to homo- and heterometallic hydrocarbyl-containing clusters is the condensation of a metal carbonyl fragment with an organometallic complex (*1c,28*). To this end metal propargyl complexes are no different. In-depth studies have shown that a wide range of transition metal propargylic complexes act as a convenient source of an allenyl ligand and a metal atom fragment on condensation with a binary metal carbonyl (*5b,29*). Such reactions require a propargyl to allenyl rearrangement, the mechanism of which is still not firmly established. The product distribution in these reactions is sensitive to various factors including substituent effects, the metal carbonyl used, and the reaction conditions employed.

3. With Group VI Metal Propargyls

The pendant acetylenic functionality in $[M(CO)_3Cp(\eta^1\text{-}CH_2C\equiv CR)]$ (**45**) (M = Mo, W; R = Me, Ph) has been used as a template for the two-step synthesis of heteronuclear μ-alkyne complexes (*30*) [Scheme 12, path (i)]. Attempts were made to extend this chemistry to the group VIII metal carbonyls. Instead of alkyne-containing products, however, this procedure proved to be an efficient route to hetero bi- and trinuclear allenyl-containing clusters [Scheme 12, paths (ii) and (iii)] (*31*). The product distribution was found to be dependent on several factors including the condensing metal carbonyl used. For instance, the reaction of $[M(CO)_3Cp(CH_2C\equiv CPh)]$ (M = Mo, W) (**45**) with diiron nonacarbonyl afforded both the bi- and trinuclear clusters $[FeM(CO)_5Cp(\mu\text{-}\eta^2,\eta^3\text{-}PhC=C=CH_2)]$ (**46**) and $[Fe_2M(CO)_8Cp(\mu\text{-}\eta^1,\eta^2,\eta^2\text{-}PhC=C=CH_2)]$ (**47**) (Scheme 12), in contrast with that of dodecacarbonyltriruthenium for which only the trinuclear $\mu\text{-}\eta^1,\eta^2,\eta^2$ analog of **47**, $[Ru_2W(CO)_8Cp(\mu\text{-}\eta^1,\eta^2,\eta^2\text{-}Ph=C=CH_2)]$ (**48**), could be isolated. The NMR characteristics of these allenyl clusters have proved to be exceptionally valuable in assigning the metal–ligand bonding types (Section V). Wojcicki and co-workers (*31*) have proposed that the propargyl to allenyl rearrangement of the terminal propargylic carbon (—CR) at iron is indicative of a reaction pathway similar to that previously documented for the [2 + 3] cycloaddition reactions described in Section VII,A (Scheme 13).

4. With Group VIII Metal Propargyls

Attempts to extend the chemistry described in the previous section to group VIII propargyl metal complexes afforded considerably different products, and in some cases no allenyl-containing carbonyl complexes could be isolated.

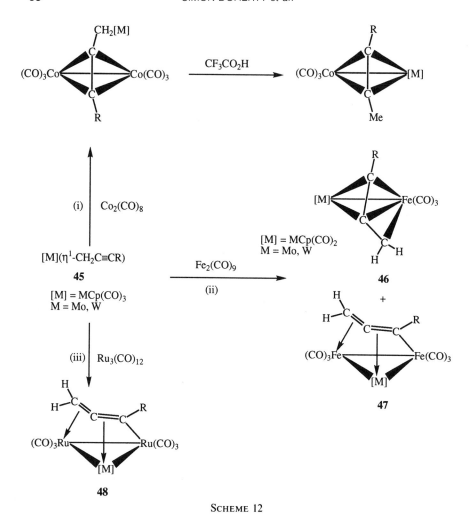

Reaction of $[Ru(CO)_2Cp(CH_2C{\equiv}CPh)]$ **(49)** with dodecacarbonyltriruthenium *(5b)* gave the homometallic cluster $[Ru_3(CO)_7Cp(\mu\text{-}\eta^1,\eta^2,\eta^2\text{-}PhC{=}C{=}CH_2)]$ **(50)** (Scheme 14) with a metal–ligand atom connectivity different from that of the mixed group VI/VIII allenyl clusters (Section III,B,3). Essentially this difference arises because either the $C_\alpha\text{-}C_\beta$ or $C_\beta\text{-}C_\gamma$ double bond of the allenyl ligand can be coordinated to the original propargyl-bound metal fragment (Section IV). The allenyl clusters prepared from group VIII metal propargyls are characterized by coordination of the external $(C_\beta\text{-}C_\gamma)$ double bond to the original propargy-

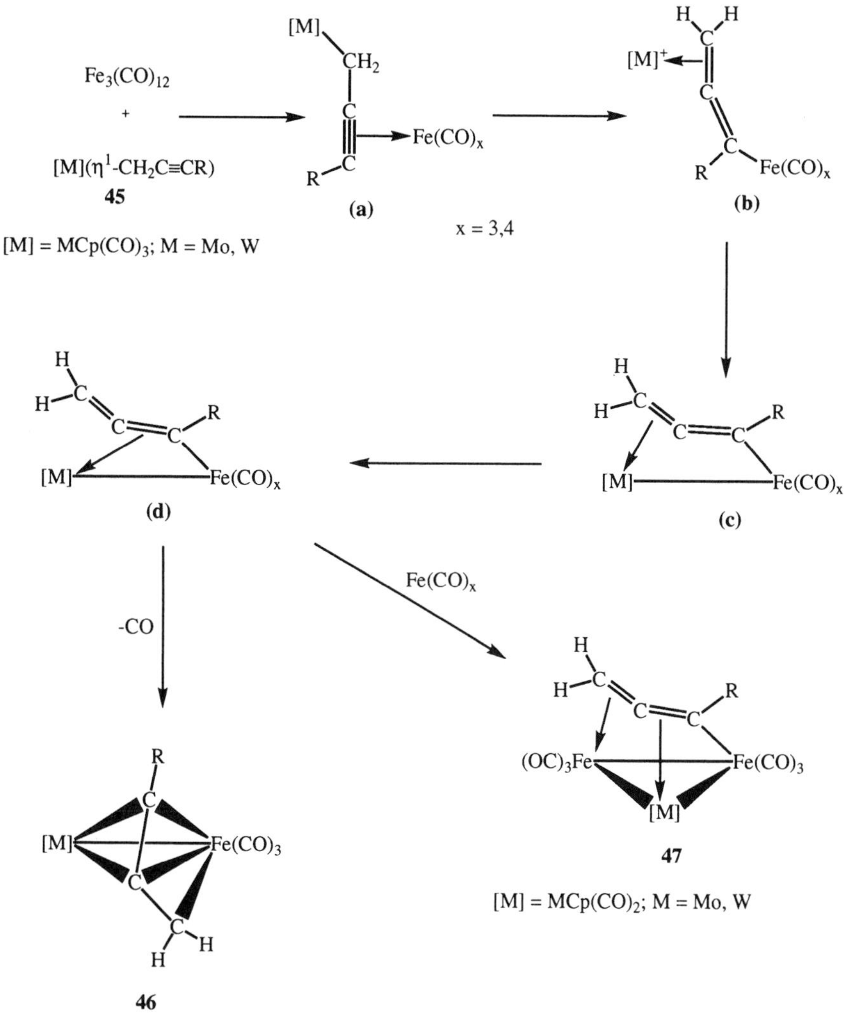

SCHEME 13

lic metal fragment, whereas those clusters formed from the condensation of group VI metal propargyls characteristically have the internal (C_α–C_β) double bond coordinated to the propargylic metal center. This metal–ligand connectivity is found for both bi- and trinuclear clusters prepared in this manner.

Several hydrocarbon-containing products were isolated from the reac-

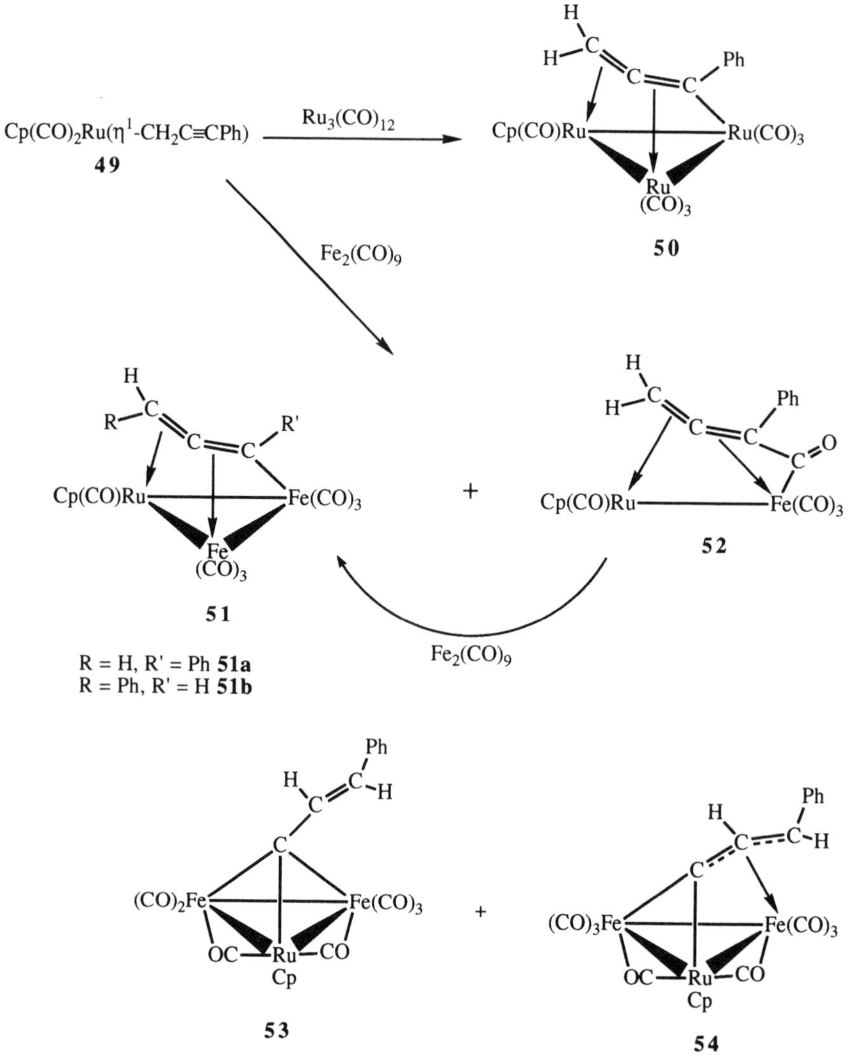

$$Cp(CO)_2Ru(\eta^1\text{-}CH_2C\equiv CPh) \xrightarrow{Ru_3(CO)_{12}} \mathbf{50}$$

49

SCHEME 14

tion of [Ru(CO)$_2$Cp(CH$_2$C$\equiv$CPh)] (**49**) with Fe$_2$(CO)$_9$ (*32*) (Scheme 14). Among them was the Fe$_2$Ru analog of **50** (**51**). However the major reaction product was identified as the binuclear allenyl (carbonyl) [RuFe-(CO)$_4$Cp{μ-η^2,η^3-C(O)C(Ph)C=CH$_2$}] (**52**). The presence of an inserted carbonyl reduces the strain otherwise present in the allenyl ligand (C–C–C angle ~145°) allowing coordination of the internal C=C (C$_\alpha$–C$_\beta$) bond to the additional condensed metal carbonyl unit. Wojcicki and co-workers demonstrated that the cluster $[Ru_2Fe(CO)_7Cp(\mu-\eta^1,\eta^2,\eta^2-PhC{=}C{=}CH_2)]$ (**51**) was the product of metal atom condensation and CO deinsertion from the allenyl(carbonyl) **52**. A similar metal–ligand atom connectivity is retained in the trinuclear species with the external C=C ($C\beta$–$C\gamma$) bond η^2-coordinated to the original metal propargyl fragment. The remaining reaction products (**53** and **54**) contained rearranged allenyl ligands (Scheme 14). The allenyl(carbonyl) bimetallic complexes described above have been utilized in the preparation of various trimetallic clusters of the type FeRuM' (Section VII,B,2) (*32*).

The condensation of **49** with various Pt(0) reagents has also been demonstrated. The outcome of the reactions depends, to a large extent, on the form of the platinum(0) reagent used. Both $\mu-\eta^1,\eta^1$, and $\mu-\eta^2,\eta^3$ dinuclear allenyl complexes are accesible (Scheme 15). An exceptionally facile conversion from the $\mu-\eta^1,\eta^2$-coordinated allenyl in $[RuPt(CO)Cp(PPh_3)_2(\mu-\eta^1,\eta^2-PhC{=}C{=}CH_2)]$ (**55**) to one $\mu-\eta^2,\eta^3$-bound in [RuPt(CO)Cp(PPh_3(μ-

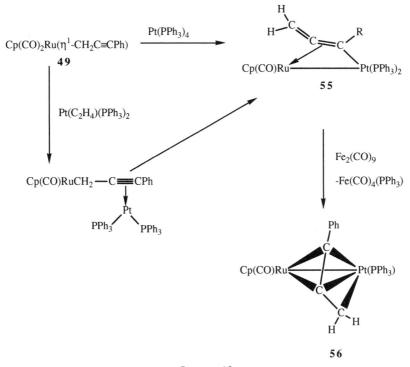

SCHEME 15

η^2,η^3-PhC=C=CH$_2$)] **(56)** is promoted by the phosphine scavenger Fe$_2$(CO)$_9$. This represents the first example of such a transformation. Reacting Pt(PPh$_3$)$_4$ with [Ru(CO)$_2$Cp(CH$_2$C≡CPh)] gave high yields of the binuclear μ-η^1,η^2-allenyl **55**, with the presence of excess phosphine preventing conversion to the μ-η^2,η^3 form (*5b*).

5. Allenyl Transfer Reagents

In contrast to the versatile procedures described in the preceeding sections for the preparation of heterometallic allenyl complexes from propargylic synthons, the analogous chromium reagents [CrCp(NO)$_2$(η^1-CH$_2$C≡CPh)] reacted with complete transfer of the propargyl moiety as an allenyl fragment (*33*), representing the first allenyl transfer reaction to be described. The homometallic allenyl-bridged cluster [Fe$_2$(CO)$_5$(NO)(μ-η^2,η^3-PhC=C=CH$_2$)] **(57)** was the major product isolated (35%). The

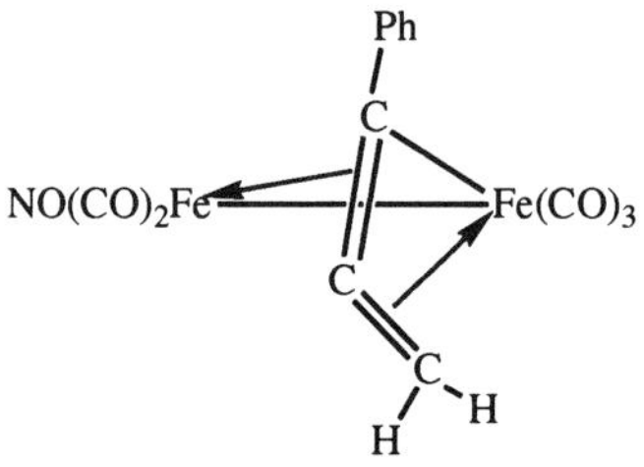

57

absence of mixed metal products in these cases was attributed to an instability of the zwitterionic intermediate.

6. Binuclear Allenylidenes

Two major developments in the preparation of bimetallic allenylidene complexes involve (i) metal fragment condensation of a monomeric allenylidene complex with a lightly stabilized metal fragment and (ii) deoxygenation of the dianionic acetylide intermediates, [M$_2$(CO)$_9${C≡C-C('Bu$_2$)O}]$^{2-}$, with COCl$_2$.

Reaction of [W(CO)$_5$(η^1-C=C=CPh$_2$)] with W(CO)$_6$ under photolytic conditions has been described by Berke *et al.* (*20a*). The binuclear allenylidene complex [{W(CO)$_5$}$_2$(μ-η^1,η^1-C=C=CPh$_2$)] **(23)** is formed [Eq. (16)]. A similar procedure proved useful for the preparation of [Fe$_2$(CO)$_8$(μ-η^1,η^1-C=C=C'Bu$_2$)] **(58)** from [Fe(CO)$_4$(η^1-C=C=C'Bu)] **(24)** and Fe(CO)$_5$ (*20b*). Furthermore, the condensation of 2 mol equiv of Fe(CO)$_4$ with **24** resulted in the isolation of a rare trinuclear allenylidene

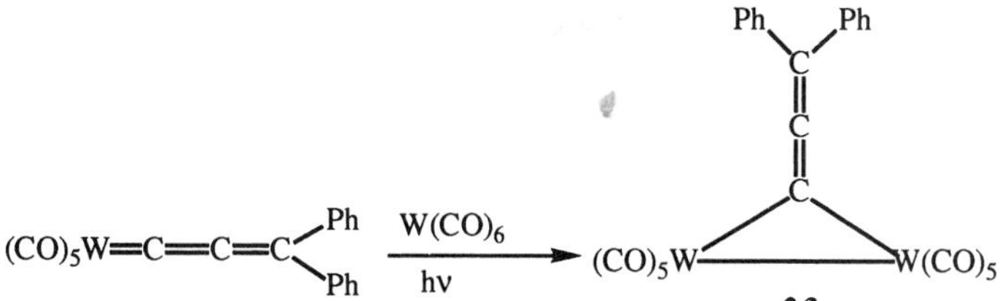

$$(16)$$

$$(CO)_5W{=}C{=}C{=}C\underset{Ph}{\overset{Ph}{\diagdown}} \quad \xrightarrow[hv]{W(CO)_6} \quad (CO)_5W\text{———}W(CO)_5$$

23

cluster, $[Fe_3(CO))_{10}(\mu_3\text{-}\eta^1,\eta^1,\eta^2\text{-}C{=}C{=}C'Bu_2)]$ (**59**). Earlier examples of metal fragment condensation included the reaction of $[Mn(CO)_2Cp(\eta^1\text{-}C{=}C{=}CPh_2)]$ with $[Mn(CO)_2Cp(Et_2O)]$ and $Fe_2(CO)_9$ to give the homo- and heterobimetallic $[\{Mn(CO)_2Cp\}_2(\mu\text{-}\eta^1,\eta^1\text{-}C{=}C{=}CPh_2)]$ (**60**) (*34a*) and $[Mn(CO)_2Cp(\mu\text{-}\eta^1\text{-},\eta^1\text{-}C{=}C{=}CPh_2)Fe(CO)_4]$ (**61**), respectively (*34b*).

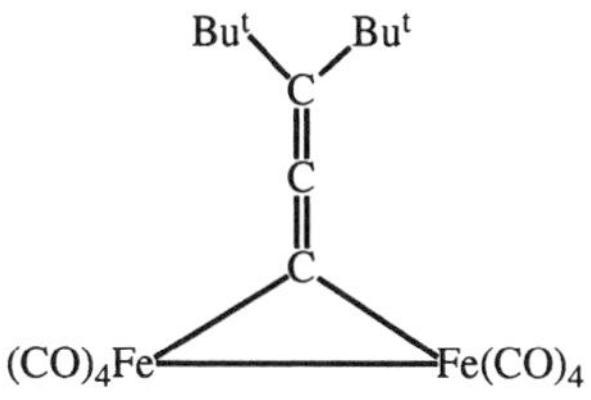

58

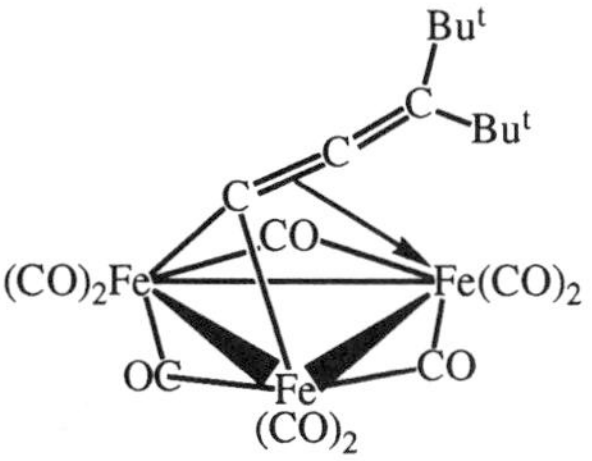

59

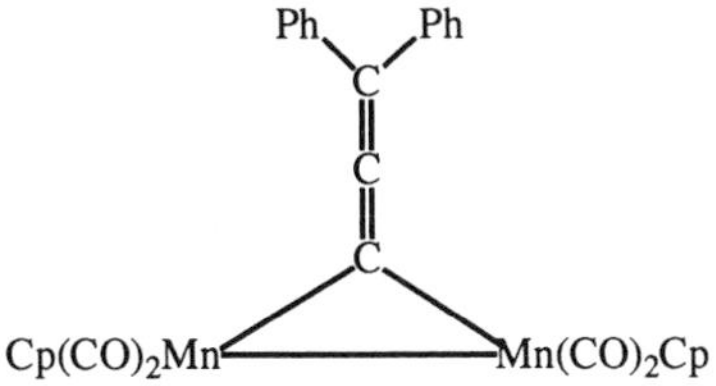

60

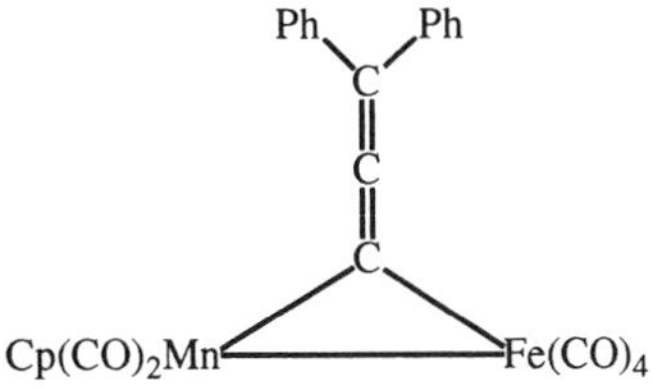

61

Procedures of type (ii) involve reacting a binuclear binary carbonyl (or lightly stabilized derivative) with the deprotonated propargyl alcohol $[C{\equiv}CCR_2O]^{2-}$. Treatment of the intermediate acetylide dianion $[M_2(CO)_9(C{\equiv}CCR_2O)]^{2-}$ with phosgene then liberates CO_2, forming $[M_2(CO)_9(\mu\text{-}\eta^1,\eta^1\text{-}C{=}C{=}C'Bu_2)]$ (M = Mn, Re). This procedure bears a remarkably close similarity to that successfully employed for the preparation of monomeric allenylidene complexes described above. For M = Mn, the $\mu\text{-}\eta^1,\eta^1$-symmetrically bridged allenylidene complex $[Mn_2(CO)_9(\mu\text{-}\eta^1,\eta^1\text{-}C{=}C{=}C'Bu_2)]$ (**62**) was identified. However, the rhenium analog represents a rare example of the η^1-coordination of an allenylidene at a single metal center of a binuclear system (*35*).

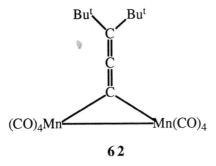

6 2

Few examples of bimetallic allenylidene complexes in which the ligand adopts the μ-η^1:η^2(4e) unsymmetrical coordination mode have been recognized. Addition of $H_2C\!\!=\!\!C(Me)C\!\!\equiv\!\!CLi$ to a tetrahydrofuran (THF) solution of $[M_2(CO)_4Cp_2]$ (M = Mo, W) at $-78°C$ led to rapid adduct formation (*36*). Protonation of the reaction mixture on an alumina column gave a separable mixture of $[M_2(CO)_4Cp_2(\mu$-η^2,η^2-$HC_2C(Me)\!\!=\!\!CH_2)]$ (**63**) and $[M_2(CO)_4Cp_2(\mu$-η^1,η^2-$C\!\!=\!\!C\!\!=\!\!CMe_2)]$ (**64**), the latter containing a highly fluxional allenylidene bridging ligand (Scheme 16). Protonation of the η^2-ethene complex $[ZrCp_2(PMe_3)(\eta^2$-$C_2H_4)]$ by a cyclopropene provides the most recent example of an unsymmetrical allenylidene (*37*). The product, $[\{ZrCp_2\ Et\}_2(\mu$-η^1:η^2-$C\!\!=\!\!C\!\!=\!\!CMe_2)]$ (**65**), contains a character-

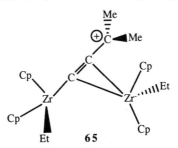

istic μ-η^1:η^2-dimethylallenylidene, postulated to be formed via an intermediate dizirconated cyclopropane which readily rearranges to give **65**.

C. Homometallic Allenyls via Carbon–Carbon Bond Formation

Because the allenyl fragment is constructed from C (carbide), CH (alkylidyne) and CH_2 (alkylidene) units, several possible routes to allenyl complexes via C–C coupling reactions can be envisaged; for instance, reaction of a metal C_1 fragment (alkylidene, alkylidyne) with a C_2 substrate

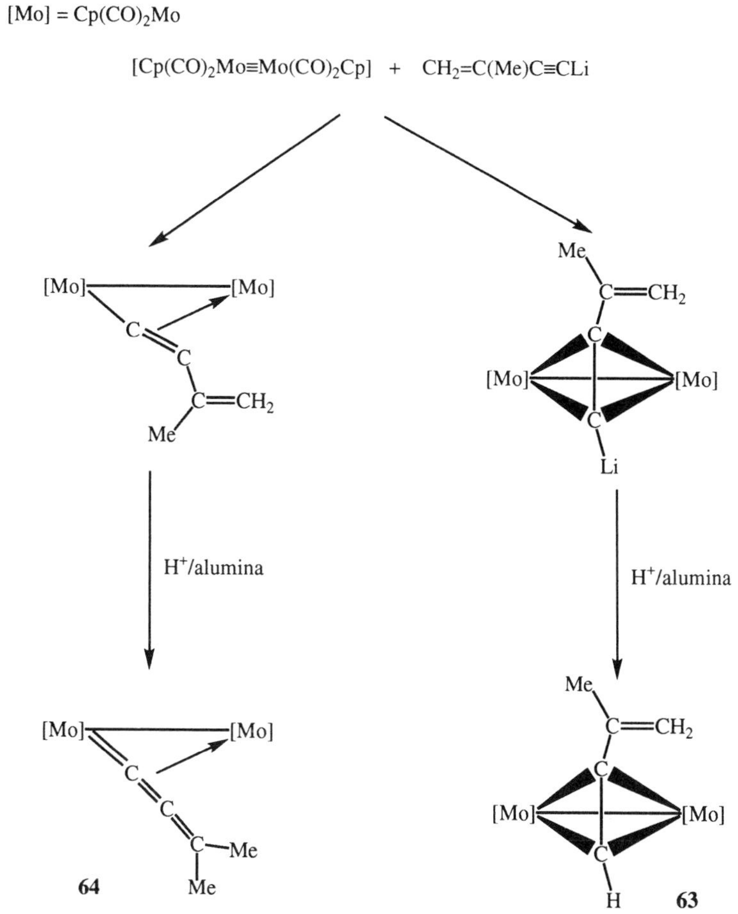

[Mo] = Cp(CO)$_2$Mo

[Cp(CO)$_2$Mo≡Mo(CO)$_2$Cp] + CH$_2$=C(Me)C≡CLi

SCHEME 16

(alkyne and 1,2 shift; acetylide) or a metal C$_2$ fragment (vinylidene, acetylide) with a C$_1$ fragment. However, these routes remain either largely unexplored or underdeveloped. The most direct route to bi- and polynuclear allenyl clusters described to date involves reaction of multisite-bound acetylide ligands with diazoalkanes. Several years ago it was established that the Cα atom of σ–π-bound acetylide complexes in both μ-η^1,η^2 binuclear and μ-η^1,η^2,η^2 trinuclear clusters are susceptible to regioselective nucleophilic attack by neutral ligands, including carbenes generated from diazoalkanes (38a). The preparation of bi- and trinuclear allenyl clusters using this methodology is considered separately.

1. *Preparation of Homobinuclear Allenyl Complexes*

A significant advancement in the development of allenyl chemistry was the observation that the μ-η^1,η^2 bridging acetylides in [Ru$_2$(CO)$_6$(μ-PPh$_2$)(μ-η^1,η^2-C≡CPh)] (**66**) reacted regiospecifically with diazomethane, nucleophilic addition at Cα dominating, to afford the binuclear phosphido-bridged allenyl complexes [Ru$_2$(CO)$_6$(μ-PPh$_2$)(μ_2-η^1,η^2-PhC=C=CH$_2$)] (**67**) (Scheme 17) (*38a*). This methodology has been successfully extended to the preparation of the osmium congener (*38b*). Two different bonding modes have been discerned for these allenyl clusters, differentiated only by their η^2-connectivity, involving either the internal or external double bonds (Section IV,B) (*38b*). A rich chemistry has been developed for both di- and trimetallic allenyl clusters (discussed further in Section VII).

Attempts to extend the synthesis of [M$_2$(CO)$_6$(μ-PPh$_2$)(μ-η^1,η^2-PhC=C=CH$_2$)] (M = Ru, Os) to the iron analog revealed that simple 1:1 addition to the μ-η^1,η^2-acetylide to afford the allenyl-bridged cluster was not the dominant reaction pathway; rather, facile double addition of methylene occurred, affording the μ-η^2,η^2-coordinated butadienylidene complex [Fe$_2$(CO)$_6$(μ-PPh$_2$){μ-η^2,η^3-H$_2$C=C('Bu)C=CH$_2$}] (**68**) [Eq. (17)]

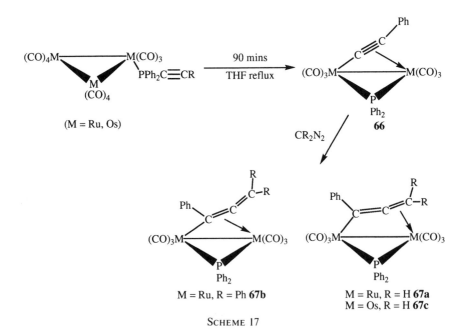

M = Ru, R = Ph **67b**

M = Ru, R = H **67a**
M = Os, R = H **67c**

Scheme 17

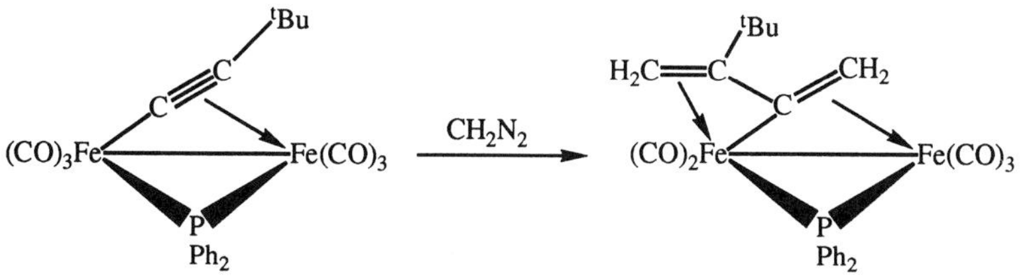

$$(39).$$ There was no evidence for an allenyl-bridged intermediate in the reaction sequence.

2. Preparation of Trimetallic Allenyl Clusters

Carbon–carbon bond forming reactions between diazoalkanes and multisite-bound acetylides can also be used to generate allenyl clusters. Thus $C\alpha$ of the μ_3-η^1,η^2,η^2-coordinated acetylide in $[Ru_3(CO)_8(\mu$-$PPh_2)(\mu_3$-η^1,η^2,η^2-$C\equiv C^iPr)]$ **69** is susceptible to attack by a variety of nucleophilic reagents $(40a)$. Reaction of **69** with diazomethane afforded $[Ru_3(CO)_8(\mu$-$PPh_2)(\mu_3$-η^1,η^2,η^2-$^iPrC\!=\!C\!=\!CH_2)]$ **(71)**. The synthesis of **71** from the open 50-electron cluster $[Ru_3(CO)_9(\mu$-$PPh_2)(\mu_3$-η^1,η^2,η^2-$C\equiv C^iPr)]$ **(69)** requires conversion to the closed 48-electron cluster $[Ru_3(CO)_8(\mu$-$PPh_2)(\mu$-η^1,η^2,η^2-$C\equiv C^iPr)]$ **(70)** prior to C–C bond formation (Scheme 18) $(40b)$.

3. Further Carbon–Carbon Bond Forming Reactions

Two-electron reduction of the bisalkylidyne cluster $[Fe_3(CO)_9(\mu_3$-$CCH_2 R)(\mu_3$-$COC_2H_5)]$ **(72)** by $[Mn(CO)_5]^-$ induces alkylidyne–alkylidyne coupling to give an intermediate μ_3-η^2-bound alkyne cluster $[Fe_3(CO)_9(\mu_3$-η^1,η^1,η^2-$RCH_2C\equiv COEt)]$ (41). Two competing reaction pathways have been identified, dependent on the alkylidyne substituents. For R = Ph or C_3H_7, ethoxide elimination always gave acetylide clusters of the formula $[Fe_3(CO)_9(\mu_3$-η^1,η^2,η^2-$CCCH_2R)]^-$ **(73)**. In contrast, reduction of the bisalkylidyne cluster with R substituents of C(O)OMe and C(O)Me gave the allenyl-bridged cluster $[Fe_3(CO)_9(\mu_3$-η^1,η^2,η^2-$(OEt)C\!=\!C\!=\!CHR)]^-$ **(74)** via hydride elimination (Scheme 19). Whether this reaction can be adapted to a general strategy for accessing allenyl clusters remains to be demonstrated.

The chemistry of the highly electrophilic cationic diiron bridging methylidyne complex $[\{Fe(CO)Cp\}_2(\mu$-$CH)(\mu$-$CO)]^+$ **(75)** has been probed extensively with an emphasis on C–C bond forming capabilities. For instance, alkenes react to give one of two possible products: (i) alkyli-

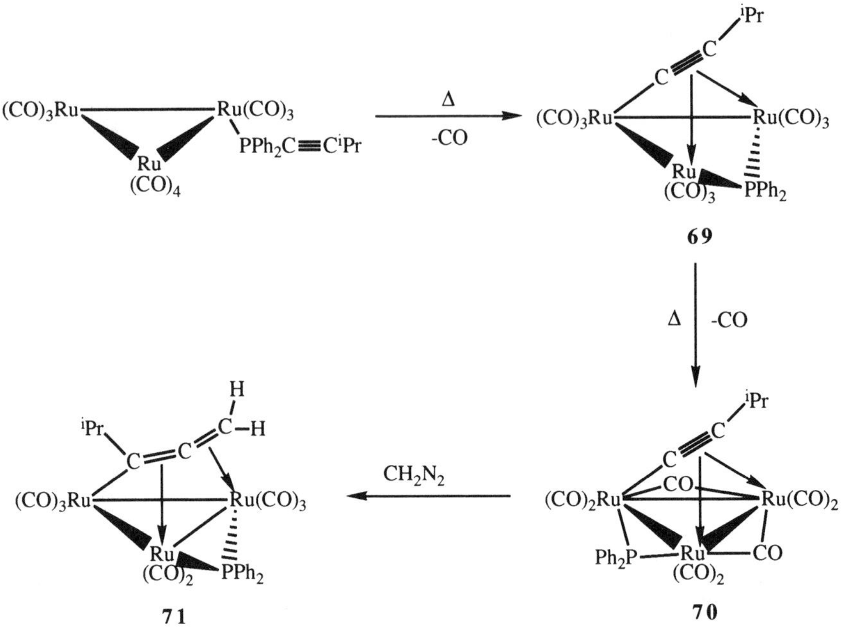

$(CO)_3Ru$⎯⎯$Ru(CO)_3$ $\quad$ $PPh_2C\equiv C^iPr$ $\xrightarrow[-CO]{\Delta}$ $(CO)_3Ru$ ... $Ru(CO)_3$

Ru $(CO)_4$

$(CO)_3$ PPh_2 ^iPr

69

Δ | $-CO$

$(CO)_2Ru$ ... $Ru(CO)_2$ $\quad$ Ph_2P—Ru—CO $(CO)_2$ ^iPr CO

70

$\xleftarrow{CH_2N_2}$

$(CO)_3Ru$ ... $Ru(CO)_3$ $\quad$ ^iPr H C=C C—H Ru $(CO)_2$ PPh_2

71

SCHEME 18

dyne-bridged complexes (**76**) via 1,3-hydride migration from methylidyne to alkene (*42a*) or (ii) μ-alkenyl complexes (**77**) via a 1,2-carbon or 1,2-hydrogen shift (pathways a and b, respectively, in Scheme 20) (*42b*). However, reaction of **75** with alkynes did not proceed as anticipated (pathways c and d in Scheme 21). Instead of complexes analogous to the alkenyl (**80**) and alkylidyne (**79**) adducts of alkene insertion, products best described as μ-alkenyl(carbonyl) [vinyl ketenes] (**78**) were isolated (*42c*). Pathway d arises from formal 1,2-carbon migration and represents a potentially powerful route to allenyl-bridged bimetallic systems. However, in the majority of cases formation of μ-η^1,η^4-vinyl ketene complexes [{Fe-(CO)Cp}$_2$(μ-CO){(μ-η^1,η^4-CHC(R)C(R')CO}] (**78**) was favored, proceeding via regiospecific insertion of the alkyne between the methylidyne carbon and a carbon monoxide ligand of **75**. In contrast, *tert*-butylacetylene reacted with **75** to afford two products, both of which involved regioselective C–C coupling. The major product [{Fe(CO)Cp}$_2$(μ-CO){μ-η^1,η^2-CH=C(H)=C(Me)=C(Me)$_2$}]$^+$ (**81**) contains a μ-η^1,η^2-dienyl ligand which arises from a 1,2-methyl migration, whereas the minor product [{Fe(CO)Cp}$_2$(μ-CO)(μ-η^1,η^2-CH=C=CH^{t}Bu]$^+$ (**82**), bearing a bridging allenyl ligand, results from a 1,2-hydride migration (Scheme 22). The μ-η^1,η^2-vinylketene complexes (**78**) described above closely resemble the

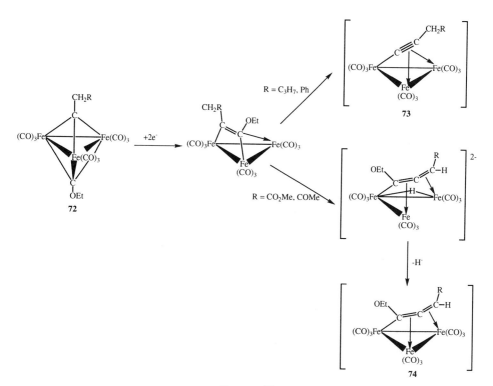

SCHEME 19

SCHEME 20

69

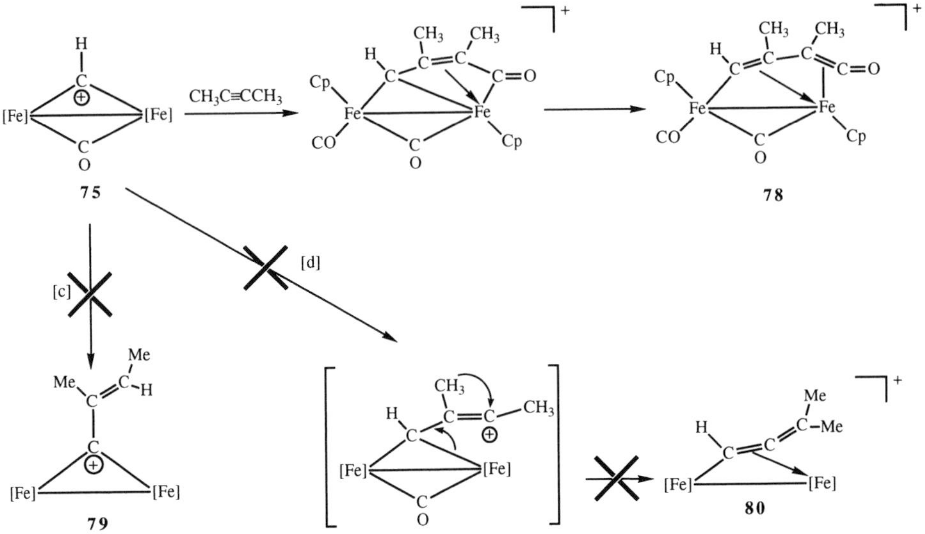

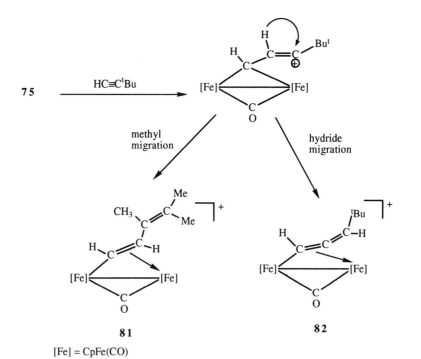

[Fe] = CpFe(CO)

SCHEME 21

[Fe] = CpFe(CO)

SCHEME 22

70

allenyl(carbonyl) dimer **52** described by Wojcicki and co-workers, for which a similar metal fragment condensation and CO deinsertion reaction would give the thermodynamically favored 1,3-dimetalloallyl isomers directly.

The vinyl ketenes have been shown to rearrange readily to vinyl carbenes on treatment with $NaBH_4$. Similar vinyl carbene bridges (e.g., **83**) have previously been prepared via elimination of HF between the μ-methylene ligand of $[\{RuCp\}_2(\mu\text{-CO})(CO)(\mu\text{-CH}_2)(\eta^2\text{-C}_2F_3R)]$ and its η^2-coordinated fluoroalkene, facilitated by the presence of intramolecular H⋯F hydrogen bonds [Eq. (18)] *(43)*.

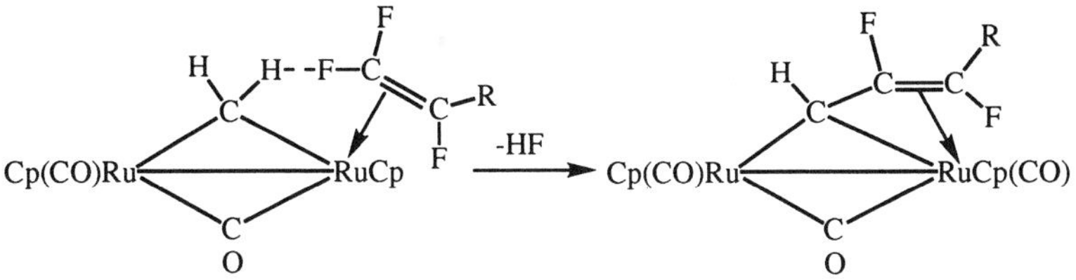

$$(18)$$

83

Addition of the imine nitrogen of $Ph_2C{=}NH$ to the electrophilic carbon of $[\{Fe(CO)Cp\}_2(\mu\text{-CO})(\mu\text{-CH})]^+$ **(75)** gave, after deprotonation, $[\{Fe(CO)Cp\}_2(\mu\text{-CO})(\mu\text{-}\eta^1,\eta^1\text{-HCN}{=}CPh_2)]$ **(84)**, a complex in which two iron

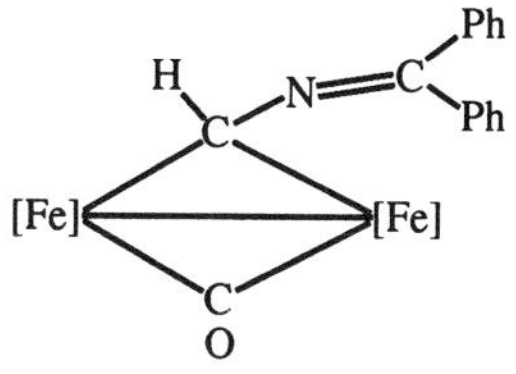

84

[Fe] = Fe(CO)Cp

atoms are bridged by a hydrocarbyl ligand most appropriately described as a $\mu\text{-}\eta^1$-heteroallenyl ligand *(44)*.

Casey, Hoel, and co-workers have reported that the ethylidene-bridged complex $[\{Fe(CO)Cp\}_2(\mu\text{-CO})(\mu\text{-}\eta^1,\eta^1\text{-C}{=}CH_2)]$ reacts with diazoalkanes, giving the cyclopropylidene complexes $[\{Fe(CO)Cp\}_2(\mu\text{-CO})(\mu\text{-}\eta^1,\eta^1\text{-C}{-}CH_2CHR)]$ which readily undergo photolytic ring opening to form the $\mu\text{-}\eta^2,\eta^2$-allene-bridged complexes $[\{Fe(CO)Cp\}_2(\mu\text{-CO})(\mu\text{-}\eta^2,\eta^2\text{-CH}_2{=}C{=}CHR)]$ **(85)**. Alternatively, acid-catalyzed ring opening affords the substituted vinylidene $[\{Fe(CO)Cp\}_2(\mu\text{-}\eta^1,\eta^1\text{-C}{=}CHCH_2R)]$ **(86)** (Scheme 23) *(45)*. These C–C bond forming processes between

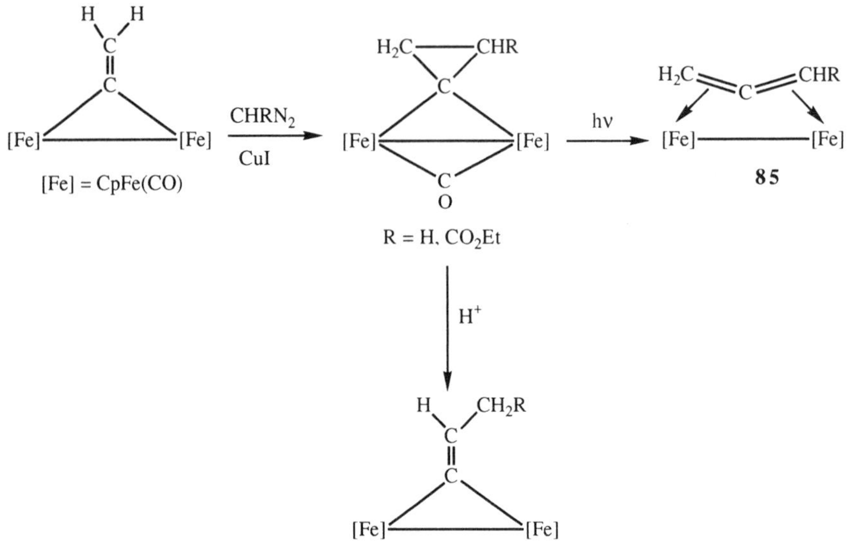

SCHEME 23

μ-bound unsaturated hydrocarbyls and diazoalkanes have precedent in the reactions with μ-η^1,η^2- and μ-η^1,η^2,η^2-acetylides described earlier (*38a,b,40a,b*). Similar C–C forming processes include the following: reaction of the lightly stabilized complex [Ru$_2$(CO)$_2$Cp$_2$(MeCN)(μ-C≡CH$_2$)] with diazomethane to afford the μ-η^2,η^2-allene-bridged [Ru$_2$(CO)$_2$Cp$_2$(μ-η^2,η^2-H$_2$C≡C≡CH$_2$)] (**87**) (*46a*) and reaction of the side-on coordinated

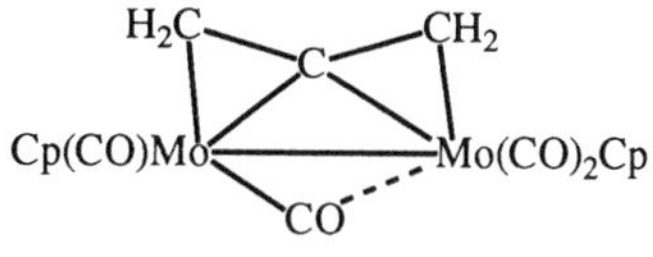

87

vinylidene in [Mo$_2$(CO)$_4$Cp$_2$(μ-η^1,η^2-C≡CH$_2$)] to give the allene-bridged dimer [Mo$_2$(CO)$_2$Cp$_2$(μ-η^2,η^2-H$_2$C≡C≡CH$_2$)] (**88**) via regiospecific C–C bond formation at C$_\alpha$ (*47*). These σ–π-bound vinylidenes are thermodynamically unstable, readily rearranging to the isomeric alkyne counterparts (**89**) via 1,2-hydrogen migration (Scheme 24) (*47,48*).

A number of other C–C bond forming reactions have been described which afford not allenyl but metalloallyl clusters. Carbyne–alkyne coupling reactions are potentially useful as sources of allenyl fragments but

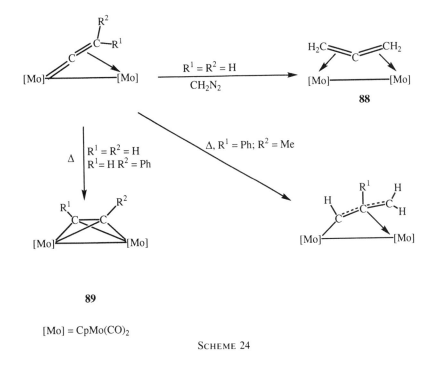

[Mo] = CpMo(CO)$_2$

SCHEME 24

require a hydrogen atom migration to generate the allenyl moiety. However, the products initially formed in such reactions are thermodynamically stable, and only a limited number of examples of the reverse rearrangement to afford allenyl products are known (49). Keister and co-workers have reported that the reaction of the heteroatom–alkylidyne stabilized cluster [Ru$_3$H$_3$(CO)$_9$(μ_3-CSEt)] with alkynes affords the allyl-containing cluster [Ru$_3$H(CO)$_9$(μ_3-η^1,η^1,η^3-SEtCCRCR')] (90) (Scheme 25) (50). A subsequent isomerization gave the vinylcarbyne (1,1-dimetalloallyl) type clusters 91. These C–C coupling reactions have been proposed as models for heterogeneously catalyzed hydrocarbon chain growth occurring at a metal surface.

D. *Carbon–Hydrogen Activation as Source of Allenyl Groups*

1. *Directly from Allene*

In principle a convenient route to allenyl-bridged clusters is via the oxidative addition of a C–H bond of allene. Unfortunately in many instances the reaction of an allene with metal clusters leads directly to

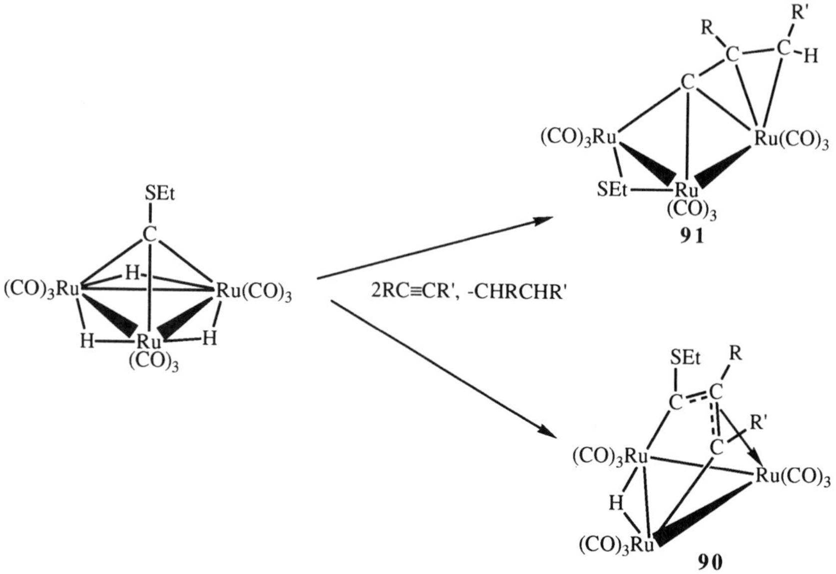

SCHEME 25

complicated products often involving C–C coupling reactions (46a–c). Thus C_6 hydrocarbyl ligands from allene dimerization have been found supporting both metal–metal bonded and nonbonded systems. Attempts to avoid carbon–carbon coupling by reacting the lightly stabilized clusters $[Os_3(CO)_{11}(MeCN)]$ or $[Os_3(CO)_{10}(MeCN)_2]$ with allene under exceptionally mild conditions have yielded the allene complexes $[Os_3(CO)_{11}(\mu-\eta^1,\eta^3-CH_2=C=CH_2)]$ (**92**) and $[Os_3(CO)_{10}(\mu_3-\eta^1,\eta^2,\eta^3-C_6H_8)]$ (**93**) respectively, the latter product arising from facile allene coupling (51). In this case the final product distribution was found to be highly dependent on the reaction conditions employed. Direct thermolysis of **92** gave binuclear $[Os_2(CO)_7(\mu-\eta^1,\eta^3-CH_2=C=CH_2)]$ (**94**), whereas irradiation with UV light afforded high yields of $[Os_3H(CO)_9(\mu_3-\eta^1,\eta^2,\eta^2-HC=C=CH_2)]$ (**95**). The 1,3-dimetallacycle $[Os_3H(CO)_9(\mu_3-\eta^1,\eta^1,\eta^1-HCCHCH)]$ (**96**) was accessible via thermal isomerization of the allenyl cluster **95** (Scheme 26).

Dickson et al. have generated an allene from the metal-induced decarboxylation of a diketene (52) by $[Rh_2(CO)Cp_2(\mu-\eta^2,\eta^2-CF_3C\equiv CCF_3)]$. Subsequent C–H bond activation led to the vinyl–allenyl-bridged dimer $[Rh_2(CO)Cp_2(\mu-\eta^1,\eta^2-CF_3C=CHCF_3)(\mu-\eta^1,\eta^2-HC=C=CH_2)]$ (**97**) (52). The reactivity of this dimer is discussed further in Section VII,B,2.

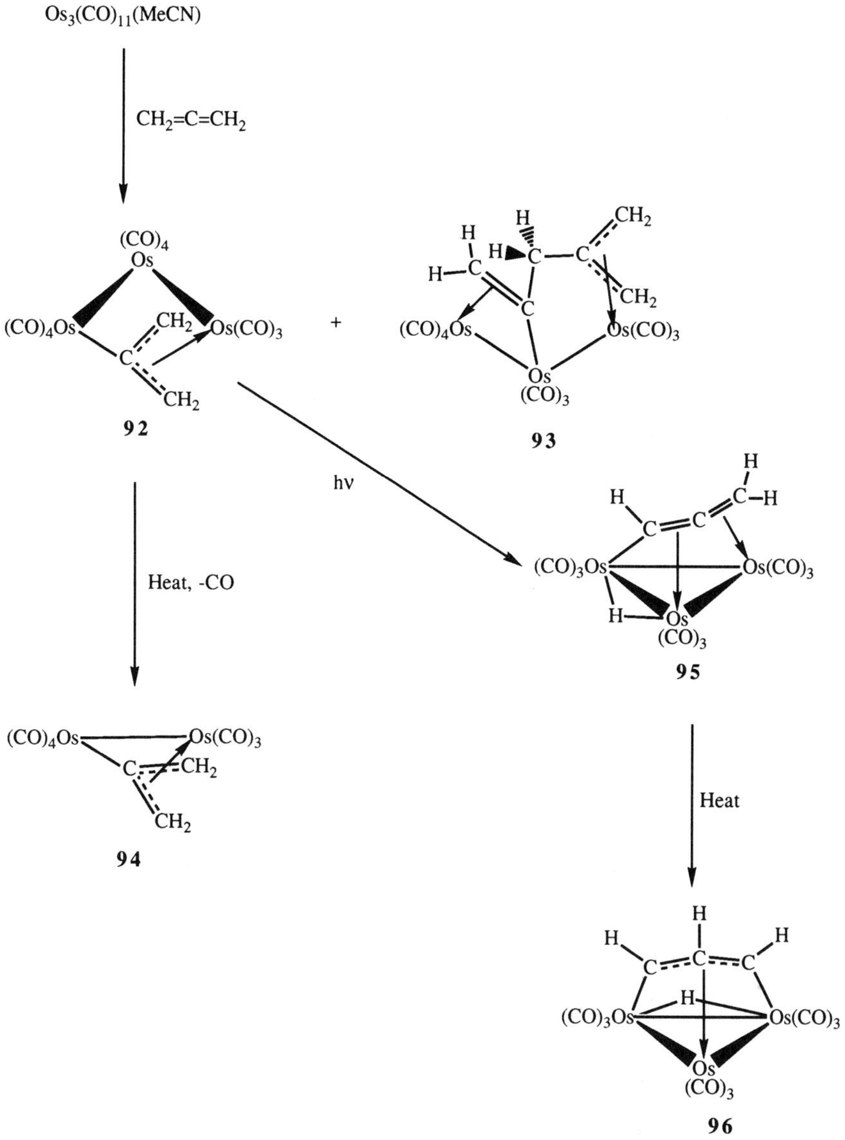

SCHEME 26

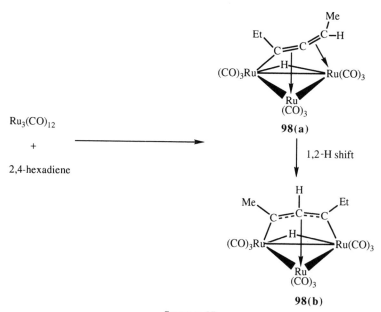

97

2. Dienes as Source of Allenyls

The earliest examples of σ–π-allenyl clusters arose from the thermal C–H activation of diene ligands. Reaction of isomers of 2,4-hexadiene with dodecacarbonyl triruthenium gave two isomers of $[Ru_3H(CO)_9(\mu_3$-$C_6H_9)]$, **98(a)** and **98(b)** (Scheme 27) (*53,a,b*). Isomer **98(a)** readily transforms to **98(b)** thermally via a facile 1,2-hydrogen shift. The structure of **98(b)**, which has been known since 1972, contains a 1,3-dimetalloallyl hydrocarbon ligand. Its isomeric counterpart was the first allenyl cluster to be structurally characterized. Gervasio and co-workers (*53c*) showed

SCHEME 27

that **98(a)** contains a μ-η^1,η^2,η^2-allenyl ligand. Alkenes and dienes containing at least three carbon atoms and with at least one hydrogen atom α to the double bond often react with metal carbonyls in a similar manner to form allenylic clusters as the dominant reaction products (*54*).

3. Allenyl Ligands from Alkynes

Alkynes containing methylene (—CH_2—) functionalities α to the triple bond readily undergo thermal C–H activation under conditions necessary to initiate reaction with group VIII metal carbonyls (*54*). The allenyl clusters obtained isomerize thermally via a 1,2-hydrogen atom shift to afford the thermodynamically more favorable 1,3-dimetalloallyl clusters (Table I) (*49,55*). A similar chemistry has been recognized for osmium clusters, albeit under more severe conditions.

The influence of heteroatom substituents on the reactivity of alkynes with group VIII metal carbonyls has been examined by Deeming and co-workers (*56a–c*) The reactivity of HC≡CCH$_2$NMe$_2$ (2-dimethylamino-

TABLE I

1,3-DIMETALLOALLYL AND ALLENYL CLUSTERS FORMED FROM Ru$_3$(CO)$_{12}$ AND
UNSATURATED HYDROCARBONS

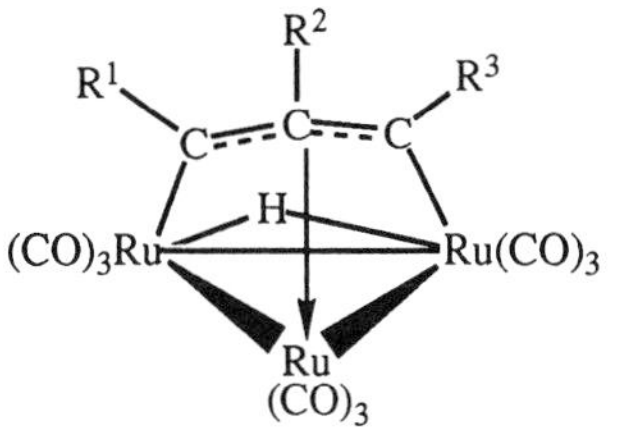

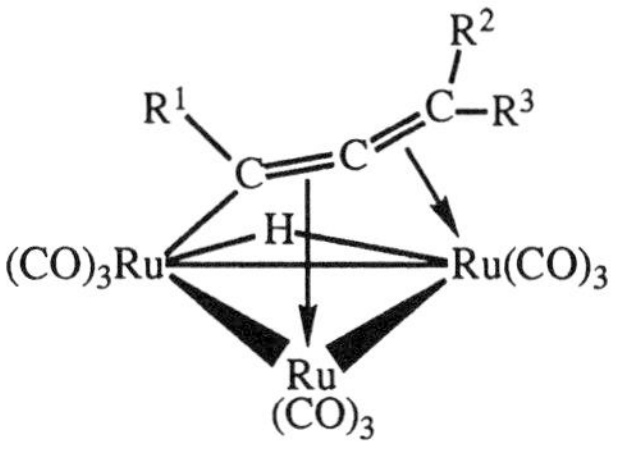

	1,3-Dimetalloallyl cluster (I) R^1, R^2, R^3	Allenyl cluster (II) R^1, R^2, R^3
Alkenes or alkynes		
Hexadiene	Me, H, Et	Et, H, Me
4-Methyl-2-pentyne	—	Me, Me, Me
Pentene, penta-1,3-dienes	Me, H, Me	Me, H, Me
Buta-1,3-diene	H, H, Me	—
1,4-Diphenylbuta-1,3-diene	Ph, H, CH$_2$Ph	—
Hex-3-yne	Et, H, Me	Me, H, Et
1,4-Dihydroxybut-2-yne	H, H, CHO	
Cyclopentadiene	R^{1}R^3 = —CH$_2$CH$_2$—, R^2 = H	

prop-1-yne) and $MeC \equiv CCH_2NMe_2$ (1-dimethylaminobut-2-yne) toward dodecacarbonyltriruthenium appears to be superficially similar to that of other nonfunctionalized acetylenes. There are several differences, however, most notably in the structures and dynamic behavior of the final reaction products.

Terminal alkynes often undergo competing reactions with metal carbonyls, coordination followed by C–H activation frequently being the most favorable reaction pathway (*57*). Deeming and co-workers suggested that 3-dimethylaminopropyne reacted in this manner but that a subsequent rapid 1,3-NMe_2 (inter- or intramolecular nucleophilic) migration occurred, affording the allenylic cluster $[Ru_3H(CO)_9(\mu_2\text{-}\eta^1,\eta^1,\eta^2\text{-}CH_2=C=CNMe_2)]$ (**99**) (Scheme 28) (*56b*). An X-ray study revealed that the substituent arrangement on the C_3 hydrocarbyl ligand was allenylic in nature. However, the metal–ligand bonding was modified by the strong π donating characteristics of the heteroatom substituent, and a zwitterionic description, $[Ru_3^{\ominus}H(CO)_9(\mu_3\text{-}\eta^1,\eta^1,\eta^2\text{-}Me_2N^{\oplus}=C\text{---}C=CH_2)]$, was con-

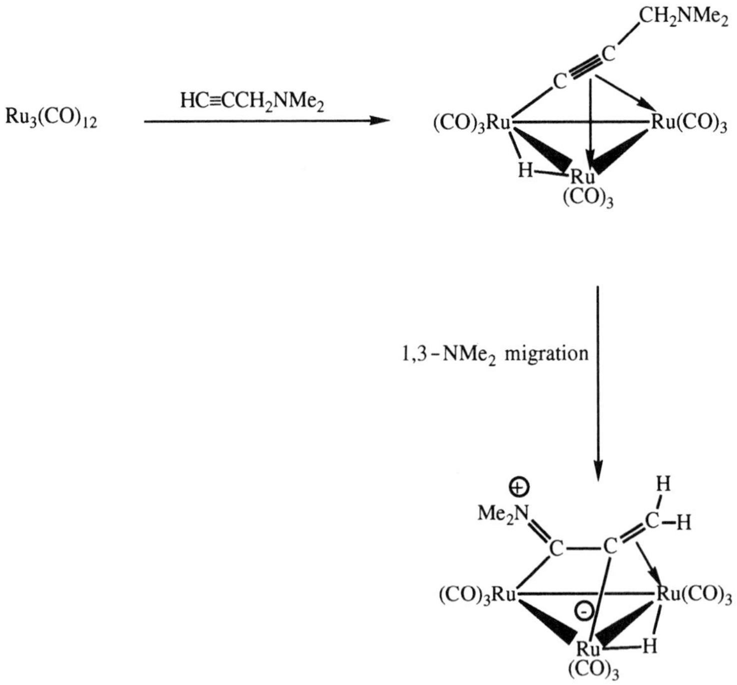

SCHEME 28

sidered more appropriate. In this case the familiar allenyl to metalloallyl isomerization was shown to be thermodynamically inaccessible, and $[Ru_3H(CO)_9(\mu_3-\eta^1,\eta^1,\eta^2-Me_2N{=}C{-}C{=}CH_2)]$ remained stable at temperatures up to 140°C. However, a PPh_3-catalyzed shift was observed at 55°C, giving $[Ru_3H(CO)_9(\mu_3-\eta^1,\eta^1,\eta^1-Me_2NCCHCH)]$ (**100**) (Scheme 29). Investigations of the reaction of $DCCCH_2NEt_2$ with $Ru_3(CO)_{12}$ gave in good yields the NEt_2 homolog of **99**. The structure has been confirmed by a single-crystal X-ray diffraction study. Although the structure corresponds to that of **99**, the deuterium atom was identified as one of the methylenic substituents, not as a bridging hydride. Nevertheless, the thermodynamic inaccessibility of the allenyl–allyl isomerization was confirmed. The reaction described above also gave low yields of the NEt_2 analog of **100**. In this complex, however, the deuterium atom bridges an Ru–Ru edge. This behavior indicates a different reaction path for the two complexes.

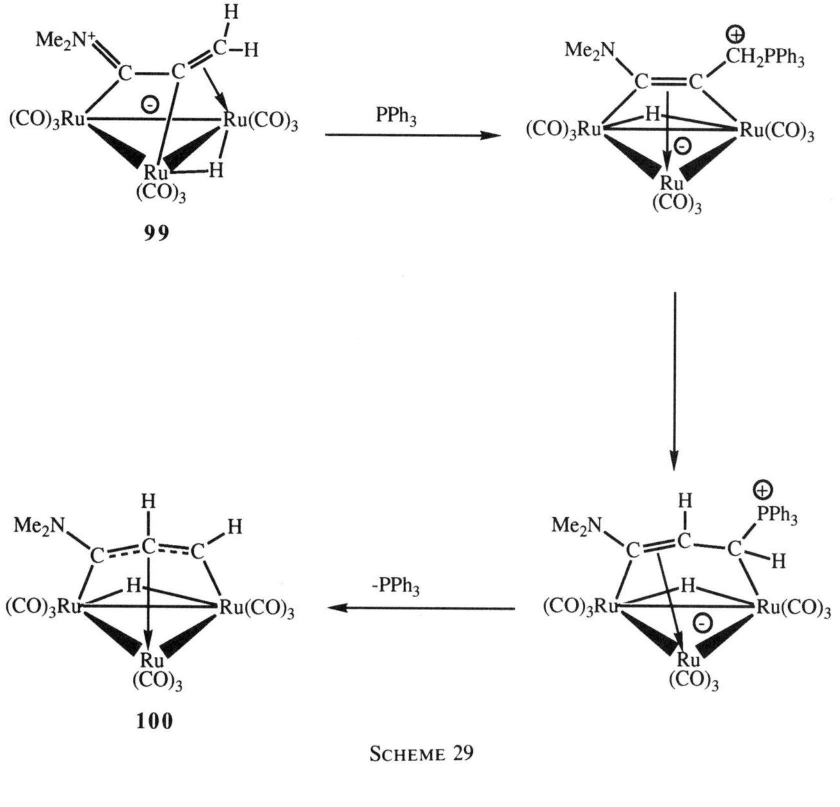

SCHEME 29

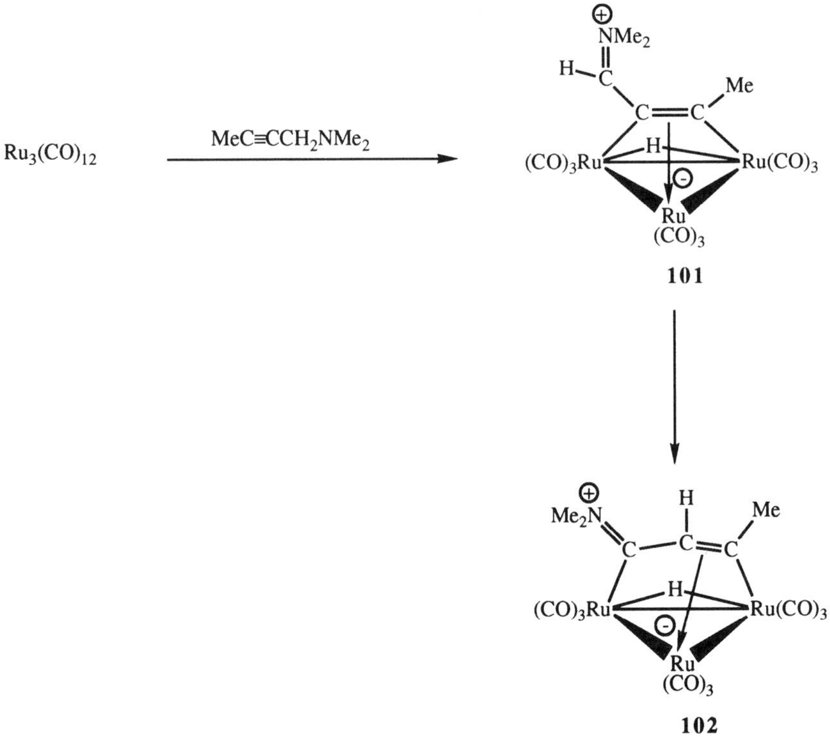

$$Ru_3(CO)_{12} \xrightarrow{\quad MeC\equiv CCH_2NMe_2 \quad} \textbf{101}$$

101

102

SCHEME 30

Reaction of 1-dimethylaminobut-2-yne with dodecacarbonyltriruthenium (*56a*) gave the allenylic cluster $[Ru_3H(CO)_9(\mu\text{-}\eta^1,\eta^1,\eta^2\text{-MeC}=C=CHNMe_2)]$ (**101**), presumably formed directly via methylenic C–H bond activation, in contrast with that proposed for the terminal aminoacetylene described above in Scheme 28. Cluster **101** readily undergoes thermal 1,2-hydrogen atom migration to afford the allylic isomer $[Ru_3H(CO)_9(\mu\text{-}\eta^1,\eta^1,\eta^2\text{-MeCCHCNMe}_2)]$ (**102**), consistent with characteristic reactivity patterns of allenylic clusters (Scheme 30). The metal–ligand bonding types in both the allenyl **101** and allylic **102** clusters are distorted from the ideal $\mu\text{-}\eta^1,\eta^2,\eta^2$ and $\mu\text{-}\eta^1,\eta^1,\eta^3$ coordination geometries. In the former case Deeming described the metal–ligand bonding as that of a modified $\mu_3\text{-}\eta^2$-alkyne type coordination with the zwitterionic formulation $[Ru_3^{\ominus}H(CO)_9(\mu\text{-}\eta^1,\eta^1,\eta^2\text{-MeC}\equiv CC=^{\oplus}NMe_2H)]$ being most appropriate. This structure displayed typical acetylenic characteristics with two σ bonds, an η^2 interaction, and a planar nitrogen atom. Skeletal modifications as a consequence of strong π donor substituents are known

(58) and may explain some of the unusual properties of these cluster types such as the high basicity of the cluster core and unexpected barriers to C–NMe_2 rotation.

The amide-substituted acetylene $HC{\equiv}CCMe_2NHCOR$ (R = C_6H_9, Ph) reacts with dodecacarbonyltriruthenium in a different manner, giving the anticipated hydrido acetylide $[Ru_3(CO)_9(\mu_3-\eta^1,\eta^2,\eta^2-C{\equiv}C-CMe_2NHCOR)]$ (59). In this case there was no evidence to suggest the presence of a modified allenylic cluster generated via an amide migration analogous to that described above.

Reaction of $Os_3H_2(CO)_{10}$ with internal alkynes gave both vinyl $[Os_3H(CO)_{10}(\mu-\eta^1,\eta^2-CR^1{=}CHR^2)]$ and alkyne $[Os_3(CO)_{10}(\mu_3-\eta^1,\eta^1,\eta^2-R^1C{\equiv}CR^2)]$ clusters. The latter cluster ($R^1 = R^2 = Me$) readily lost carbon monoxide with transfer of a hydrogen atom to the metal framework, forming $[Os_3H(CO)_9(\mu_3-\eta^1,\eta^2,\eta^2-MeC{=}C{=}CH_2)]$ (103) [Eq. (19)]. This

(19)

103

cluster was subsequently transformed to the allylic isomer on heating in toluene (156°C) (60).

In the presence of the radical substitution promoter $[Fe_2(CO)_4(\mu-SR)_2(PPh_3)_2]$, tertiary amines react with transition metal carbonyl clusters under exceptionally mild conditions. Modified allenyl and allylic clusters similar to those described earlier have been isolated from such reactions. Two distinct types of products have been isolated: (i) those involving the elimination of an alkyl group and (ii) those involving C–C coupling reactions. The product formation described in Scheme 31 was preceded by amine coordination, C–H activation, C–N cleavage, carbene–amine complex formation, transamination, and C–C coupling (61a,b). Such processes are of interest in the area of hydrodenitrification (61c).

E. Higher Nuclearity Allenyl Clusters

An allenyl ligand coordinated to the basal plane of a square pyramidal metal atom framework has been reported (62). Reaction of $[PPh_4]$ $[Fe_2(CO)_7(\mu-CH{=}CH_2)]$ with $[RhCl(CO)_2]_2$ in the presence of $TlBF_4$

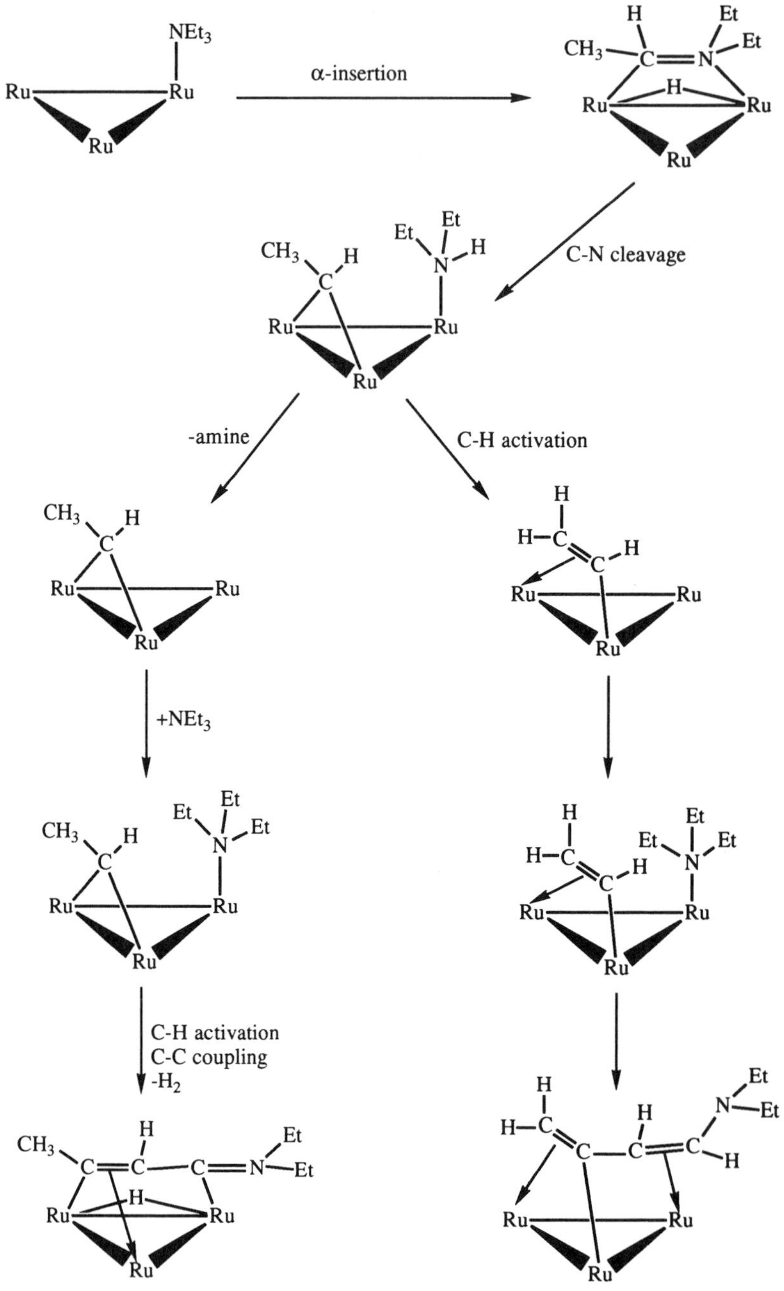

NEt3
Ru
α-insertion
H
Et
CH3
C
N
Et
H
Ru
C-N cleavage
-amine
C-H activation
+NEt3
C-H activation
C-C coupling
-H2

SCHEME 31

gave the pentanuclear cluster $[Fe_3Rh_2(CO)_{10}(\mu\text{-}CO)_3(\mu_4\text{-}\eta^1,\eta^2,\eta^2,\eta^2\text{-}CH_3C{=}C{=}CH_2)]^-$ **(104)** [Eq. (20)]. The rhodium atom is at the apex of

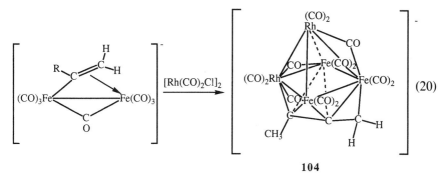

$$(20)$$

104

the cluster. The 3-methyl-3-allenyl ligand was assumed to result from the coupling of an ethylidyne and a vinylidene ligand, the latter derived from the initial vinyl ligand. The allenyl ligand displays an unusual μ_4-$\eta^1,\eta^2,\eta^2,\eta^2$ coordination to the Fe$_3$Rh base.

IV

STRUCTURAL CHARACTERISTICS OF ALLENYL AND RELATED COMPLEXES

A. Mononuclear Complexes

A description of both the allenyl and propargylic tautomers is included here because there is convincing evidence that the former often contains a significant contribution from a propargylic resonance structure (see Section V).

Both terminal allenyl and propargyl complexes are well known. However, structural data for both types of complexes remain scarce (Table II).

TABLE II

STRUCTURAL CHARACTERISTICS OF MONONUCLEAR η^1- AND η^3-PROPARGYL AND ALLENYL COMPLEXES

Compound (ref.)	CH$_2$–C (Å)	C–CR (Å)	∠Cβ (°)
$[Mo(C_6Me_6)(CO)_2(\eta^3\text{-}CH_2{-}C{\equiv}CH)]^+$ (63)	1.380(4)	1.236(4)	150.9(3)
$[Pt(PPh_3)_2(\eta^3\text{-}CH_2{-}C{\equiv}CPh)]^+$ (26c)	1.39(2)	1.23(1)	152(1)
$[ZrCp_2(Me)(\eta^3\text{-}CH_2C{\equiv}CPh)]$ (19)	1.344(5)	1.259(4)	155.4(3)
$[W(CO)_2Cp\{P(OMe)_3\}(\eta^1\text{-}CH_2{-}C{\equiv}CH)]$ (9c)	1.428(10)	1.174(11)	178.9(7)

Lin and co-workers have structurally characterized the tungsten propargyl complex [W(CO)$_2$Cp{P(OMe)$_3$}{η^1-CH$_2$C≡CH}] [2(I)a] (9c). In this compound the pendant acetylenic functionality retains triple bond character [C≡C length 1.174(11) Å] as well as a near-linear arrangement of carbon atoms [Cβ angle 178.9(7)°]. The role of the pendant propargylic triple bonds in template-promoted metal fragment condensations and the formation of allenylic clusters has already been discussed (Section III,B). As an alkyne-functionalized alkyl group, this organic fragment possesses the capacity to coordinate several metal centers (structures A–D in Scheme 32) as does the allenylic analog. For such complexes structural assignments have usually been based solely on spectroscopic characteristics. The spectroscopic properties (^{1}H, ^{13}CNMR; IR) of σ- bound allenyl and propargylic complexes differ considerably and provide a diagnostic tool for their differentiation (Section V).

Several mononuclear η^3-complexed propargyl ligands have been reported by Casey and Yi (13), Krivykh et al. (63), and Wojcicki and co-workers (26c). Casey described the hydrocarbyl ligand in [Re(η^5-C$_5$Me$_5$)(CO)$_2$(η^3-CH$_2$C≡CH$_3$)]$^+$ as a hybrid of both the propargylic and allenylic resonance structures [8(I) and 8(II) in Scheme 1], on the basis of spectroscopic evidence. For [Mo(η^6-C$_6$Me$_6$)(CO)$_2$(η^3-CH$_2$C≡CH)]$^+$ (105), prepared by Krivykh et al., the coordination mode was confirmed by single-crystal X-ray crystallography as was that in [Pt(PPh$_3$)$_2$(η^3-

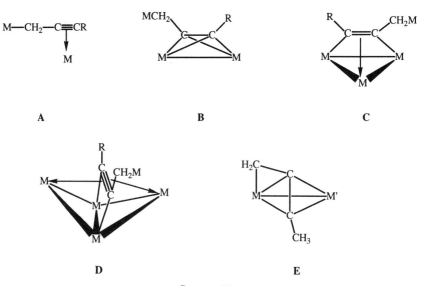

SCHEME 32

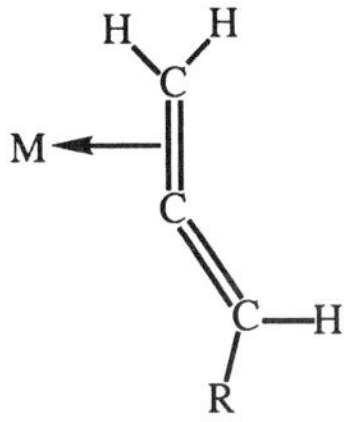

F

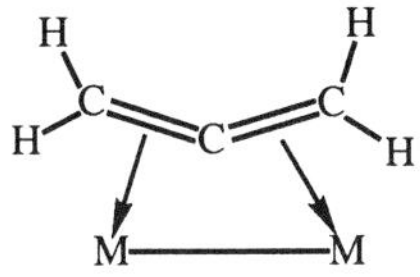

G

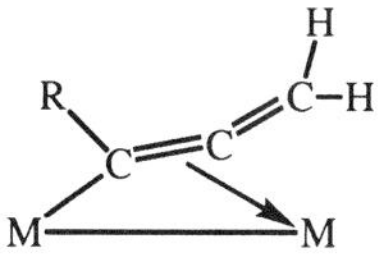

H

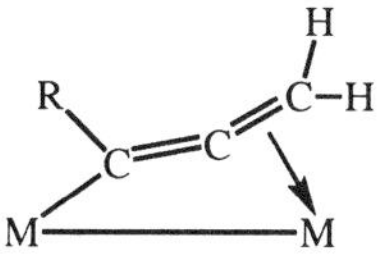

I

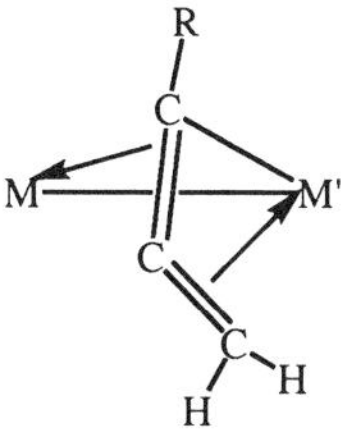

J

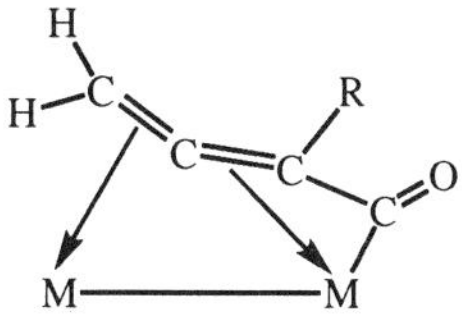

K

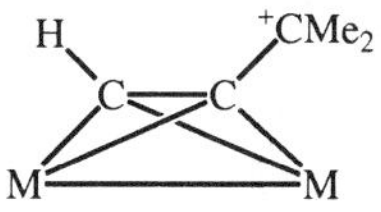

L

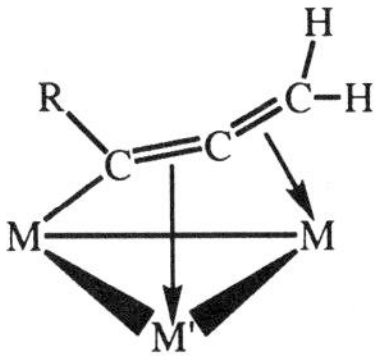

M(a)

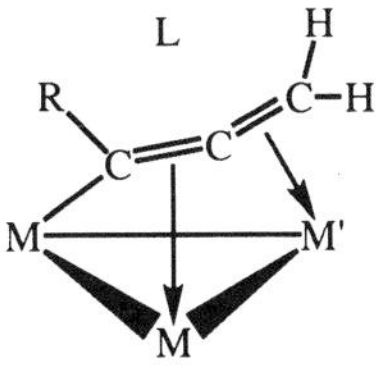

M(b)

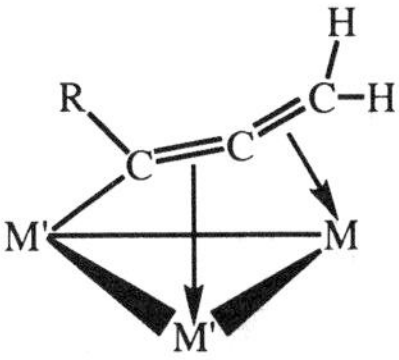

M(c)

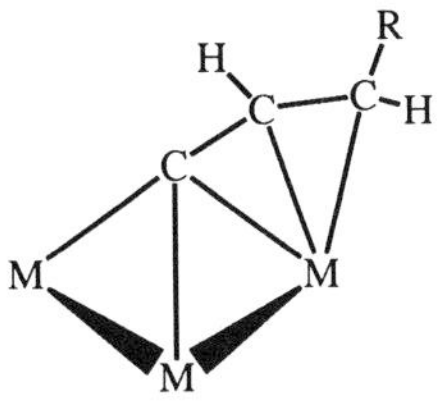

N

SCHEME 32 (*Continued*)

$CH_2\!=\!C\!=\!CPh)]^+$ (**39b**). The latter two complexes display very similar structural characteristics. Both complexes contain hydrocarbyl ligands in which $d(CH_2\text{–}C)$ exceeds $d(C\text{–}CR)$. For instance, in the molybdenum complex these distances are 1.380(2) Å and 1.263(4) Å, whereas those reported by Wojcicki are 1.39(2) and 1.23(1) Å. Such data were interpreted in terms of a large contribution from the η^3-propargyl resonance structure [**105(I)** and **39(I)**), respectively]. However, values of $^1J_{CH}$ indi-

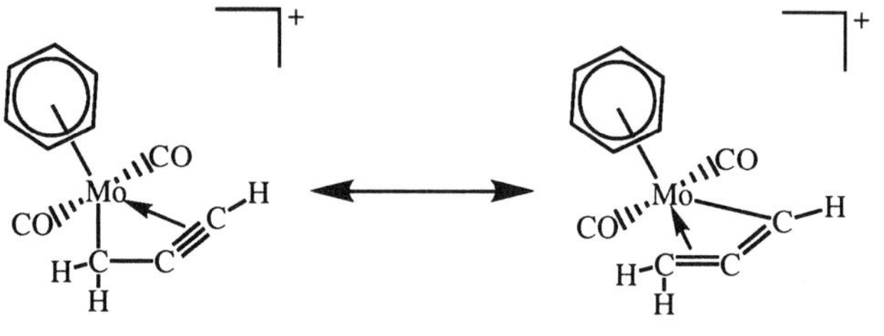

cate that the η^3-allenyl representations [**105(II)** and **39(II)**] are also important. It appears that these structural types have quite distinctive NMR characteristics (*vide infra*).

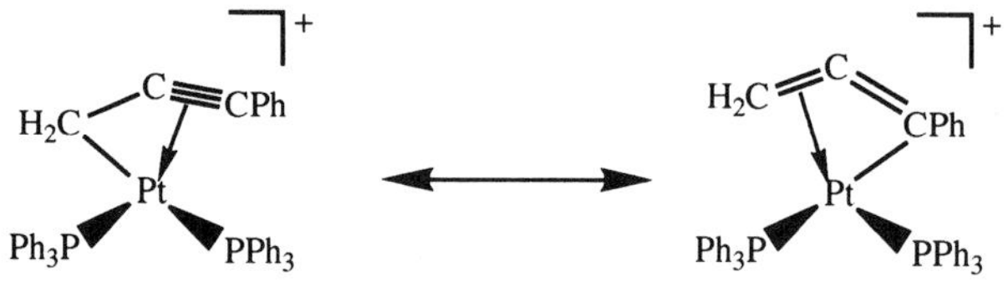

Unusual bonding was expected for the allenyl ligand of the Zr(IV) complex [$Zr(CH_3)Cp_2(\eta^3\text{-}PhC\!=\!C\!=\!CH_2)$] (**16**), and a single-crystal analysis established a rare example of an η^3-coordinated allenyl fragment (*19*). The structure of **16** consists of a normal bent metallocene fragment attached to a C_3 hydrocarbyl moiety. The bonding mode of the latter includes contributions from the η^3-allenyl and η^3-propargyl resonance forms, similar to the description used by Casey and Yi (*13*). Complex **16** has the following NMR and structural characteristics: a large $^1J_{CH}$ for the methylenic protons (167 Hz), allenylic bond lengths of 1.259(4) ($C\alpha\text{–}C\beta$) and 1.344(5) Å ($C\beta\text{–}C\gamma$), and angles summing to 359° at C_γ. These data support a greater contribution from resonance structure **16(II)**. As yet there are no reports of reactivity to confirm either allenylic or propargylic properties.

Osella and co-workers have prepared and characterized a novel example of a $\mu\text{-}\eta^1,\eta^2,\eta^2$-coordinated propargyl-bridged binuclear complex of

16(I)   16(II)

type **E**, via dehydroxylation of alkyne mono- and diol ligands (*27*). The close similarity between structures **A** and **E** is apparent, with metal–metal bond formation and coordination of the remaining orthogonal π orbital converting **A** to **E** (Scheme 32).

B. Binuclear Allenyl and Propargyl Complexes

Scheme 32 illustrates the known bonding types for binuclear allene- and allenyl-bridged clusters. The availability of two orthogonal π orbitals for metal–ligand bonding enables this fragment to adopt a range of structural types. The coordinating ability of the allenyl unit is not dissimilar to that of an alkyne; however, the presence of a hinge (=C=) carbon atom allows greater structural versatility. There are only a few examples of allenes with a single η^2-interaction (*9b,64a*) (type **F**). However, Wojcicki (*5b*) has suggested that the η^2-metalloallene (R = M) exists as an intermediate in the condensation reactions of propargylic metal complexes (Scheme 13). There are several reports of allenes adopting the μ-η^2,η^2-bridging mode (type **G**) across both bonded and nonbonded metals (*64b,c*). Allene coordinated in this manner is characterized by an axis skewed with respect to the metal–metal bond, enabling an optimization of its π-bonds for bridging. In complexes of this type the central carbon is distinctly bent and lies in closer proximity to the metal atoms than the terminal atoms. The allene unit appears to rehybridize and no longer contains orthogonal p orbitals. As a result the hydrogen atoms tend toward coplanarity.

Allenyl ligands bridging two metal atoms exist in one of two common structural types: μ-η^1,η^2 (**H**) and μ-η^2,η^3 (**J**). Both **H** and **J** type complexes have been isolated from the reaction of metal propargyls and metal carbonyls. Unfortunately, factors governing formation of a particular structural type are not well understood. The facile transformation of a μ-η^1,η^2-bound allenyl to a μ-η^2,η^3-coordination type by ligand abstraction has been described (*5b*). Triphenylphosphine abstraction from [Ru(CO)Cp(μ-η^1,η^2-PhC=C=CH$_2$)Pt(PPh$_3$)$_2$] by the phosphine scavenger Fe$_2$(CO)$_9$ provides the vacant coordination site necessary to accommodate the additional η^2 interaction. The μ-η^1,η^2-bonding mode has been described for several homonuclear allenyl complexes prepared either via the reaction of a propargylic halide with a dimeric metal carbonylate (*23*) or by nucleophilic attack of a carbene at C_α of a coordinated acetylide (*38*). Dickson *et al.* isolated the binuclear allenyl rhodium complex [Rh$_2$(CO)Cp$_2$(μ-η^1,η^2-CF$_3$C=CHCF$_3$)(μ-η^1,η^2-HC=C=CH$_2$)] of type **H** from the action of allene with the alkyne-bridged complex [Rh$_2$(CO)Cp$_2$(μ-η^2,η^2-CF$_3$C≡CCF$_3$)] and they showed that this cluster was capable of a further C–H oxidative addition to afford the tetranuclear compound [{Rh$_2$(CO)Cp$_2$(μ-η^1,η^2-CF$_3$C=CHCF$_3$)}$_2${μ-η^1,η^1,η^2,η^2-HC=C=CH}] (**106**) bridged by a doubly coordinating allenyl fragment (*52*).

Two forms of the μ-η^1,η^2-bridging allenyl ligand have been identified. The most common form for both homo- and heterometallic dimers is that in which the internal double bond (C_α–C_β) forms the η^2 interaction (type **H**). However, the previously unknown bonding type **I** in which the external double bond (C_β–C_γ) is coordinated in an η^2 manner to the adjacent metal center has been established (*38b*). The bimetallic compounds [Ru$_2$(CO)$_6$(μ-PPh$_2$)(μ-η^1,η^2-PhC=C=CPh$_2$)] and [Ru$_2$(CO)$_6$(μ-PPh$_2$)(μ-η^1,η^2-PhC=C=CH$_2$)] are of structural types **H** and **I** respectively, and provide an informative comparison. These types can be distinguished by the nomenclature μ-η^1,$\eta^2_{\alpha,\beta}$ and μ-η^1,$\eta^2_{\beta,\gamma}$ which specifies the internal (α,β) or terminal (β,γ) carbon atoms π-bound. The most striking differences in the structures are the angles about the central allenyl carbon, C_β. The latter clusters are characterized by a near-linear arrangement of carbon atoms (C_β angle ~172°), the former by a more acute angle (C_β angle ~145°). Table III contains some structural parameters for various μ-η^1,η^2-allenyl-bridged clusters of types **H** and **I**. Vastly disparate ^{13}C NMR

TABLE III

STRUCTURAL CHARACTERISTICS OF μ-η^1,η^2- AND
μ-η^2,η^3-ALLENYL-BRIDGED COMPLEXES

Compound (ref.)	Type	$C\alpha$–$C\beta$ (Å)	$C\beta$–$C\gamma$ (Å)	$\angle C\beta$ (°)
[Ru$_2$(CO)$_6$(μ-PPh$_2$)(μ-PhC=C=CPh$_2$)] (*38a*)	H	1.382(11)	1.350(12)	144.0(4)
[Ru$_2$(CO)$_6$(μ-PPh$_2$)(μ-PhC=C=CH$_2$)] (*38b*)	I	1.296(4)	1.355(5)	172.3
[Os$_2$(CO)$_6$(μ-PPh$_2$)(μ-PhC=C=CH$_2$)] (*38b*)	I	1.271(15)	1.353(16)	170.6(11)
[Fe$_2$(CO)$_6$(μ-SCMe$_3$)(μ-HC=C=CH$_2$)] (*23*)	H	1.363(9)	1.335(9)	156.7(7)
[Mo$_2$(CO)$_4$Cp$_2$(μ-HC=C=CH$_2$)]$^+$ (*25*)	J	1.35(3)	1.35(3)	139(2)
[RuFe(CO)$_4$Cp{μ-C(O)C(Ph)=C=CH$_2$}] (*31*)	K	1.37(2)	1.40(2)	146(1)
[WFe(CO)$_5$Cp(μ-C$_6$H$_4$MeC=C=CH$_2$)] (*31*)	J	1.372(8)	1.390(9)	130.2(6)
(second molecule in the asymmetric unit)		1.377(8)	1.390(9)	127.6(6)
[W$_2$(CO)$_4$Cp$_2$(μ-CH=C=CMe$_2$)]$^+$ (*65*)	L	1.536(19)	1.415(21)	136.1(14)

spectroscopic properties have been identified for each cluster type (see Section V).

Condensation reactions of group VI metal propargyls proceed, in all cases, with transfer of the terminal propargylic —CR fragment to the incoming metal carbonyl with formation of a σ M–C bond. Of the two possible isomeric μ-η^2,η^3-coordination types only that of **J** has been observed, that is, with the condensed metal fragment σ-bound to —CR and η^2-coordinated to the external (C_β–C_γ) allenyl double bond. The angular requirements imposed on this ligand as a result of these interactions result in an exceptionally acute angle at C_β.

Binuclear complexes containing allenyl(carbonyl) bridges (**K**) have been isolated from the condensation reactions of group VIII metal propargyls with $Fe_2(CO)_9$. These molecules differ from those of type **J** by the insertion of an additional carbonyl fragment into the hydrocarbon bridge. The effect on the metal–ligand connectivity is to cause a reversal of the η^2 interactions. The metal–ligand connectivity now involves a similar M–CR σ bond (albeit with an inserted CO) to the incoming metal fragment. However, this same metal is now η^2-coordinated to the internal (C_α–C_β) allenylic double bond, whereas the external double bond coordinates to the original propargylic metal center. The preference for this bonding mode over the analogous μ-η^2,η^3 type has been attributed to reduced angular strain, as a result of the extended hydrocarbon chain of the allenyl (carbonyl). We have drawn a comparison between these coordination types here because both types of binuclear complexes have been used successfully in further cluster expansion reactions, retaining their inherent η^2 connectivity present in the starting materials.

An example of bonding mode **L** was prepared via protonation of the μ-η^1,η^2-(4e)-allenylidene complex $[M_2(CO)_4 Cp_2(\mu$-η^1,η^2-C=C=CMe$_2)]$ (M = Mo, W) with $HBF_4\cdot Et_2O$ (65). Attack at C_α gave $[M_2(CO)_4 Cp_2(\mu$-η^2,η^2-HC$_2$CMe$_2)]^+$ (**107**) most accurately described as a propargylic car-

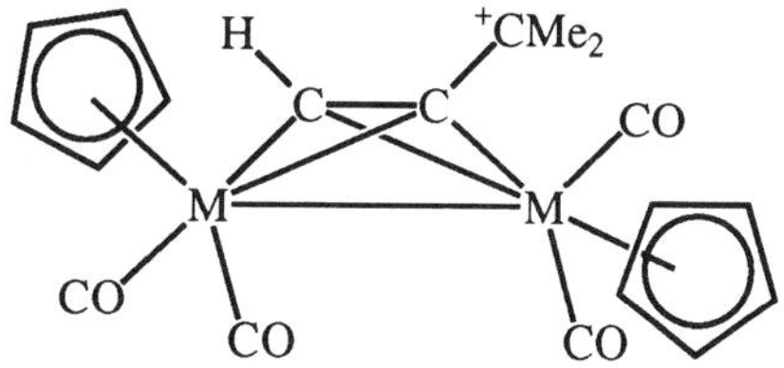

107

M= Mo, W

bocation, the partial delocalization of charge being reflected in the elongated η^2-C–C bond [1.536(19) Å].

Although the spectroscopic characteristics of allenyl-bridged complexes make an important contribution to their structural identification (Section V) single-crystal X-ray analysis is essential for unequivocal identification. Unfortunately there are only a handful of structurally characterized allenyl clusters. The formation of heterometallic allenyl complexes of type **H** and **J** from the condensation of a metal propargyl complex and a metal fragment requires the rehybridization of C_γ from sp^3 to sp^2. However, some structural and spectroscopic features indicate that they retain some of the original propargylic character. For instance, the C–C bond lengths in [Fe(CO)$_3$(μ-η^2,η^3-C$_6$H$_4$MeC=C=CH$_2$)W(CO)$_2$Cp] (**46**) support the allenylic formulation [C_α–C_β length 1.375(8) Å,C_β–C_γ length 1.390(9)Å] as does the trigonal environment about C_γ and its sp^2 hybridization ($^1J_{CH}$ 164 Hz). However, the angle about the central allenyl carbon atom is grossly distorted from 180°, suggesting a contribution to the structure by the propargylic resonance form.

The wide range of angles at C_β reflects the versatile coordinating ability of this ligand. For instance the μ-η^2,η^3-coordination types are typified by exceptionally acute angles at C_β (as low as 128.9°), whereas the μ-η^1,η^2-allenyl and μ-η^2,η^3-allenyl(carbonyl) bonding types **H, I,** and **K** contain a more linear arrangement of carbon atoms (145° < C_β angle < 172°). The C–C bond lengths of the η^2 interactions in compound types **H** and **I** are similar [C_α–C_β length 1.382(11) Å, C_β–C_γ length 1.355(5) Å for [Ru$_2$-(CO)$_6$(μ-PPh$_2$)(μ-η^1,$\eta^2_{\alpha,\beta}$-PhC=C=CPh$_2$)] (type **H**) and [Ru$_2$(CO)$_6$(μ-PPh$_2$)(μ-η^1,$\eta^2_{\beta,\gamma}$-PhC=C=CH$_2$)] (type **I**) respectively; the uncoordinated bonds, although longer, also have comparable lengths in both cluster types.

C. Trinuclear and Tetranuclear Allenyl Clusters

Type **M** is by far the most common coordination mode for cluster-bound allenyl fragments. Several homo- and heterometallic derivatives have been crystallographically characterized, and Table IV contains a selection of the more salient structural features. Deeming et al. have described several clusters in which severe structural distortions of the μ-allenyl ligand arise from the presence of heteroatom substituents with strong π donor character (56). These structural types are discussed further below. Both homo- and heterometallic clusters with μ-η^1,η^2,η^2-coordinated allenyls have been prepared. The C–C bond distances in these cluster types are similar to those found in the bimetallic μ-η^2,η^3-bridged

TABLE IV

STRUCTURAL CHARACTERISTICS OF μ-η^1,η^2,η^2-TRINUCLEAR ALLENYL AND MODIFIED ALLENYL CLUSTERS

Compound (ref.)	$C\alpha$–$C\beta$ (Å)	$C\beta$–$C\gamma$ (Å)	$\angle C\beta$ (°)
[Ru$_3$H(CO)(μ-EtC=C=CHMe)] (53)	1.365(8)	1.369(8)	142.3(6)
[Ru$_3$(CO)$_8$(μ-PPh$_2$)(μ-^{i}PrC=C=CH$_2$)] (40b)	1.352(8)	1.386(8)	143.7(3)
[Ru$_3$(CO)$_7$(μ-PPh$_2$)(μ-CH$_2$)(μ-^{i}PrC=C=CH$_2$)] (40b)	1.340(9)	1.371(10)	146.8(4)
[Fe$_2$W(CO)$_8$Cp(μ-C$_6$H$_4$MeC=C=CH$_2$)] (31)	1.371(5)	1.385(6)	144.3(3)
[Ru$_2$W(CO)$_8$Cp(μ-PhC=C=CH$_2$)] (31)	1.40(2)	1.37(2)	138(2)
[RuFe$_2$(CO)$_8$Cp(μ-PhC=C=CH$_2$)] (31)	1.372(7)	1.409(8)	148.4(5)
[Fe$_3$(CO)$_9${μ-OEtC=C=C(O)Me}]$^-$ (41)	1.378(4)	1.382(4)	140.6(2)
[Ru$_3$(CO)$_9$(μ-NEt$_2$C=C=CH$_2$)] (56b)	1.439(6)	1.394(6)	127.4(4)
[Ru$_3$H(CO)$_9$(MeC=C=CHNEt$_2$)] (56a)	1.385(5)	1.435(5)	131.9(3)

complexes. For example, C_α–C_β lengths of 1.372(8) and 1.371(5) Å and C_β–C_γ lengths of 1.390(9) and 1.385(6) Å are similar in [Fe(CO)$_3$(μ-η^2,η^3-C$_6$H$_4$MeC=C=CH$_2$)W(CO)$_2$Cp] (**46**) and [Fe$_2$W(CO)$_8$Cp(μ-η^1,η^2,η^2-PhC=C=CH$_2$)] (**47**), respectively. The most striking structural difference between these coordination modes is the angle at C_β. For ligands bridging three metal centers there is far less strain associated with the C$_3$ hydrocarbyl ligand, and the angle about this central carbon atom more closely resembles those found in the μ-η^1,η^2-allenyl (**H**) and the less strained μ-η^2,η^3-allenyl(carbonyl) complex (**K**).

The condensation products formed from the μ-η^1,η^2-binuclear allenyl systems retain their regiospecificity of η^2 interaction, namely, that to the propargylic metal center involves the internal double bond (C$_\alpha$–C$_\beta$). These trinuclear allenyl clusters appear to exist in a single isomeric form. In a similar manner the cluster expansion reactions of binuclear allenyl (carbonyl) complexes appear to proceed with retention of metal ligand η^2-interactions, with the external double bond retaining its interaction with the original propargylic metal atom. These clusters show no evidence of a dynamic "wagging" motion (Section VI) previously described for homonuclear allenyl clusters. Heterometallic allenyl clusters containing at least two different metals possess the capacity to exhibit isomerism as a consequence of the metal–hydrocarbon connectivity. However, cluster formation appears to occur with exceptional stereospecificity, a single isomer dominating in each instance.

Keister and co-workers isolated a cluster containing a hydrocarbyl fragment of type **N**, isomeric with the allenyl ligand (50). The cluster was identified as an intermediate in the carbon chain growth reactions of [Ru$_3$H$_3$(CO)$_9$(μ_3-CSEt)]. However the carbon–carbon bond lengths for

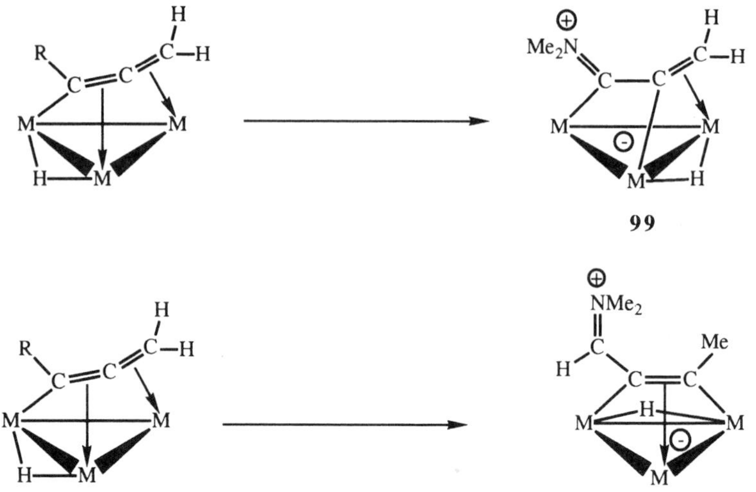

SCHEME 33

this bonding type are longer than those of normal allenyl ligands [1.400(8) and 1.429(8) Å], and the acute angles at the central carbon [116.6(6)°] closely resemble those of an allyl fragment.

Several hydrocarbyl-containing clusters were isolated from the reaction of $[Ru(CO)_2Cp(CH_2C\equiv CPh)]$ with $Fe_2(CO)_9$. Two of these, a binuclear allenyl(carbonyl) (52) (type K) and an allenyl-bridged trinuclear complex (51) (type M) have already been described. The remaining cluster compounds isolated contained rearranged allenyl ligands of an unprecedented form (Scheme 14). It was proposed that the isomeric allenyl ligand of cluster 51 arose from hydrogen scrambling, possibly via a hydridic intermediate. Clusters 53 and 54 containing the alkenyl-substituted alkylidyne and the allyl-substituted alkylidene (dimetalloallyl) ligands both result from hydrogen migration in the μ-η^1,η^2,η^2-allenyl cluster. Such clusters may well represent possible precursors to the isomeric allenyl cluster 51 (Scheme 14).

Unusual heteroatom-substituted allenyl clusters were isolated from the thermal reaction of amino-substituted alkynes and $Ru_3(CO)_{12}$. The most striking features of these clusters are the modified metal–ligand bonding interactions, a direct result of the π-donating capabilities of the heteroatom substituents. Scheme 33 describes the normal and modified bonding for these structural types; $[Ru_3(CO)_9(\mu_3$-η^1,η^1,η^2-$NMe_2C=C=CH_2)]$ (99) and $[Ru_3H(CO)_9\{\mu_3$-η^1,η^1,η^2-$CH_3C=C=C(NMe_2)H\}]$ (101) contain hydrocarbyl fragments in which the substituent positions indicate a formal ressemblance to allenyl ligands ($R^1C=C=CR^2R^3$). Table IV lists some of the more important structural features. A zwitterionic description $[Ru_3^{\ominus}H(CO)_9(\mu_3\text{-}\eta^1,\eta^1,\eta^2\text{-}Me_2N^+\!\!=\!\!C\!\!-\!\!C\!\!=\!\!CH_2)]$ of **99**, formally derived form an allenyl group by heteroatom substitution at C_α, was considered most appropriate. The allenyl bonding mode **98(a)** (Scheme 27) typified by $[Ru_3H(CO)_9(\mu_3\text{-}\eta^1,\eta^2,\eta^2\text{-}EtC\!\!=\!\!C\!\!=\!\!CHMe)]$ **(53)** contains identical C–C bond lengths for $C_\alpha\text{–}C_\beta$ and $C_\beta\text{–}C_\gamma$ [1.369(8) and 1.365(8) Å, respectively]. However, a structural clarification of **99** revealed rather different C–C bond lengths [1.439(5) and 1.394(6) Å]. The angle at the central allenyl carbon is also much smaller [127.4(4)°] than that in typical $\mu_3\text{-}\eta^1,\eta^2,\eta^2$-allenyl ligands [142.3(6)°].

A different type of structural distortion arises on heteroatom substitution at C_γ. The modified $\mu_3\text{-}\eta^1,\eta^1,\eta^2$-allenyl metal–ligand bonding in $[Ru_3H(CO)_9(\mu_3\text{-}\eta^1,\eta^1,\eta^2\text{-}MeC\!\!\equiv\!\!CCNMe_2H)]$ **(101)** is best described as a zwitterionic $\mu_3\text{-}\eta^2$-alkyne bridge containing two Ru–C σ bonds and an η^2 interaction. The C–C bond lengths in **101** are significantly different from those in a conventional allenyl cluster [1.385(5) and 1.435(5) Å]. Despite the modification in metal to ligand bonding, **101** exhibits a reactivity characteristic of a "normal" $\mu_3\text{-}\eta^1,\eta^2,\eta^2$-allenyl bridged cluster, that is, it readily undergoes thermal 1,2-hydrogen atom migration to give a 1,3-dimetalloallyl cluster, albeit with modified bonding.

V

NMR CHARACTERISTICS OF ALLENYL COMPOUNDS

The importance of ^{13}C and ^{1}H NMR spectroscopy as powerful methods for the assignment of coordination modes for hydrocarbyls in the absence of X-ray structure analysis cannot be overemphasized. The chemical shifts of ^{13}C nuclei in hydrocarbyl ligands are often diagnostic of a specific bonding type (66a,b). In this section we summarize some of the more important characteristics and trends in the ^{13}C and ^{1}H NMR data of propargylic and allenylic complexes. Conventional labeling of the allenyl carbon atoms as depicted in **VII** (Fig. 1) will be used throughout, and that for propargyl carbon atoms is shown in **VIII**: the metal-bound carbon is C_α, the central carbon is C_β, and the terminal carbon is labeled C_γ independent of the structural type.

A. Mononuclear Allenyl and Propargyl Complexes

During early pioneering work in this area of organometallic chemistry (12), the long-range couplings $^4J_{HH}$ were recognized as being diagnostic of the tautomeric bonding mode adopted. For instance, the prototypical allenylic fragment ($—CH{=}C{=}CH_2$) displays a larger four-bond coupling ($^4J_{HH}$ 6–7 Hz) than the propargylic counterpart ($^4J_{HH}$ 2–3 Hz). Lin and co-workers have demonstrated that $^4J_{HH}$ is diagnostic, providing spectroscopic data for both tautomeric forms of the tungsten complex [W(CO)$_3$ Cp(η^1-CH$_2$CCH)] [**2(I)**, propargyl; **2(II)**, allenyl] (*9a,c*). The allenylic isomer was reported to have a $^4J_{HH}$ of 6.5 Hz, greater than in the propargylic isomer ($^4J_{HH}$ 2.8 Hz). These ideas were reinforced by the value of 6.4 Hz for $^4J_{HH}$ in [ReCp(NO)(PPh$_3$)(η^1-CH=C=CH$_2$)] (*9b*). Table V contains ^{1}H NMR spectroscopic data for a range of allenyl and propargylic complexes.

Usually for a given structural type protons attached to the carbons directly bonded to the metal resonate to low field of the remaining protons in the hydrocarbon ligand. In other words, for the allenyl and propargyl fragments δHC_α exceeds δHC_γ (Table V).

With more ^{1}H NMR specroscopic data available for these complexes, further trends between the proton chemical shifts and structural types are becoming apparent. For example, in tautomerically related complexes, the methylene protons of η^1-bound allenyl ligands (type **VII**) appear downfield of propargylic counterparts ($1 \leq \Delta\delta \leq 2.5$ ppm), allowing a

TABLE V

SELECTED ^{1}H NMR DATA FOR ALLENYL AND PROPARGYL COMPLEXES

Compound (ref.)	δ CH (ppm)	δ CH$_2$ (ppm)	$^4J_{HH}$(Hz)
[Ru(CO)$_2$Cp(η^1-CH$_2$—C≡CPh)] (*10*)	—	2.17	—
[Ru(CO)$_2$Cp(η^1-CH=C=CH$_2$)] (*10*)	5.37	3.99	6.4
[ReCp(NO)(PPh$_3$)(η^1-CH$_2$—C≡CH)] (*9b*)	2.15	2.47/3.01	2.7
[ReCp(NO)(PPh$_3$)(η^1-CH=C=CH$_2$)] (*9b*)	6.59	3.98/4.05	6.4
[W(CO)$_3$Cp(η^1-CH$_2$—C≡CH)] (*9a*)	2.18	1.90	2.8
[W(CO)$_3$ Cp(η^1-CH=C=CH$_2$)] (*9a*)	5.47	4.06	6.7
[W(CO)$_2$Cp{P(OMe)$_3$}(η^1-CH$_2$—C≡CH)] (*9c*)	2.08	1.75	—
[W(CO)$_2$Cp{P(OMe)$_3$}(η^1-CH=C=CH$_2$)] (*9c*)	6.01	4.26	—
[Re(CO) Cp(η^3-CH$_2$—C≡CMe)] (*13*)	—	4.38/3.32	2.7
[Zr(Cp)$_2$(η^1-CH$_2$C≡CPh)(η^3-CH$_2$C≡CPh)] (*19*)	—	2.93	—
[Zr(CH$_3$)(Cp)$_2$(η^3-PhC=C=CH$_2$)] (*19*)	—	3.37	—
[Pt(PMe$_3$)$_3$(η^1-PhC=C=CH$_2$)]$^+$ (*26c*)	—	4.26	—
[PtBr(PPh$_3$)$_2$(η^1-CH$_2$—C≡CPh)] (*12*)	—	1.39	—
[Pt(PPh$_3$)$_2$(η^3-CH$_2$—C≡CPh)]$^+$ (*26c*)	—	2.74	—
[Pt(PPh$_3$)$_2$(η^3-CH$_2$—C≡CH)]$^+$ (*26b*)	4.60	2.91	—
[Pt(CO)(PPh$_3$)$_2$(η^1-CH$_2$—C≡CPh)]$^+$ (*26c*)	—	1.98	—
[Pt(CO)(PPh$_3$)$_2$\{η^1-C(Ph)=C=CH$_2$\}]$^+$ (*26c*)	—	3.65	—
[Mo(η^6-C$_6$Me$_6$)(CO)$_2$(η^3-CH$_2$—C≡CH)]$^+$ (*63*)	4.02	3.27	—

TABLE VI

SELECTED ^{13}C{^{1}H}NMR DATA FOR MONONUCLEAR ALLENYL AND
PROPARGYL COMPLEXES

Compound (ref.)	δCR (ppm)	δ—C— (ppm)	δCH$_2$ (ppm)	$^1J_{CH_2}$ (Hz)
[Ru(CO)$_2$Cp(η^1-CH$_2$—C≡CPh)] (10)	101.2	81.6	−26.4	142.6
[Ru(CO)$_2$Cp(η^1-CH=C=CH$_2$)] (10)	58.4	206.1	63.1	167.0
[W(CO)$_3$Cp(η^1-CH$_2$—C≡CH)] (9a)	68.4	92.0	−33.3	—
[W(CO)$_3$Cp(η^1-CH=C=CH$_2$)] (9a)	48.6	209.1	62.2	—
[W(CO)$_2$Cp{P(OMe)$_3$}(η^1-CH$_2$—C≡CH)] (9c)	65.8	95.5	−33.0	—
[ReCp(NO)(PPh$_3$)(η^1-CH$_2$—C≡CH)] (9b)	66.5	85.4	1.5	—
[ReCp(NO)(PPh$_3$)(η^1-CH=C=CH$_2$)] (9b)	64.3	207.1	60.7	—
[Re(CO)Cp(η^3-CH$_2$—C≡CMe)] (13)	76.6	56.7	29.0	—
[Zr(Cp)$_2$(η-CH$_2$—C≡CPh)$_2$] (19)	103.5	128.4	61.7	158.0
[Zr(CH$_3$)(Cp)$_2$(η^3-PhC=C=CH$_2$)] (19)	114.1	120.5	55.5	167.0
[PtBr(PPh$_3$)$_2$(η^1-CH$_2$—C≡CPh)] (26c)			−5.5	140
[Pt(PMe$_3$)$_3$(η^1-PhC=C=CH$_2$)]$^+$ (26c)	101.6	202.4	69.9	167.0
[Pt(PPh$_3$)$_2$(η^3-CH$_2$—C≡CPh)]$^+$ (26c)	102.1	97.3	48.3	170
[Pt(PPh$_3$)$_2$(η^3-CH$_2$—C≡CH)]$^+$ (26b)	90.6	101.4	51.9	171
[Pt(PPh$_3$)$_2$ (CO){η^1-C(Ph)=C=CH$_2$}]$^+$ (26c)	101.5	203.1	72.9	—
[Pt(CO)(PPh$_3$)$_2$(η^1-CH$_2$—C≡CPh)]$^+$ (26c)	92.2	86.1	7.1	—
[Mo(C$_6$Me$_6$)(CO)$_2$(η^3-CH$_2$—C≡CH)]$^+$ (63)	74.5	69.9	39.6	166.6

ready and reliable distinction between these η^1-hydrocarbyl bonding types. It appears that σ-bound allenylic methylene protons resonate within the region δ 3.9–4.3 ppm, whereas those associated with the propargylic tautomer clearly lie at much higher field between δ 1.75 and 3.0 ppm. Similarly large chemical shift differences ($\Delta\delta$) between the methyne (CH) protons of bonding types VII (5.37 < δCH < 6.01) and VIII (2.08 < δCH < 2.18) (see Table V) are apparent, reinforcing structural assignments based solely on spectroscopic evidence. Abnormal ^{1}H NMR chemical shifts for ligands expected to be allenylic in nature are often a reliable indication of the presence of an unusual bonding mode, such as η^3-coordination (vide infra), or a significant contribution from an alternative resonance structure. For instance, the methyne protons of η^3-propargyl/allenyl ligands lie midway between these two extremes (4.0 < δCH < 4.06), and although data are presently somewhat limited this has proved to be a reliable structural probe.

Table VI lists several of the characteristic ^{13}C NMR parameters used for structural elucidation of allenyl and propargyl complexes. The distinctive ^{13}C chemical shifts (δC) and associated $^1J_{CH}$ values of allenyl complexes have been used to distinguish different bonding types. The allenylic ^{13}C methylene signal appears downfield of the corresponding resonance of the propargyl tautomer. For instance, C_γ of $[W(CO)_3Cp(\eta^1\text{-}CH{=}C{=}CH_2)]$ (δ 62.2 ppm) lies to low field of its corresponding propargylic C_α resonance located at δ $-$ 33.3 ppm: for $[ReCp(NO)(PPh_3)(\eta^1\text{-}CH{=}C{=}CH_2)]$ (δC_γ 60.7 ppm) a similar spectroscopic relationship exists with $[ReCp(NO)(PPh_3)(\eta^1\text{-}CH_2C{\equiv}CH)]$ (δC_γ 1.5 ppm).

Contrasting spectroscopic properties of the two isomeric forms of $[Pt(CO)(PPh_3)_2(PhC_3H_2)]$ have been reported. Vastly disparate ^{13}C chemical shifts for the methylene and quaternary carbon atoms were recognized (propargyl: CH_2, 7.1; $\equiv C$, 86.1; CPh, 92.2; and allenyl: CH_2, 72.9; $=C=$, 203.1; CPh, 101.5 ppm). The quarternary carbon atoms of the η^1-allenylic complexes invariably resonate at extremely low field (δ 200 ppm) distinct from those of the propargylic tautomer which appear at much higher field ($\Delta\delta \geq 120$ ppm). This low-field signal of the central η^1-allenyl carbon atom signifies a high degree of allenic character for C_β (66a) [note that δ C_β for free allene is at 213.5 ppm (67)]. The values of $^1J_{CH}$ often reflect the different states of hybridization of the carbon atoms in these complexes, while remaining a reliable indication of the extent of contributions from competing resonance structures.

The carbon atom chemical shifts for the majority of σ-bound allenylic and propargylic complexes obey a general order of upfield movement ($\delta C_\beta > \delta C_\alpha$, δC_γ) similar to that previously established for dinuclear $\mu\text{-}\eta^1,\eta^2\text{-}$ and $\mu\text{-}\eta^2,\eta^3$-bound allenyl complexes (66a,b). Independent of structural type, the highest field signal often corresponds to that carbon directly σ-bonded to the metal, although there are exceptions to this generalization.

Both Green and co-workers (15) and Templeton and co-workers (16a) have reported unusual ^{13}C chemical shifts for the carbon atoms of η^3(3e)-bound allenylic ligands prepared by deprotonation of cationic η^2-alkyne complexes. A low-field signal (δ 253.5 ppm) in the ^{13}C NMR spectrum of $[MoCp\{P(OMe)_3\}\{\eta^2\text{-}C(Ph)C{=}CHPh\}]$ (10) was suggested as evidence supporting the presence of an $Mo{=}C$ double bond possessing a high degree of carbenic character. Higher field shifts (δ 151.6 and 109.0 ppm) were assigned to C_β and C_γ, respectively. Templeton and co-workers reported (16a) spectroscopic properties similar to those noted by Green and co-workers, with the complex $[W(CO)(dppe)(dtc)(\eta^3\text{-}MeOC{=}C{=}CHPh)]$ (11) also displaying a characteristic low-field shift for C_α (δ $\sim$250 ppm).

The spectroscopic properties of η^3-propargyl complexes reinforce the importance of ^{13}C NMR spectroscopy in establishing the extent of contribution from competing resonance structures. One report (13) acknowledged that the formally propargylic complex $[Re(CO)Cp(\eta^3\text{-}CH_2C{\equiv}CH)]$ [8(I)] contained a significant contribution from allenylic $[Re(CO)Cp(\eta^3\text{-}$

$CH=C=CH_2)$] [**8(II)**], a deduction based solely on its spectroscopic characteristics, namely, the large value of $^1J_{CH}$ associated with the methylenic resonance. Similar deductions by Chen established the existence of competing resonance forms for [(Pt(PPh$_3$)$_2$(η^3-CH$_2$CCH)]$^+$ (**39a**) (*26b*).

The nonrepresentative ^{13}C chemical shifts of the η^3-allenylic ligand in [Zr(CH$_3$)Cp$_2$(η^3-PhC=C=CH$_2$)] (**16**) reinforce the importance of ^{13}C NMR spectroscopy as a structural probe. The shifts C$_\alpha$ (δ 114.1 ppm) and C$_\beta$ (δ 120.5 ppm) appear at low field of the region normal for η^1- and η^3-bound propargyl carbon atoms, whereas C$_\beta$ lies far upfield of the region normally associated with η^1-allenylic ligands (δ ~200 ppm). These features suggest a severe reduction in allenic character. However, this upfield shift is not within the range expected for η^3-propargyl coordination. The structure of **16** was crystallographically established; the sum of the bond angles about the methylenic carbon [359(6)°], the value of $^1J_{CH}$ for these protons (167 Hz), and the bond distances [d(CH$_2$–C) 1.344(5) Å; d(C–CH) 1.259(4) Å] support the importance of the allenic resonance structure. In addition, the methylenic proton chemical shift of δ 3.37 ppm is unusually downfield of that found for η^1-propargylic complexes but at too high a field to be η^1-allenylic in nature. Thus, this resonance appears in the region that has previously been assigned to methylenic protons of propargylic ligands adopting an η^3-coordination mode (Table V).

B. Binuclear Allenyl Complexes

Tables VII and VIII contain a selection of ^{1}H and ^{13}C NMR spectroscopic data available for a range of allenyl-containing bimetallic complexes. In many instances the methylenic protons of μ-η^1,η^2- and μ-η^2,η^3-allenyl ligands exhibit magnetic nonequivalence, and variable temperature NMR spectroscopy is often an effective probe for studying

TABLE VII

Selected ^{1}H NMR Data for Allenyl-Bridged Binuclear Complexes

Compound (ref.)	δH_α (ppm)	δH_β (ppm)	δH_γ (ppm)
[Ru$_2$(CO)$_6$(μ-PPh$_2$)(μ-PhC=C=CH$_2$)] (*38*)	—	2.11	1.46
[Fe$_2$(CO)$_6$(μ-SCMe$_3$)(μ-HC=C=CH$_2$)] (*23b*)	7.39	5.39	—
[Fe$_2$(CO)$_6$(μ-SEt)(μ-HC=C=CH$_2$)] (*23b*)	7.39	5.30	—
[Fe$_2$(CO)$_6$(μ-SCMe$_3$)(μ-HC=C=CH$_2$)] (*23b*)	—	5.20	—
[Mo$_2$(CO)$_4$Cp$_2$(μ-CH=C=CH$_2$)]$^+$ (*25*)	6.76	5.33	5.05
[MoFe(CO)$_5$Cp(μ-PhC=C=CH$_2$)] (*31b*)	—	4.73	3.95
[WFe(CO)$_5$Cp(μ-MeC=C=CH$_2$)] (*31b*)	—	5.10	3.65

TABLE VIII

SELECTED ^{13}C NMR SPECTROSCOPIC DATA FOR ALLENYL-BRIDGED
BINUCLEAR COMPLEXES

Compound (ref.)	δC_α (ppm)	δC_β (ppm)	δC_γ (ppm)	$^1JC_\alpha H$ (Hz)	$^1JC_\gamma H$ (Hz)
[Ru$_2$(CO)$_6$(μ-PPh$_2$)(μ-PhC=C=CH$_2$)] (38b)	140.2	99.2	1.0	—	—
[Fe$_2$(CO)$_6$(μ-SCMe$_3$)(μ-HC=C=CH$_2$)] (23b)	113.2	177.7	95.5	162.0	166.0
[Fe$_2$(CO)$_6$(μ-SEt)(μ-HC=C=CH$_2$)] (23b)	114.8	176.6	93.4	162.0	167.0
[Fe$_2$(CO)$_6$(μ-SCMe$_3$)(μ-EtC=C=CH$_2$)] (23b)	148.6	177.1	94.7	—	163.0
[Fe$_2$(CO)$_6$(μ-SCMe$_3$)(μ-HC=C=CMe$_2$)] (23b)	119.6	171.8	115.4	166.0	—
[MoFe(CO)$_5$Cp(μ-MeC=C=CH$_2$)] (31b)	93.08	92.07	67.24	—	—
[MoFe(CO)$_5$Cp(μ-PhC=C=CH$_2$)] (31b)	83.1	120.1	70.00	—	—
[WFe(CO)$_5$Cp(μ-MeC=C=CH$_2$)] (31b)	84.58	112.3	67.88	—	164.0
[WFe(CO)$_5$Cp(μ-PhC=C=CH$_2$)] (31b)	80.5	111.76	70.53	—	—
[RuFe(CO)$_4$Cp{μ-C(O)C(Ph)C=CH$_2$}] (32)	47.0	201.8	15.8	—	162.6

exchange mechanisms (Section VI). The ^{1}H NMR spectra of μ-η^1,η^2-bound allenyl ligands are characterized either by a single high-field signal or two weakly coupled resonances in the region δ 3.65–5.5 ppm associated with the methylenic protons and a vinylic (CH) signal located at low field (δ 6.5–7.5 ppm). The geminal coupling ($^2J_{HH}$) can vary within the range 0.7–11.0 Hz.

^{13}C NMR spectroscopy is also a very useful predictive tool (66a,b). The chemical shift values of the C$_3$ allenylic carbon atoms cover a wide range, both for homo- and heterometallic complexes (δC_α 80.5–148.6 ppm, δC_β 92.07–177.7 ppm, δC_γ 1.00–115.4 ppm). However, the chemical shifts of these carbon atoms generally move upfield in the order $\delta C_\beta > \delta C_\alpha > \delta C_\gamma$, there being very few exceptions to this ordering. Again, the magnitude of $^1J_{CH}$ in these complexes attests to the sp^2 character of C$_\gamma$ ($^1J_{CH}$ 160–175 Hz). As might be expected the central allenylic carbon atom chemical shift appears to be sensitive to the angle at C$_\beta$. Substantiating this is the fact that complexes containing the μ-η^2,η^3-allenyl ligands (type J), (Scheme 32) display higher field shifts for this carbon than do the μ-η^1,η^2 counterparts. Further evidence in support of this correlation is provided by a comparison of the ^{13}C NMR data of the dinuclear μ-η^2,η^3- and trinuclear μ-η^1,η^2,η^2-allenyl ligands (Section V,C) for which there is a dramatic difference in the angles at C$_\beta$. The chemical shifts in μ-η^2,η^3 complexes have C$_\beta$ typically in the region δ 92–120 ppm, whereas those of trinuclear μ-η^1,η^2,η^2 complexes appear at δ 142–162 ppm, indicating greater allenic character.

As an example of the value of ^{1}H and ^{13}C NMR studies in this area the anomalously high-field chemical shift of C$_\gamma$ in [Ru$_2$(CO)$_6$(μ-PPh$_2$){μ-

$\eta^1,\eta^2_{\beta,\gamma}$-C(Ph)=C=CH$_2$}] (**67**) led to a reinvestigation of the structure of this molecule, which was previously assumed to have the μ-$\eta^1,\eta^2_{\alpha\beta}$-coordination mode established by X-ray crystallography for [Ru$_2$(CO)$_6$(μ-PPh$_2$)(μ-$\eta^1,\eta^2_{\alpha\beta}$-C(Ph)=C=CPh$_2$)] (*38a*). An X-ray structural analysis (*38b*) of **67** revealed an unprecedented μ-η^1,η^2 bonding interaction involving the external (C$_\beta$–C$_\gamma$) double bond of the allenyl ligand. The high-field shift associated with C$_\gamma$ is consistent with a degree of M–C σ bonding and shielding of the C$_\gamma$ nucleus by the metal.

C. Trinuclear Allenyl Clusters

Table IX contains ^{13}C NMR data for both homo- and heteronuclear allenyl clusters. The ^{13}C NMR chemical shifts of these cluster-bound hydrocarbyl ligands differ substantially from those of the binuclear analogs. However, there appears to be a correlation between bonding type and the number of metal atoms involved. Thus C$_\gamma$ of a μ_3-η^1,η^2,η^2-bound allenyl ligand resonates at high field, δ 20.1–38.1 ppm, of the corresponding carbon in a μ-η^2,η^3-coordination environment, δ 67.24–70.53 ppm ($\Delta\delta$ ~40–50 ppm). The central allenyl carbon, however, appears at lower field [in the range δ 142.9–161.59 ppm ($\Delta\delta$ 30–40 ppm)] than in binuclear μ-η^2,η^3 counterparts (δ 92.07–120.1 ppm). This shift correlates with an increase in the angle at C$_\beta$, possibly indicating greater allenic character for the former carbon atoms. The remaining carbon atom C$_\alpha$ also displays a low-field shift relative to that of a μ-η^2,η^3-coordinated ligand ($\Delta\delta$ 40 ppm). These trends reflect changes in δ(^{13}C) for other cluster-bound hydrocarbyls including acetylides.

In summary, there is a wide range of chemical shift values for C$_\alpha$, C$_\beta$, and C$_\gamma$ of the allenyl ligands. However, the ^{13}C chemical shifts of individual carbon atoms of the allenyl group are sensitive to whether they are (i) coordinated or uncoordinated (e.g., η^1 versus η^3) (ii) the specific bonding

TABLE IX
^{13}C{^{1}H} NMR Data for Selected Trinuclear Allenylic Clusters

Compound (ref.)	δCα (ppm)	δCβ (ppm)	δCγ (ppm)
[Ru$_3$(CO)$_8$(μ-PPh$_2$)(μ-^{i}PrC=C=CH$_2$)] (*40b*)	172.0	142.7	20.1
[Fe$_3$(CO)$_9${μ-OEtC=C=CHC(O)OMe}]$^-$ (*41*)	200.1	146.1	27.7
[Fe$_3$(CO)$_9$(μ-OEtC=C=CHCOMe)]$^-$ (*41*)	201.9	145.3	38.1
[Fe$_2$W(CO)$_8$(μ-PhC=C=CH$_2$)] (*31b*)	—	155.86	27.8
[Fe$_2$W(CO)$_8$(μ-C$_6$H$_4$Me—C=C=CH$_2$)] (*31b*)	136.36	156.68	26.99
[Ru$_2$W(CO)$_7$(μ-PhC=C=CH$_2$)] (*31b*)	114.09	161.59	26.99

mode (e.g., μ-η^1,$\eta^2_{\alpha,\beta}$ versus μ-η^1,$\eta^2_{\beta,\gamma}$), and (iii) the number of metals involved (e.g., μ_2-η^2,η^3 in a binuclear complex where all three carbons are bound versus μ_3-η^1,η^2,η^2 in a trinuclear system where all three carbons are attached). Although there are clearly some pitfalls in assigning bonding modes soley on the basis of chemical shifts, the general ordering of $\delta(^{13}C)$ is relatively consistent with $\delta c_\gamma \ll \delta c_\beta \leq \delta c_\alpha$.

VI

LIGAND DYNAMICS: SOLUTION NMR STUDIES

^{13}C and ^1H NMR spectroscopy has been extensively used for probing the ligand dynamics of metal-bound C_2 hydrocarbyl fragments (68), and many of these processes are now well understood. However, there have been relatively few reports on the fluxional characteristics of C_3 allenylic hydrocarbyl units.

[ZrCp$_2$(η-CH$_2$CCPh)$_2$] (17), prepared by Wojcicki and co-workers (19), is the first complex reported to contain both η^1- and η^3-allenyl/propargyl ligands. At ambient temperatures both ^1H and ^{13}C NMR spectroscopy indicated an apparent equivalence of the two C_3 hydrocarbyl ligands. The ^1H NMR spectrum of 17 is temperature dependent. At 180 K the peak originally assigned to the time-averaged methylene resonance (δ 2.80 ppm) appeared as two separate signals, one centered at δ 3.30 the other at δ 1.90 ppm. The latter was readily assigned to the methylene protons of an η^1-coordinated propargyl ligand. The former signal, however, could be attributed neither to the methylene protons of an η^1-allenyl ligand (see Section V,A) nor to those of an η^1-propargyl. However, the resonance lies within the range now established as characteristic of the unusual η^3-coordination type (Section V). These spectroscopic characteristics suggested that the ground state structure of 17 was [ZrCp$_2$(η^1-CH$_2$C≡CPh)(η^3-PhC=C=CH$_2$)] with a low-energy barrier to the interconversion process (illustrated in Scheme 34). Previous examples of related η^1-η^3 hydrocarbyl interconversions include those of the allyl type ligand and have been extensively documented (69).

Templeton demonstrated that the two isomers of [W(CO)(dppe)(dtc) (η^2-MeOC=C=CHPh)] interconvert (T_c ~120°) on the NMR time scale. Molecular orbital (MO) calculations on the model complex [WH$_4$(CO)(η^2-HOC=C=CH$_2$)] provided the rotational profile shown in Fig. 2 characterized by a low barrier to rotation of the allenyl fragment with respect to the metal (rotation about the z axis) and a higher barrier for rotation about the C_β–C_γ double bond. The former process possesses two energy minima

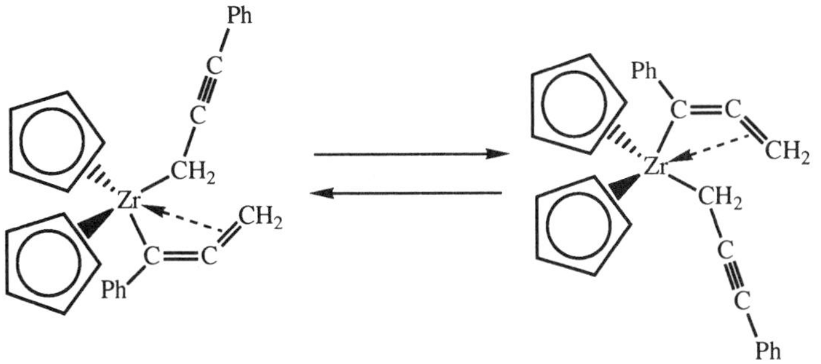

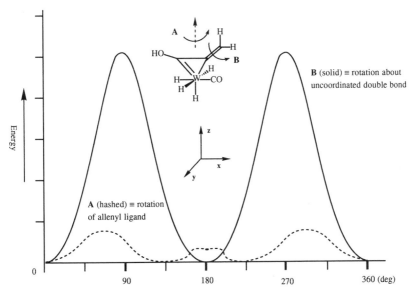

SCHEME 34

for rotation of the allenyl, one at 130° and the other at 230°. Both the NMR characteristics and the MO calculations supported an isomerization mechanism via restricted rotation of the allenyl fragment rather than rotation about C_β –C_γ (*16a*).

Rapid rotation of the allenyl fragment in [MoCp{P(OMe)$_3$}{η^2-C(Ph)C═CHPh}] was reported earlier by Green and co-workers (*15*). The observation of two AB quartets in the low-temperature ^{31}P NMR spectrum of **10** was attributed to isomers **10a** and **10b** which differ by virtue of

FIG. 2. Rotational profile for the model complex WH$_4$(CO)(η^2-HOC═C═CH$_2$) (see text). Adapted from Gamble *et al.* (*16a*).

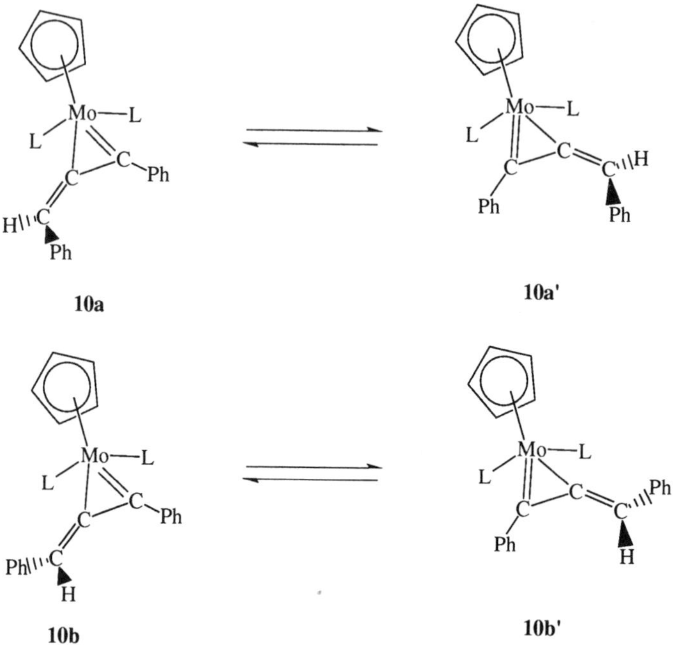

10a

10a'

10b

10b'

SCHEME 35

the disposition of the C_γ substituents with respect to the molybdenum atom. The two isomers interconvert with their mirror images via a windshield wipe motion, namely, exchange of **10a** with **10a'** and **10b** with **10b'**, affording two sharp singlets at the fast exchange limit (Scheme 35).

One of the first fluxional processes of an allenyl ligand to be investigated by variable-temperature ^{1}H and ^{13}C NMR spectroscopy was that of [Ru$_3$H(CO)$_9$(μ-η^1,η^2,η^2-CH$_3$C=C=CMe$_a$Me$_b$)] (**98**) (70). The high-temperature equivalence of the methyl groups and a 2-fold symmetry plane at the metal cluster led to the proposal of a "wagging" motion for the organic moiety. Such a process implies an oscillation of the η^2-interactions between the two metal centers coupled with edge hopping of the hydride as shown in Scheme 36.

Detailed ^{1}H and ^{13}C variable-temperature NMR studies of [Ru$_2$(CO)$_6$(μ-η^1,η^2-PhC=C=CH$_2$)] (**67**) have been carried out as part of a study of the ligand dynamics of a μ-η^1,η^2-allenyl ligand (38a). The molecule is static on the NMR time scale at room temperature, with the NMR characteristics being fully consistent with the solid state structure. Variable-temperature ^{13}C NMR studies revealed trigonal rotation of a single Ru(CO)$_3$ unit as the only process accessible by NMR spectroscopy in the temperature range 30–90°C. Within the same temperature range ^{1}H NMR spectra revealed exchange of the methylenic protons. Several plausible mechanisms were

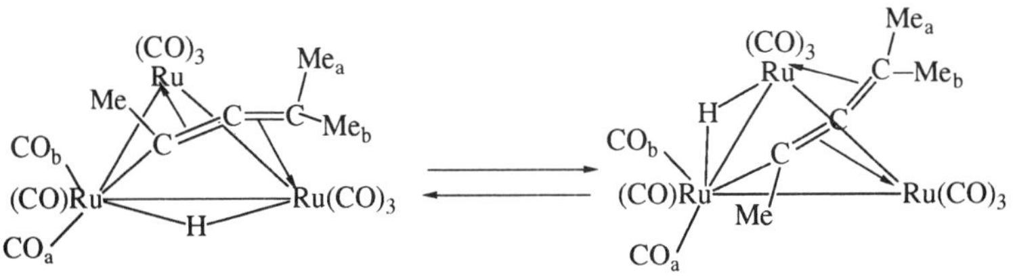

98c   98d

SCHEME 36

examined to account for these observations, including (i) $\sigma-\pi$ windshield wiper type exchange, similar to that well established for acetylides and vinyls, (71), (ii) phosphido bridge exchange (72), (iii) base-catalyzed isomerization via nucleophilic attack at C_β generating an intermediate alkylidene and leading to free rotation about C_α–C_β (73), and (iv) rotation about C_β–C_γ via a zwitterionic transition state. A suitable model for the transition state associated with this process is provided by the structurally characterized cluster $[Fe_2(CO)_6(\mu\text{-}PPh_2)(\mu\text{-}\eta^2\text{-}C\{CNMe(CH_2)_2NMe\}CPh)]$ (108) for which restricted rotation about C_β–C_γ would explain methyl resonance exchange (74).

Seyferth and co-workers described the preparation and spectroscopic characteristics of a series of $\mu\text{-}\eta^1,\eta^2$-allenyl-bridged complexes $[Fe_2\text{-}(CO)_6(\mu\text{-}SR)(\mu\text{-}\eta^1,\eta^2\text{-}HC{=}C{=}CH_2)]$ (34) for which the methylene protons appeared equivalent at room temperature (23). A close structural similarity between the allenyl bonding mode and that of a bridging $\sigma-\pi$-vinyl was noted and led to the suggestion that the allenyl and vinyl ligands underwent a similar fluxional process (Scheme 37).

The cationic allenyl-bridged complex $[Mo_2(CO)_4Cp_2(\mu\text{-}\eta^2,\eta^3\text{-}HC{=}C{=}CH_2]^+BF_4^-$ (36) is also dynamic (25). Here the methylene protons remained distinct, but exchange of the cyclopentadienyl groups was observed (25). The temperature dependence of the ^1H NMR spectra was rationalized by a process (Scheme 38) which interconverts enantiomers 36a and 36b (75).

The possibility that a similar exchange process occurs in the cationic cluster $[Fe_2(CO)_3Cp_2(\mu\text{-}\eta^2,\eta^3\text{-}HC{=}C{=}C^tBuH)][PF_6]$ has not been ruled out (42c). Unfortunately the asymmetry of the molecule precluded an investigation of ligand dynamics, the cyclopentadienyl rings remaining nonequivalent even in the presence of an allenyl isomerization. The fluxional μ-allenyl isomerization (Scheme 39) shows a transition state closely resembling that proposed by Curtis (25). However, the environments of the cyclopentadienyl rings remain distinct.

Unusual ligand dynamics in the modified allenyl cluster $[Ru_3H(CO)_9(\mu\text{-}\eta^1,\eta^1,\eta^2\text{-}Me_2NC{=}C{=}CH_aH_b)]$ (99) were described by Aime et al. (56b).

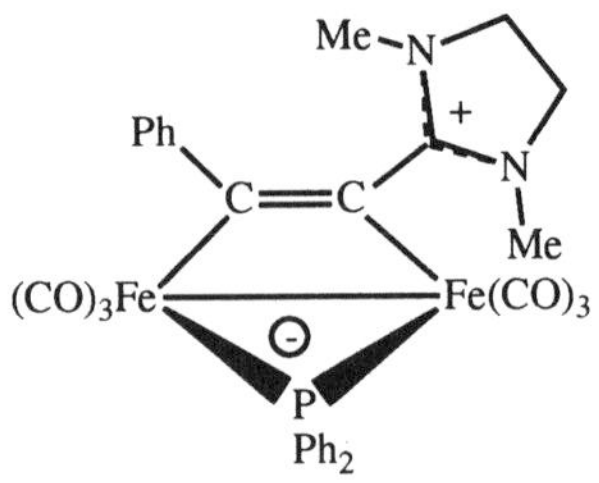

108

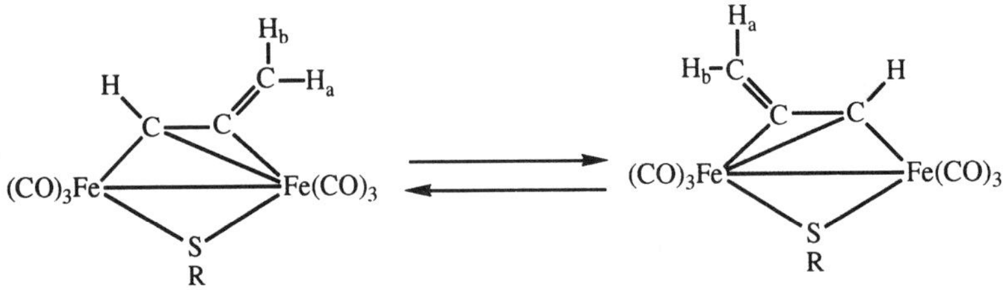

SCHEME 37

Simple σ-π-vinyl type motion as shown for $[Os_3H(CO)_{10}(\mu\text{-}\eta^1,\eta^2\text{-}CH{=}CH_2)]$ (*71a*) and $[Os_3H(CO)_9(\mu_3\text{-}\eta^1,\eta^2\text{-}PEt_2CH{=}CH_2)]$ (*76*) was not sufficient to explain the observed NMR averaging of the methylenic protons in **99,** since such a process would maintain the cis and trans relation-

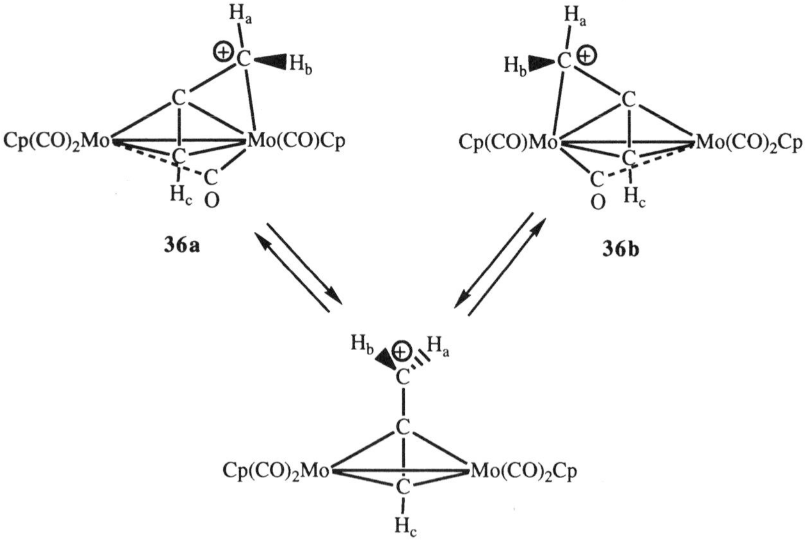

SCHEME 38

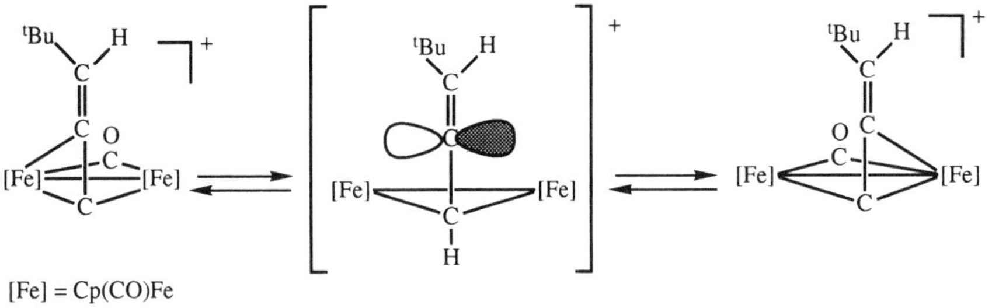

$[Fe] = Cp(CO)Fe$

SCHEME 39

ship of H_a and H_b with respect to the Ru–C σ bond. Processes X and Y (Scheme 40) were proposed to account for the observed NMR averaging. A minor geometric perturbation of **99** would transform the modified hydrocarbyl ligand into the more familiar μ-η^1,η^2,η^2-bonding type and coupled with the low-energy process (Y) would exchange the methylenic environments (Scheme 40). The latter process has precedence in the ligand dynamics of $[Ru_3H(CO)_9(\mu\text{-}\eta^1,\eta^2,\eta^2\text{-}MeC{=}C{=}CMe_aMe_b)]$ **(98c,d)** (Scheme 36) described in the early studies of Rosenberg and coworkers *(70)*.

The allenyl cluster $[Ru_3H(CO)_9(\mu\text{-}\eta^1,\eta^1,\eta^2\text{-}MeCCC(H)NMe_2)]$ **(101)**, with a hydrocarbyl ligand which closely resembles a zwitterionic alkyne,

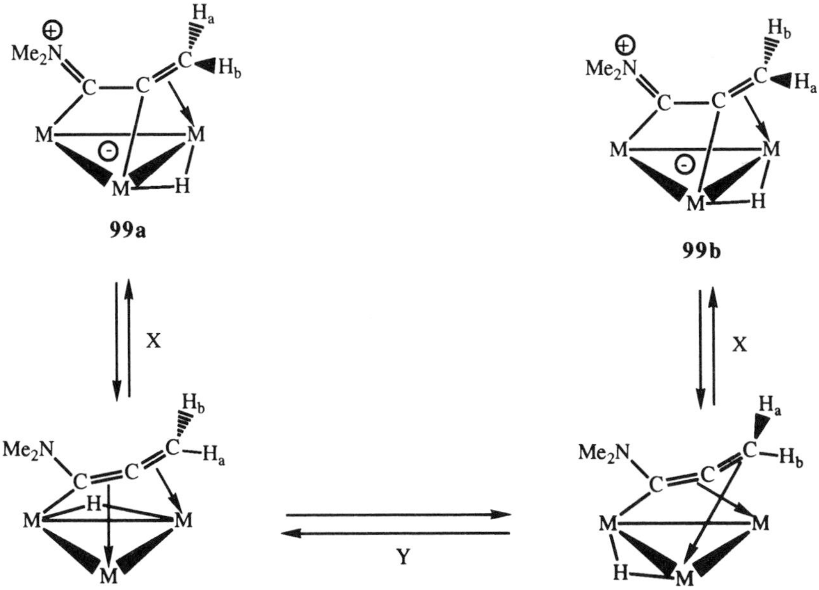

SCHEME 40

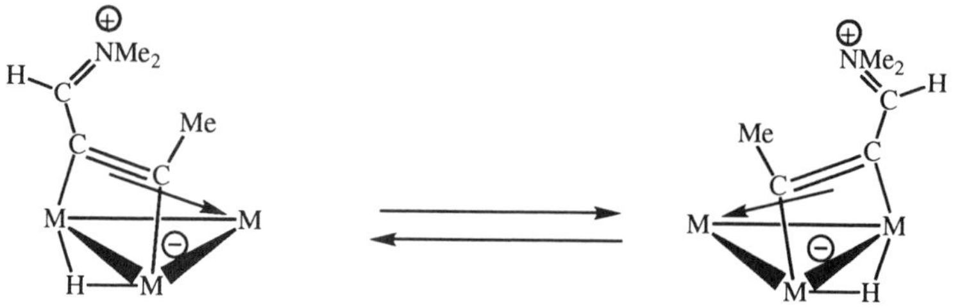

101a   101b

SCHEME 41

is dynamic in solution. An alkyne oscillation and hydride migration in conjunction with rapid carbonyl exchange interconvert enantiomers **101a** and **101b,** generating a time-averaged mirror plane (Scheme 41). Total carbonyl scrambling occurs at elevated temperature (>90°C), requiring a further hydrocarbyl oscillation similar to that described in Scheme 40. A combination of these processes would permute all possible orientations of the hydrocarbyl ligand on the Ru$_3$ framework and account for the NMR equivalence of all carbonyls (*56a*). These dynamic processes closely resemble the familiar alkyne fluxionality for which the zwitterionic resonance representation of the allenyl clusters bears a remarkable similarity (*68*).

Finally, in this section we briefly mention the dynamic characteristics of unsymmetrically bridging allenylidene (allenediyl) ligands. Green and co-workers were the first to discover evidence for allenylidene ligand exchange in binuclear complexes (*36*). During investigations into the relationship between $\mu\text{-}\eta^1,\eta^1\text{-}(2e)$- and $\mu\text{-}\eta^1,\eta^2\text{-}(4e)$-vinylidene bridging arrangements, they described a site-averaging exchange process in the dimer [{M(CO)$_2$Cp}$_2${$\mu\text{-}\eta^1,\eta^2$-C=CMe$_2$}] (M = Mo, W), involving a high-energy but accessible $\mu\text{-}\eta^1,\eta^1\text{-}(2e)$ intermediate as shown in Scheme 42. A similar site exchange process may be relevant to [{Mo(CO)$_2$Cp}$_2${$\mu\text{-}\eta^1,\eta^2$-C=C=CMe$_2$}] (**64**). Convincing spectroscopic evidence was provided for such a process.

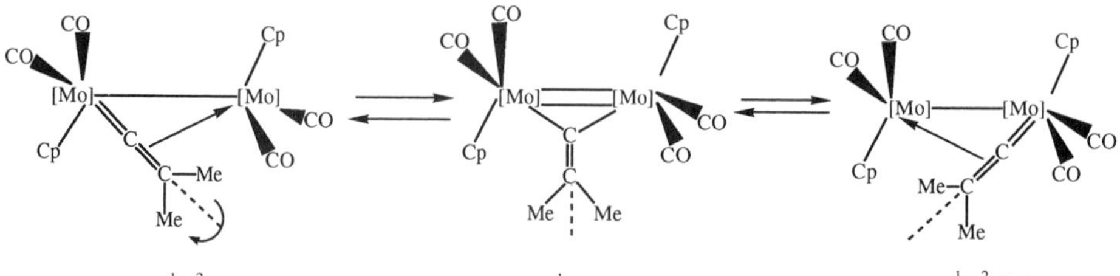

$\eta^1,\eta^2\text{-}(4e)$ $\eta^1\text{-}(2e)$ $\eta^1,\eta^2\text{-}(4e)$

SCHEME 42

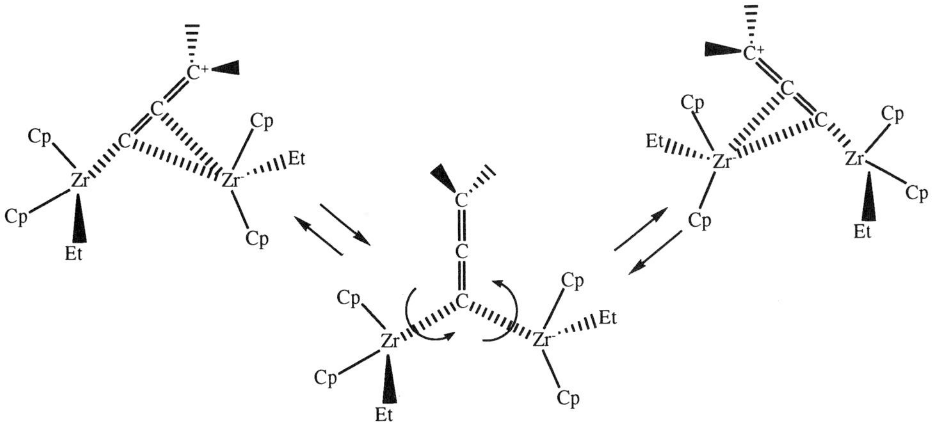

SCHEME 43

In contrast to the reports of Green and co-workers, Krüger and co-workers (37) described an exceptionally facile site exchange in the early transition metal allenylidene-bridged complex [{ZrEtCp$_2$}$_2${μ-η^1,η^2-C=C=CMe$_2$}] (**65**). Krüger *et al.* suggested that the μ-η^1,η^2-site exchange occurred rapidly, incorporating a higher energy μ-η^1,η^1-symmetrically bridged allenylidene intermediate (Scheme 43); however, the only available evidence for such a process was a slight broadening of the methyl carbon resonances of the ethyl substituents of the product at the lowest accessible temperatures (37).

VII

REACTIVITY

A. Reactivity of σ-Bound Propargyl and Allenyl Ligands

Early investigations into the chemistry of σ-bound transition metal propargyl complexes concentrated on their reactivity toward charged and uncharged electrophiles. In the former case η^2-bound allene complexes [ML$_n$(η^2-CH$_2$=C=CH$_2$)]$^+$ were formed via protonation [Eq. (21)]

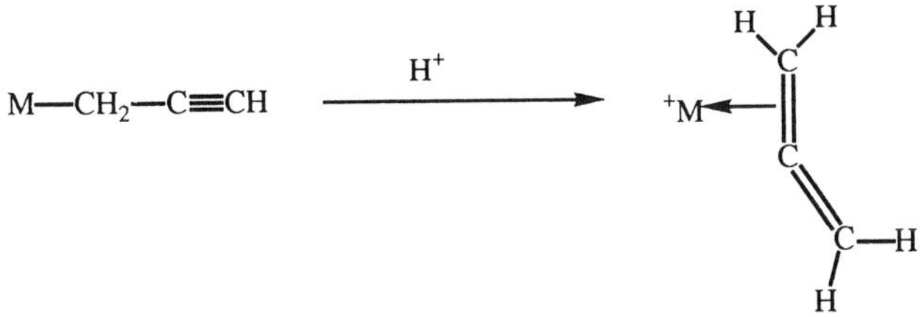

$$\text{(21)}$$

(*6a,b,11*), whereas the latter reaction was typified by [2 + 3] cycloaddition reactions (Scheme 44) (*7*). These [2 + 3] cycloaddition reactions offer
organic chemists an alternative approach to the synthesis of five-membered ring heterocycles provided that the metal fragment can be cleaved
from the ring in high yield. Despite the potential of the latter process in
synthesis, only relatively recently has there been success in its application, most notably via the research of Welker and co-workers (*77*). The
cationic allene complexes formed by the addition of a proton to a propargyl ligand react further with a variety of nucleophiles to afford metal
alkenyl complexes [Eq. (22)] (*6a,78*). The [2 + 3] cycloaddition reactions

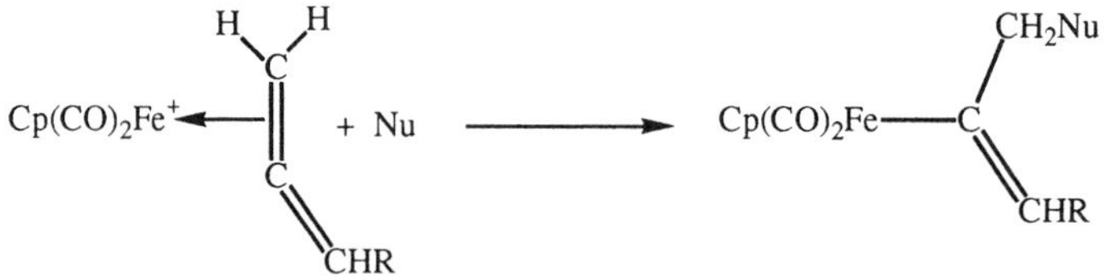

$$Cp(CO)_2Fe^+ \longleftarrow \underset{\underset{CHR}{\overset{\displaystyle C}{\Vert}}}{\overset{\displaystyle C\overset{H}{\underset{H}{\diagup}}}{}} + Nu \longrightarrow Cp(CO)_2Fe\underset{CHR}{\overset{CH_2Nu}{\diagup}}C \qquad (22)$$

yield cyclic alkenyl products, presumably via a process closely related to
that described for the successive addition of proton and nucleophile, that
is, a two-step mechanism as shown in Scheme 44 with a zwitterionic
intermediate. Propargylic metal complexes may react with metal carbonyls in a similar manner, affording heteronuclear bi- and trinuclear allenyl
clusters (Section III,B,3).

1. *Nucleophiles*

The successful preparation and isolation of both tautomers of
[W(CO)$_3$Cp(η^1-C$_3$H$_3$)] [**2(I)**, propargyl; **2(II)**, allenyl] enabled Lin and
co-workers (*9a,c*) to demonstrate the distinct reactivities of these ligand
types (*79*). In both instances facile regiospecific C–C bond formation was
observed (Scheme 45). Reactivities differed in the regiospecificity of addition, with the propargylic form undergoing C–C bond formation at C$_\beta$ via
a postulated η^2-allene intermediate in contrast to the allenylic complex in
which C–C coupling occurred at C$_\alpha$ (*9c*). However, the latter reaction

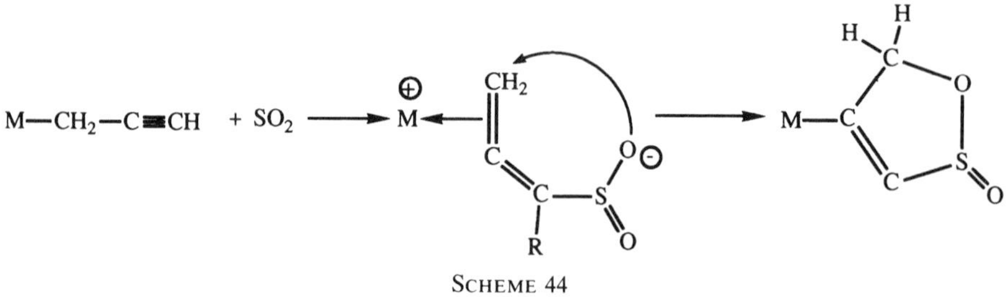

$$M{-}CH_2{-}C{\equiv}CH + SO_2 \longrightarrow M \longleftarrow \overset{\overset{\displaystyle \oplus}{}}{\underset{\underset{\underset{R}{|}}{C}{-}S\overset{\displaystyle O}{\underset{\displaystyle O}{\Vert}}}{\overset{\displaystyle CH_2}{\underset{\displaystyle C}{\Vert}}}{}\overset{O}{\underset{\displaystyle \ominus}{}} \longrightarrow M{-}\underset{\underset{C}{\Vert}}{\overset{\displaystyle C}{}}\overset{\overset{\displaystyle H}{\underset{\displaystyle C}{\diagdown}}}{\underset{\underset{\displaystyle S}{\diagdown}}{}}\overset{O}{\underset{O}{}}$$

SCHEME 44

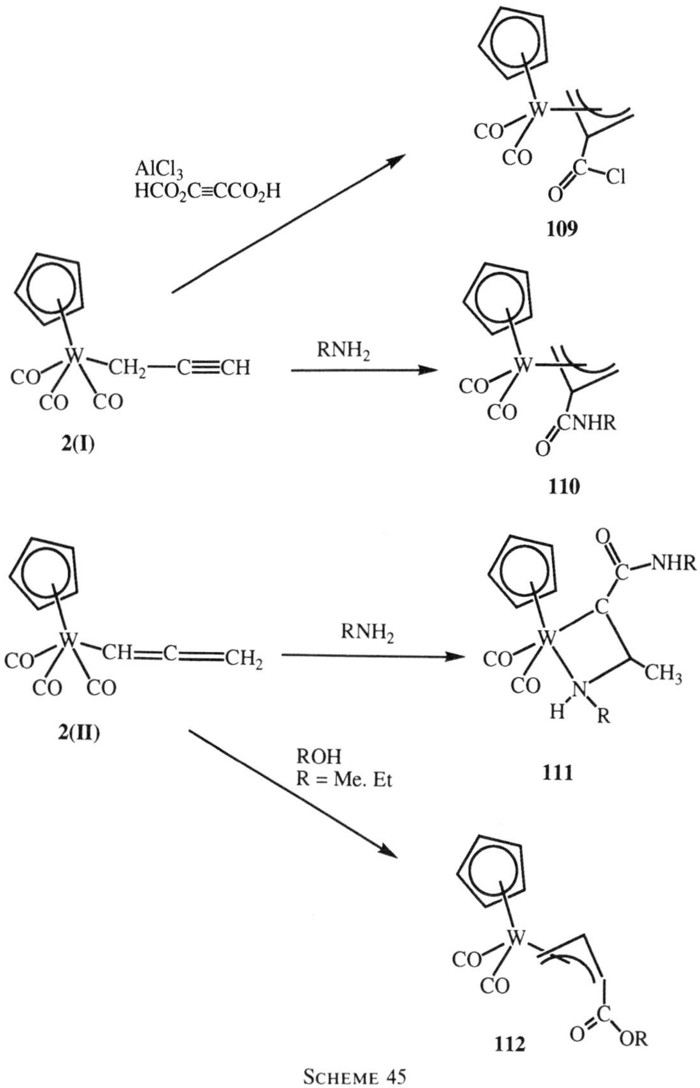

SCHEME 45

probably involved nucleophilic attack by amine at C_β of the allenyl ligand followed by coupling at C_α with a carbamoyl group formed by amine attack at a CO group. In separate studies the phosphite-substituted propargyl derivative [W(CO)$_2$Cp{P(OMe)$_3$}(η^1-CH$_2$C≡CH)] [2(I)a] gave the expected [2 + 3] cycloaddition product [W(CO)$_2$Cp{P(OMe)$_3$}{C$_3$H$_3$-C$_2$(CN)$_2$}] (113a) on treatment with tetracyanoethylene (TCNE) (Scheme 46). When treated with Os$_3$H$_2$(CO)$_{10}$ its allenylic tautomer [W(CO)$_2$Cp-{P(OMe)$_3$}(η^1-HC=C=CH$_2$)] (2(II)a] underwent an intermolecular C–C

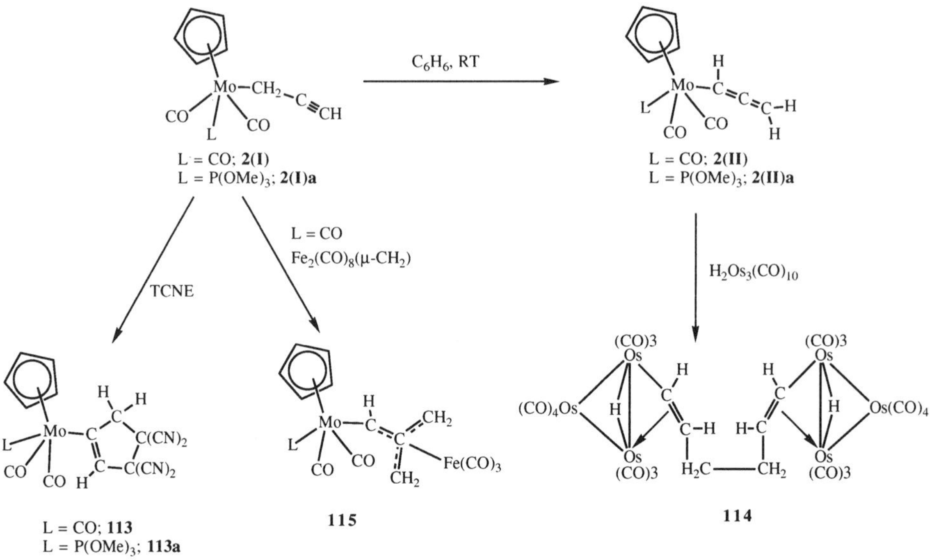

SCHEME 46

coupling reaction to give $[Os_6H_2(CO)_{20}(C_6H_8)]$ (**114**) containing a hexa-1,5-dienyl ligand linking two Os_3 clusters in a μ-η^1,η^2 manner. Other C–C bond forming reactions of allenyl clusters involve the generation of $[W(CO)_3Cp\{\mu_2$-η^1,η^3-CH—C(CH$_2$)$_2\}$Fe(CO)$_3]$ (**115**), a trimethylene-methane iron complex, isolated from the reaction of $[W(CO)_3Cp(\eta^1$-HC≡C=CH$_2$)]$ and the methylene-bridged diiron complex $[Fe_2(CO)_8(\mu$-CH$_2$)]$ (Scheme 46) (*9a*).

The central allenylic carbon atom in $[PtBr(PPh_3)_2(\eta^1$-CH=C=CH$_2$)]$ (**41**) has electrophilic character. Regioselective nucleophilic attack of amine at C_β gave the cationic complex $[Pt(PPh_3)_2\{\eta^3$-CH$_2$C(=NR R^1)CH$_2\}]$Br (**116a**) (Scheme 47) (*26a*) via a subsequent 1,3-hydrogen atom migration. This reactivity toward primary and secondary amines bears a remarkably close similarity to the behavior of $[Ru_2(CO)_6$ $(\mu$-PPh$_2)(\mu$-η^1,η^2-PhC=C=CH$_2$)]$ with similar nucleophilic substrates (Section VII,B,3), and it is becoming clear that both η^1- and μ-η^1,η^2-allenyl groups have a propensity for nucleophilic attack at C_β. Reaction of **41** with N_2H_4·H_2O initially gave an azatrimethylenemethane adduct $[Pt(PPh_3)_2\{\eta^3$-CH$_2$C(=NHNH$_2$)CH$_2\}]$ (**117**). Further hydrogen atom migration and ring closure produced the cationic metallopyrazoline complex $[Pt(PPh_3)_2\{CH_2C(Me)=NNH_2\}]$Br (**118**) which readily transformed to the neutral species $[PtBr(PPh_3)\{CH_2C(Me)=NNH_2\}]$ (**119**) or to $[Pt(PPh_3)\{CH_2C(Me)NNH\}]$ (**120**) on treatment with base. In contrast to

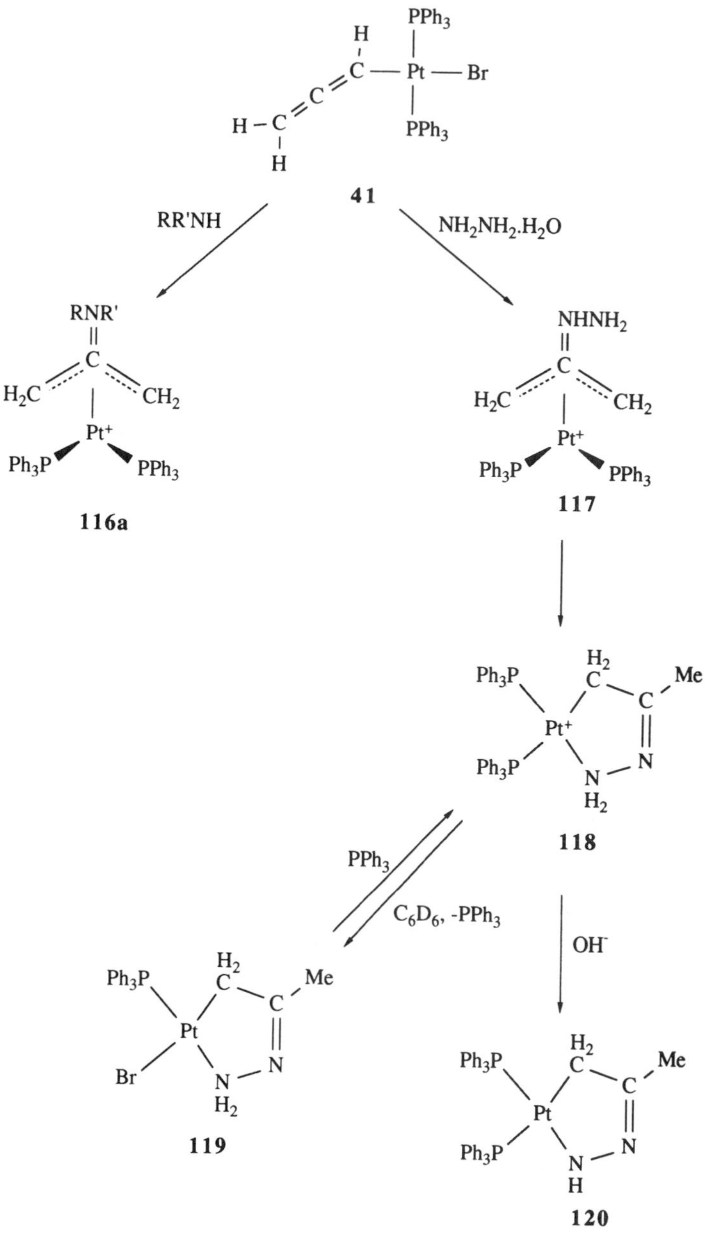

SCHEME 47

these studies, however, Elsevier and co-workers have observed the facile insertion of carbon monoxide and isocyanide ligands into the M–C_α bond (M = Pt^{II}, Pt^{II}) of σ-allenyl complexes (25d,e). There was no evidence for attack at C_β in any of the reactions reported. The proposed mechanism for the formation of the metallovinylketenimine complexes in nonpolar solvents involved initial coordination of RNC to the metal center forming a five-coordinate intermediate, migratory insertion into the M–C allenyl bond, and a subsequent 1,3-metal shift.

Only relatively recently have mononuclear η^3-allenyl and η^3-propargyl complexes been described. $trans$-[$PtBr(PPh_3)_2(\eta^1$-CH=C=CH_2)] (41) was successfully converted to [$Pt(PPh_3)_2(\eta^3$-CH_2C≡CH)]$^+$ (39a) via halide abstraction using silver salts [Eq. (14)]. The chemistry of 39a toward soft nucleophiles revealed a high propensity for nucleophilic attack at C_β, affording a range of β-substituted η^3-allyl-containing complexes (Scheme 48) (26b). This chemistry was extended further by Wojcicki and co-workers (26c) providing (i) more versatile routes to both η^1- and η^3-propargyl and -allenyl complexes, (ii) evidence for the intermediacy of an η^3-propargyl/allenyl complex in η^1-propargyl/allenyl tautomerizations, and (iii) a facile route to various β-heterosubstituted η^3-allyl complexes via reaction with various nucleophiles (Scheme 49).

The η^3-propargyl ligand in [$Re(\eta^5$-C_5Me_5)(CO)(η^3-CH_2C≡CH)] (8) is exceptionally electrophilic, undergoing nucleophilic attack exclusively and regiospecifically at the central carbon atom to afford rhenacyclobutene derivatives (Scheme 50) (13). For example, treatment of 8 with excess PMe_3 gave the phosphine-substituted complex [$Re(\eta^5$-C_5Me_5)(CO)_2(\eta^2$-MeC—C(PMe_3)—CH_2)]$^+$ (127), whereas the anionic nucleophiles C≡CMe$^-$ and CH(CO_2Et)_2$^-$ gave the neutral rhenacyclobutene products [$Re(\eta^5$-C_5Me_5)(CO)_2(\eta^2$-MeC—C(R)—CH_2)] [R = C≡CMe, 128; R = CH(CO_2Et)_2, 129].

2. Electrophiles

Templeton and co-workers, on the other hand, have demonstrated that neutral η^2-allenyl complexes [$W(CO)Tp'(\eta^2$-PhC=C=CMeH)] (12) are highly nucleophilic in character. Thus, deprotonation of [$W(CO)Tp'(\eta^2$-PhC≡CCH_2Me)]$^+$ [Tp' = hydridotris(3,5-dimethylpyrazolylborate)] followed by reaction with benzylbromide gave a single diastereoisomer of the η^2-alkyne complex [$W(CO)Tp'(\eta^2$-PhC≡CCBzMeH)]. Conversely *in situ* generation of [$W(CO)Tp'(\eta^2$-PhC=C=CBzH)] from [$W(CO)Tp'(\eta^2$-PhC≡CCBzH_2)]$^+$ and subsequent methylation gave the opposite diastereoisomer.

As expected the site of attack on σ-bound allenyl ligands by electrophiles is not C_β. Protonation of the allenyl complex [$ReCp(NO)(PPh_3)(\eta^1$-CH=C=CH_2)] with $HBF_4 \cdot Et_2O$ occurred regiospe-

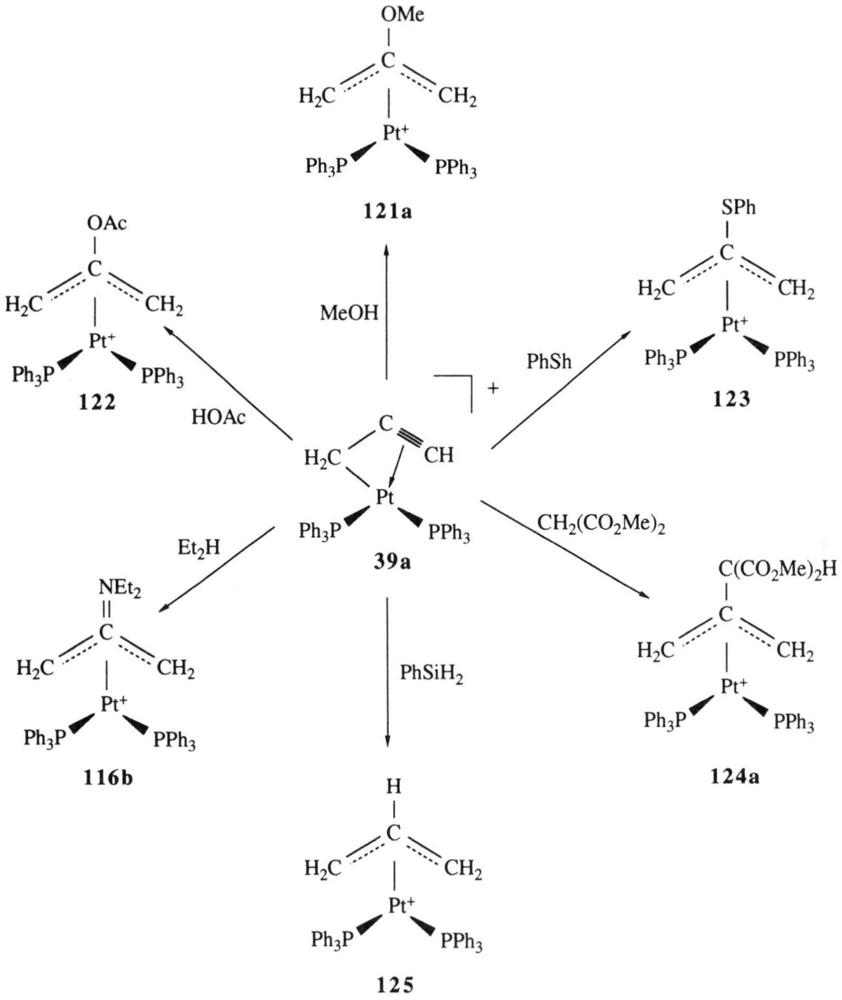

OMe
|
C
H₂C — — C — — CH₂
|
Pt⁺
Ph₃P   PPh₃

121a

OAc
|
C
H₂C — — C — — CH₂
|
Pt⁺
Ph₃P   PPh₃
122

MeOH

HOAc

SPh
|
C
H₂C — — C — — CH₂
|
Pt⁺
Ph₃P   PPh₃
123

PhSh

C
H₂C — — C — — CH
|
Pt
Ph₃P   PPh₃
39a

CH₂(CO₂Me)₂

Et₂H

NEt₂
‖
C
H₂C — — C — — CH₂
|
Pt⁺
Ph₃P   PPh₃
116b

PhSiH₂

C(CO₂Me)₂H
|
C
H₂C — — C — — CH₂
|
Pt⁺
Ph₃P   PPh₃
124a

H
|
C
H₂C — — C — — CH₂
|
Pt⁺
Ph₃P   PPh₃
125

SCHEME 48

cifically at C_γ, giving the cationic alkyne complex [ReCp(NO)(PPh₃)(η^2-HC≡CMe)]⁺ (*9b*).

B. Reaction of Allenyl Ligands Bound to Two Metal Centers

1. *Isomerization Reactions*

Wojcicki and co-workers have successfully demonstrated transformation of the μ-η^1,η^2-coordination type into a μ-η^2,η^3-bound ligand utilizing the full coordinating capacity of the unsaturated hydrocarbyl (*5b*). The

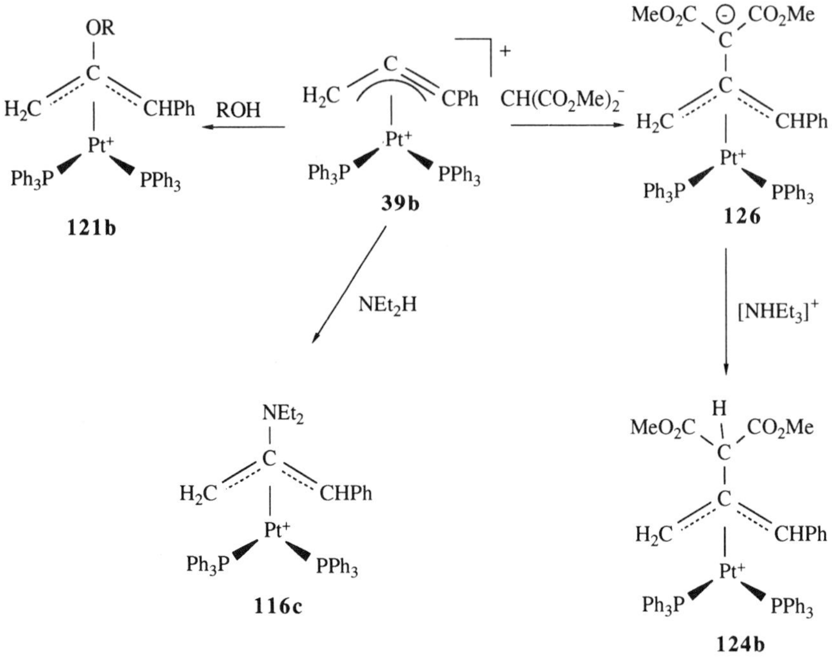

SCHEME 49

"isomerization" was initiated by $Fe_2(CO)_9$ which abstracted triphenyl-phosphine from $[Ru(CO)Cp(\mu\text{-}\eta^1,\eta^2\text{-}PhC{=}C{=}CH_2)Pt(PPh_3)_2]$, providing the vacant coordination site necessary for an extra η^2 interaction (Scheme 51).

The dominant isomerization pathway for the trimeric allenyl clusters $[M_3H(CO)_9(\mu_3\text{-}\eta^1,\eta^2,\eta^2\text{-}RC{=}C{=}CHR)]$ is via a thermal 1,2-hydrogen migration to afford the 1,3-dimetalloallyl clusters $[M_3H(CO)_9(\mu_3\text{-}\eta^1,\eta^1,\eta^3\text{-}RCCHCR)]$ (53c). Reaction of alkynes and dienes with metal carbonyls $M_3(CO)_2$ (M = Ru, Os) often occurs with C–H activation adjacent to the unsaturated carbon–carbon bond (54); however, the final products isolated are often of the 1,3-dimetalloallyl type, the intermediate allenyl complex formed being thermodynamically unstable under the reaction conditions necessary to initiate C–H activation.

Several hydrocarbyl-containing clusters were identified in the catalytic reaction mixture of $Ru_3(CO)_{12}$ and 1-pentene (80). After 50% conversion of 1-pentene to an equilibrium mixture of cis- and trans-2-pentene, the allenyl $[Ru_3H(CO)_9(\mu_3\text{-}\eta^1,\eta^2,\eta^2\text{-}CH_3C{=}C{=}CHMe)]$ and allyl $[Ru_3H(CO)_9(\mu_3\text{-}\eta^1,\eta^1,\eta^3\text{-}CH_3CCHCH_3)]$ clusters were detected, as were $Ru_4H_2(CO)_{13}$ and $Ru_4H_4(CO)_{12}$, the latter arising from dehydrogenation of 1-pentene. These clusters were identified spectroscopically as lower

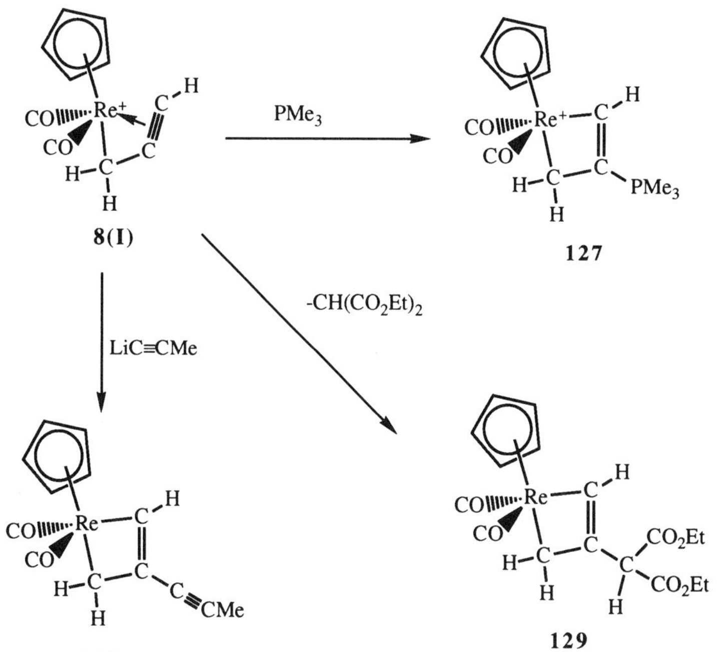

SCHEME 50

homologs of those previously characterized from the thermal reaction of $Ru_3(CO)_{12}$ with hexadienes (*53c*). The hydrogenated by-products from the reaction were shown to be catalytically active toward the isomerization of pentenes; π-allyl and π-allenyl species may be intermediates when $Ru_4H_4(CO)_{12}$ and $Ru_4H_4(CO)_{12-x}L_x$ (L = tertiary phosphine) are used as catalysts (*81*).

[NiOs$_3$(μ-H)$_3$(CO)$_9$Cp] has been established as the catalytically active species in the hydrogenation of dienes and in the selective hydrogenation of double and triple bonds (*82*), providing a rare example of cluster cataly-

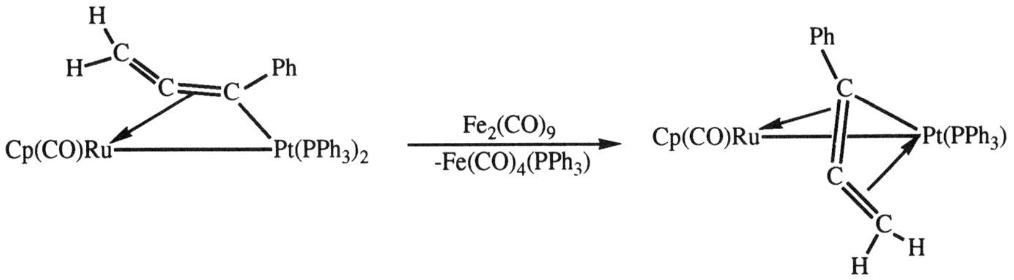

SCHEME 51

sis. The diene- and allyl-containing clusters [NiOs$_3$(μ-H)$_3$(CO)$_8$Cp(diene)] (**130**) and [Os$_3$H(CO)$_9$(μ-η^1,η^2,η^2-MeCHCMe)] (**131**) were isolated from the reaction mixture. Formation of the allyl-bridged cluster dominated over prolonged reaction times and promoted the isomerization pathway, consistent with previous reports on the catalytic activity of other allyl-containing clusters.

2. Cluster Expansion Reactions

The allenyl- and allenyl(carbonyl)-bridged dimers have proved to be versatile reagents in cluster expansion reactions, affording various higher nuclearity allenyl clusters (Section III,B) (*32*). Unfortunately at present these reactions are not well understood, and control over product distribution remains a fundamental difficulty. Clusters containing the rearranged allenyl ligands alkenyl(alkylidyne) and allyl(alkylidene) have been identified, as have carbonyl-coupled products (Scheme 13). Activation of a C$_\gamma$–H bond of the coordinated allenyl unit in [Rh$_2$(CO)Cp$_2$(μ-η^1,η^2-CF$_3$C=CHCF$_3$)(μ-η^1,η^2-HC=C=CH$_2$)] (**97**) with a second equivalent of [Rh$_2$(CO)Cp$_2$(μ-η^2,η^2-CF$_3$C≡CCF$_3$)] afforded the dimer of dimers [Rh$_2$(CO)Cp$_2$(μ-η^1,η^2-CF$_3$C = CHCF$_3$)]$_2$(μ-η^1,η^1,η^2,η^2-HC = C = CH)] (**106**) (*52*).

3. Nucleophilic Addition

There are a limited number of examples in which ligand substitution/addition at the metal cluster is the favored reaction pathway, leaving the C$_3$ hydrocarbon intact. These include the regiospecific addition of dppm to the iron atom of the heteronuclear allenyl cluster [Ru(CO)Cp{μ-η^2,η^3-C(O)C(Ph)C=CH$_2$}Fe(CO)$_3$] (**52**) to give [Ru(CO)Cp{μ-η^2,η^3-C(O)C(Ph)C=CH$_2$}Fe(CO)$_2$(η^1-dppm)] (**132**). The uncoordinated end of the ligand can be used as an effective template for subsequent condensation reactions with the lightly stabilized fragment Pt(PPh$_3$)$_2$ (Scheme 52) (*5b*). The binuclear [W(CO)$_2$Cp(μ-η^2,η^3-C$_6$H$_4$MeC=C=CH$_2$)Fe(CO)$_3$] (**46**) only reacted with PPh$_3$ in the presence of excess trimethylamine oxide, giving [W(CO)$_2$Cp(μ-η^2,η^3-C$_6$H$_4$MeC=C=CH$_2$)Fe(CO)(PPh$_3$)$_2$] (**134**) (*31*).

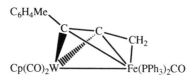

134

The chemistry of the homobimetallic phosphido-bridged allenyl compounds appears to be dominated by reactivity at the hydrocarbyl ligand, being similar to that previously described for the acetylide-bridged complexes. The binuclear $[Ru_2(CO)_6(\mu\text{-}PPh_2)(\mu\text{-}\eta^1,\eta^2_{\beta,\gamma}\text{-}PhC{=}C{=}CH_2)]$ (67) reacts with carbon, phosphorus, and nitrogen nucleophiles regiospecifically at C_β to afford the novel zwitterionic five-membered dimetallacyclopentanes $[Ru_2(CO)_6(\mu\text{-}PPh_2)\{\mu\text{-}\eta^1,\eta^1\text{-}PhC{=}C(PR_2R^1)\text{-}CH_2\}]$ (135) $(PR_2R^1$ addition) and dimetallacyclopentane $[Ru_2(CO)_6(\mu\text{-}PPh_2)\{\mu\text{-}\eta^1,\eta^1\text{-}PhHC{-}C({=}NRH){-}CH_2\}]$ (136) (N–H oxidative addition) (Scheme 53) (83). The product of isocyanide addition, $[Ru_2(CO)_6(\mu\text{-}PPh_2)\{\mu\text{-}\eta^1,\eta^1\text{-}PhC{=}C(CN^tBu)CH_2\}]$ (137), reacted smoothly with both ethylamine and ethanethiol to afford unique amidinium hydrocarbon-containing products, $[Ru_2(CO)_6(\mu\text{-}PPh_2)\{\mu\text{-}\eta^1,\eta^1\text{-}PhC{=}C\{C(EtNH)(NH^tBu)\}{-}CH_2\}]$ (138) and $[Ru_2(CO)_6(\mu\text{-}PPh_2)\{\mu\text{-}\eta^1,\eta^1\text{-}PhC{=}C\{C(EtS){=}(NH^tBu)\}{-}CH_2\}]$ (139). Reactions of this type are important to our fundamental understanding of how metal coordination modifies and directs hydrocarbyl elaboration at metal centers. The reactivity of the $\mu\text{-}\eta^1,\eta^2$-allenyl ligand described above bears a remarkably close similarity to that previously described for the $\mu\text{-}\eta^1,\eta^2$-bound acetylide ligands in $[M_2(CO)_6(\mu\text{-}PPh_2)(\mu\text{-}\eta^1,\eta^2\text{-}C{\equiv}CR)]$ (M =

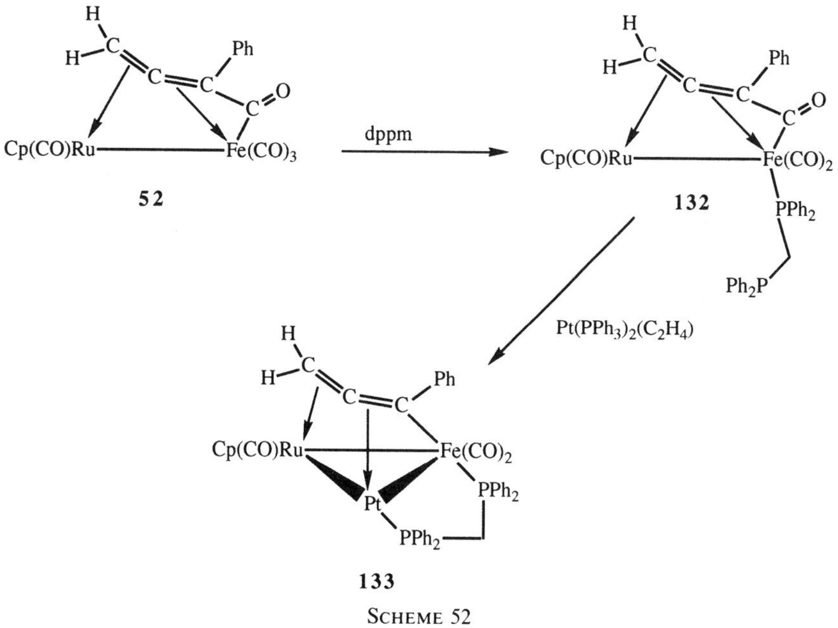

SCHEME 52

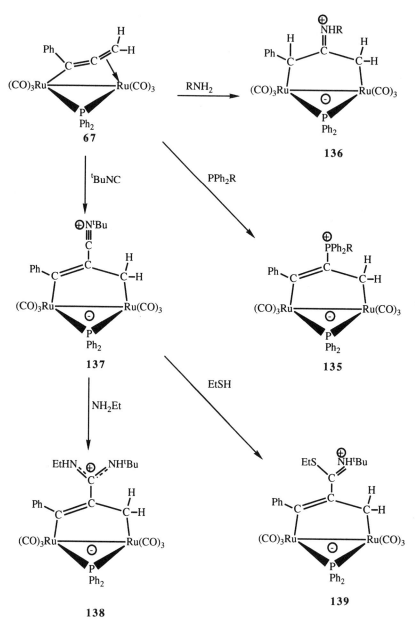

SCHEME 53

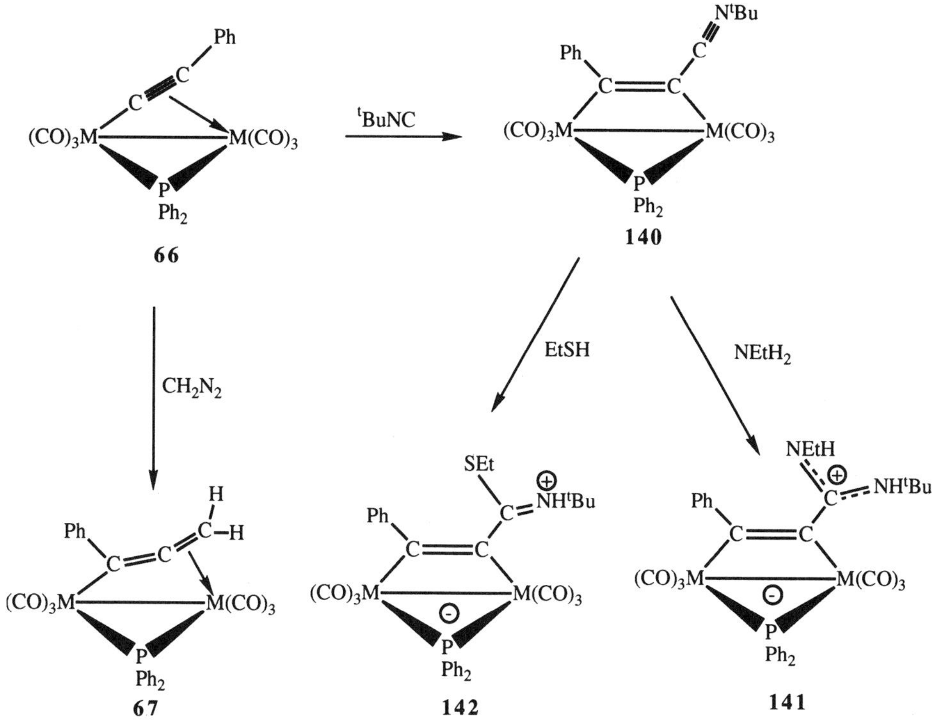

SCHEME 54

Ru, Os) (Scheme 54) (84). Product distributions in the latter class of reactions are consistent with overall orbital control of the reaction pathway.

In contrast to the binuclear iron group allenyl compounds the cationic allenyl dimer $[Mo_2(CO)_4Cp_2(\mu\text{-}\eta^2,\eta^3\text{-}HC\!=\!C\!=\!CH_2)]^+$ reacts with nucleophiles regiospecifically at C_γ to afford the alkyne-bridged dimers $[Mo_2(CO)_4Cp_2(\mu_2\text{-}\eta^2,\eta^2\text{-}HC\!\equiv\!CH_2Nu)]$ (25). In this case reactivity is consistent with the electronic structure of the molecule, the observed regiospecificity being orbital controlled as well as favored by μ-alkyne adduct formation.

4. Reactions with Alkynes

A complex reaction sequence involving carbon–carbon coupling has been established in the reaction of $[Ru_2(CO)_6(\mu\text{-}PPh_2)(\mu\text{-}\eta^1,\eta^2_{\beta,\gamma}\text{-}PhC\!=\!C\!=\!CH_2)]$ (67) and diphenylacetylene (Scheme 55) (85a). The final reaction product, $[Ru_2(CO)_4(\mu\text{-}PPh_2)\{\mu\text{-}\eta^2,\eta^5\text{-}C_5MePh_2(C_6H_4)(O)\}]_2$ (143), tetranuclear in nature, contains a novel cyclopentadienyl ring de-

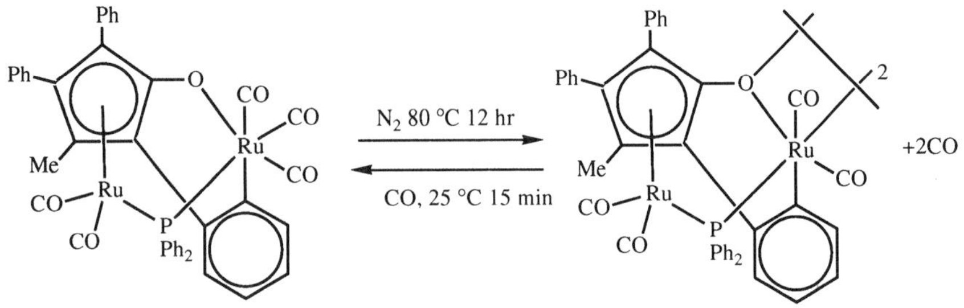

144   **143**

rived from one molecule of diphenylacetylene, a carbonyl ligand, and the original allenyl fragment (PhC=C=CH$_2$), the latter providing two ring substituents and two ring carbon atoms via hydrogen migration. The parent acetylide [Ru$_2$(CO)$_6$(μ-PPh$_2$)(μ-η^1,η^2-C≡CPh)] (**66**) reacts with diphenylacetylene in a similar manner to give the binuclear complex [Ru$_2$(CO)$_5$(μ-PPh$_2$){μ-η^2,η^5-C$_5$HPh$_2$(C$_6$H$_4$)(O)}] (**144a**), again containing a functionalized cyclopentadienyl ring (*85b*). The difference between the

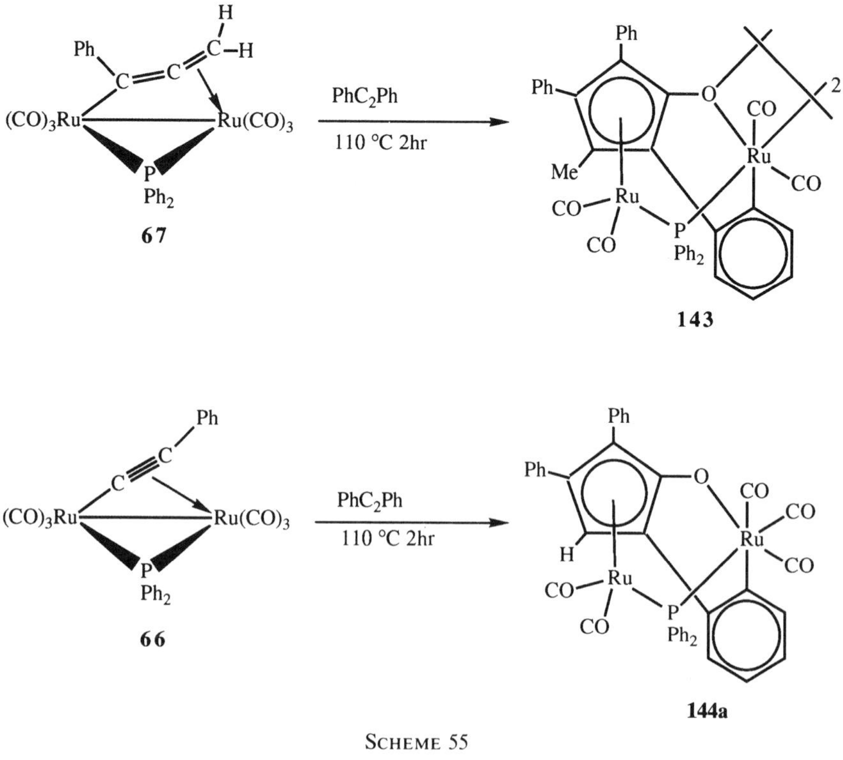

67   **143**

66   **144a**

Scheme 55

two structures lies in the cyclopentadienyl methyl substituent in **143** which arises from C_γ of the allenyl fragment in **67**.

The tetranuclear **143** exists in equilibrium with its binuclear derivative **144,** the position of the equilibrium being controlled by the presence of excess carbon monoxide. The reactivity of binuclear **67** toward alkynes remains distinct from that of its trinuclear analog (Section VII,C,3).

C. Reactions of Cluster-Bound Allenyl Groups

1. Adduct Formation

Early studies by Deeming and co-workers demonstrated the regiospecific nature of nucleophilic attack at C_γ of $\mu\text{-}\eta^1,\eta^2,\eta^2$-coordinated allenyl ligands (86). A 1:1 adduct formation between $[Os_3H(CO)_9(\mu_3\text{-}\eta^1,\eta^2,\eta^2\text{-}MeC{=}C{=}CH_2)]$ **(98)** and the tertiary phosphine PMe_2Ph gave $[Os_3H(CO)_9(\mu_3\text{-}\eta^1,\eta^1,\eta^2\text{-}MeC{\equiv}CCH_2PMe_2Ph)]$ **(145)** (Scheme 56). These adducts are best described as zwitterionic alkyne clusters in which the negative charge is delocalized onto the cluster framework. The cationic allenyl complex $[Mo_2(CO)_4Cp_2(\mu_2\text{-}\eta^2,\eta^3\text{-}CH{=}C{=}CH_2)]^+$ forms similar $\mu\text{-}\eta^2$-alkyne ligands as described in Section VII,B,3 above (25).

The 48-electron phosphido-bridged allenyl cluster $[Ru_3(CO)_8(\mu\text{-}PPh_2)(\mu_3\text{-}\eta^1,\eta^2,\eta^2\text{-}{}^iPrC{=}C{=}CH_2)]$ **(71)** reacts readily with phosphites, the dominant reaction path involving nucleophilic attack at the metal center (87). This associative reaction gives the 50-electron cluster $[Ru_3(CO)_8\{P(OMe)_3\}(\mu_3\text{-}\eta^1,\eta^2,\eta^2\text{-}{}^iPrC{=}C{=}CH_2)]$ **(146),** characterized by an open metal framework with a phosphido ligand bridging the nonbonded Ru–Ru vector (Scheme 57). Similar reactivity was established for carbon monoxide, enabling a facile interconversion between electron-precise **71** and electron-rich $[Ru_3(CO)_9(\mu\text{-}PPh_2)(\mu_3\text{-}\eta^1,\eta^2,\eta^2\text{-}{}^iPrC{=}C{=}CH_2)]$ **(147).** These reactions represent rare examples in which the allenyl ligand remains intact. Analogous reactivity has previously been described for the acetylide-bridged cluster $[Ru_3(CO)_9(\mu\text{-}PPh_2)(\mu_3\text{-}\eta^1,\eta^2,\eta^2\text{-}C{\equiv}C^iPr)]$ **(69),** a molecule capable of reversible carbon monoxide uptake (40a).

Cluster **71** readily activates dihydrogen at room temperature, convert-

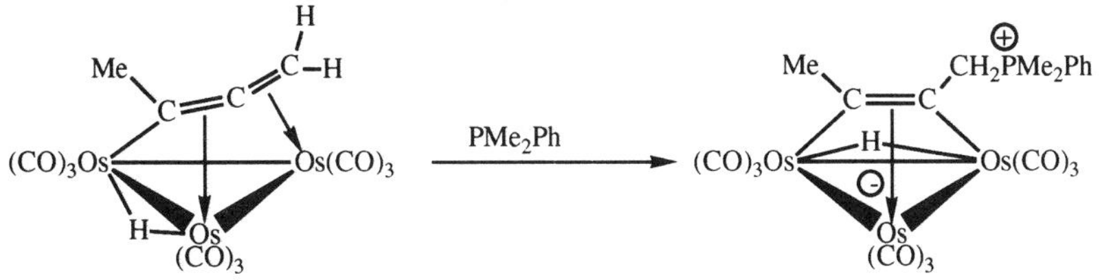

SCHEME 56

ing the μ-η^1,η^2,η^2-allenyl ligand to the hydride acetylene complex [Ru$_3$H(CO)$_8$(μ_3-η^1,η^1,η^2-CH$_3$C≡C^{i}Pr)] (148) (Scheme 57) (87). Such facile activation of dihydrogen was attributed to the availability of a vacant coordination site for the associative reaction, as described above. Valle and co-workers have reported similar allenyl hydrogenation reactions (80); however, more forcing conditions were necessary for successful hydrogenation of the allenyl ligand in [Ru$_3$H(CO)$_9$(μ_3-η^1,η^2,η^2-CH$_3$C=C=CMeH)] (type 98). The reduction of 98 was performed in refluxing cyclohexane to give the dihydride alkyne complex [Ru$_3$H$_2$(CO)$_9$(μ_3-η^1,η^1,η^2-EtC≡CMe)].

2. Carbon–Carbon Bond Forming Reactions

The preparation of [Ru$_3$(CO)$_8$(μ-PPh$_2$)(μ_3-η^1,η^2,η^2-^{i}PrC=C=CH$_2$)] (71) involved nucleophilic attack by diazomethane at C$_\alpha$ of the μ_3-η^1,η^2,η^2-bound acetylide of the electron-precise cluster [Ru$_3$(CO)$_8$(μ-PPh$_2$)(μ_3-η^1,η^2,η^2-C≡C^{i}Pr)] (70) (40a,b). It was subsequently discovered that 71 possessed a rich and diverse chemistry. Addition of excess diazomethane to [Ru$_3$(CO)$_8$(μ-PPh$_2$)(μ_3-η^1,η^2,η^2-^{i}PrC=C=CH$_2$)] generated the μ-methylene-bridged cluster [Ru$_3$(CO)$_7$(μ-PPh$_2$(μ-CH$_2$)(μ_3-η^1,η^2,η^2-^{i}PrC=C=CH$_2$)] (150), which undergoes a manifold of interesting reactions. On standing at room temperature facile cleavage of the μ-methylenic C–H bond and insertion of the methyne unit into the C$_\alpha$–Ru σ bond

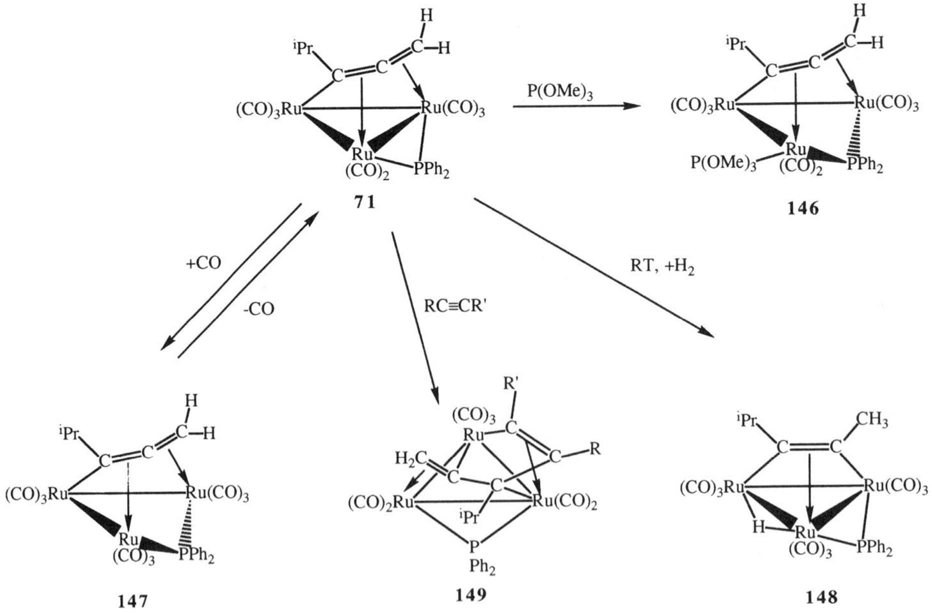

SCHEME 57

gave a rare example of an M_3 η^4-bound butadienyl ligand in $[Ru_3H(CO)_7(\mu\text{-}PPh_2)(\mu_3\text{-}\eta^1,\eta^2,\eta^3\text{-}C(H)C(^iPr)C{=}CH_2)]$ (**151**) (Scheme 58) (*88*).

Reaction of **150** with triphenylphosphine occurred with complete transfer of the methylene fragment to C_α of the allenyl ligand, affording the η^2-coordinated 2-isopropyl-1,3-butadien-3-yl ligated cluster $[Ru_3(CO)_7(\mu\text{-}PPh_2)(PPh_3)(\mu\text{-}\eta^1,\eta^2,\eta^2\text{-}CH_2C(^iPr)C{=}CH_2)]$ (**152**). This butadienyl ligand is similar to that formed in the double addition of CH_2 to the acetylide in $[Fe_2(CO)_6(\mu\text{-}PPh_2)(\mu\text{-}\eta^1,\eta^2\text{-}C{\equiv}CBu^t)]$ described in Section III,C,1.

Reaction of the methylene-bridged cluster with CO gave the 50-electron cluster $[Ru_3(CO)_9(\mu\text{-}PPh_2)(\mu_3\text{-}\eta^1,\eta^2,\eta^2\text{-}{}^iPrC{=}C{=}CH_2)]$ (**147**) described above. The same reaction carried out in the presence of methanol gave the organic by-product methylacetate. This was taken as evidence for the insertion of carbon monoxide into the Ru–methylene bond, forming a bridging ketenyl intermediate with subsequently expelled the organic product.

The reactivity of $[Ru_3(CO)_8(\mu\text{-}PPh_2)(\mu_3\text{-}\eta^1,\eta^2,\eta^2\text{-}{}^iPrC{=}C{=}CH_2)]$ (**71**) demonstrates carbon–carbon bond forming reactions under exceptionally mild conditions, underlining the potential utility of cluster-mediated transformations: addition of a single methylene unit to an acetylide forming an

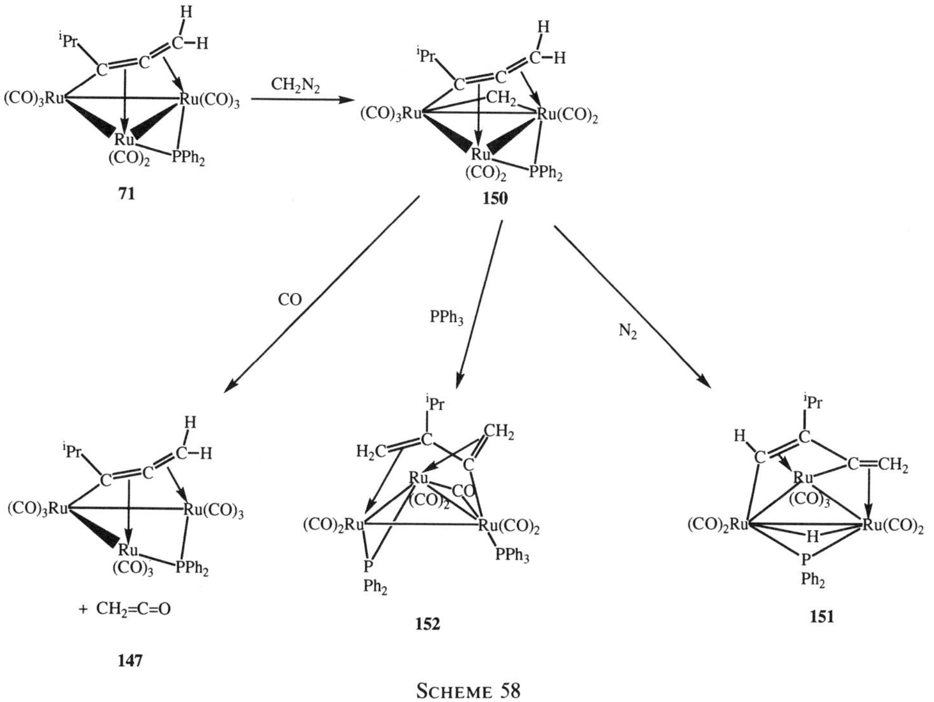

SCHEME 58

allenyl bridge and transfer of a methylene or methyne fragment to an allenyl group. Such transformations represent potential models for processes occurring on metal surfaces and in heterogeneous catalysis.

3. Reactivity Toward Alkynes

The carbon chain growth reactions of the allenyl-bridged cluster **71** were extended further by insertion of alkynes into the C_α–Ru bond and loss of CO to give a five-carbon hydrocarbon chain blanketing a closed 48-electron M_3 framework (Scheme 57). (88). The product $[Ru_3(CO)_7(\mu\text{-}PPh_2)(\mu_3\text{-}\eta^2,\eta^2,\eta^3\text{-}CH_2{=}C{-}C^iPr{-}CRCR')]$ (**149**) contains a metallocyclic ring comprising one ruthenium, two acetylenic carbons, and two carbons of the original allenyl fragment. The remaining methylenic substituent of the allenyl unit forms a metallocycle ring substituent. Alkyne insertion reactions of the dimetallaallyl-bridged clusters $[Ru_3H(CO)_9(\mu\text{-}\eta^1,\eta^1,\eta^3\text{-}CRCHCR')]$ are well established, affording closo-pentagonal bipyramidal clusters via hydride and allyl-hydrogen migration (89). Similar clusters containing metallocyclopentadiene rings are common products from alkyne coupling reactions of metal carbonyls (49).

VIII

CONCLUDING REMARKS

There has been a growing interest in the organometallic chemistry of allenyl-containing complexes. Although mononuclear propargyl complexes have proved to be versatile reagents in the synthesis of heterometallic allenyl clusters and nucleophilic attack of diazoalkanes at C_α carbon atoms of $\mu_3\text{-}\eta^1,\eta^2,\eta^2\text{-}$ and $\mu_2\text{-}\eta^1,\eta^2\text{-}$coordinated acetylides is an effective procedure for the preparation of homometallic allenyl bridged clusters, the development of other rational synthetic approaches to these compounds is pivotal to expanding the chemistry of this C_3 hydrocarbyl fragment. While there are at present no specific applications of mononuclear allenyl complexes in organic synthesis and only a limited number for tautomeric propargyl counterparts, with growing evidence for a rich and diverse reactivity of both of these species such applications in synthesis and catalysis will soon emerge.

The reactivity of the multisite-bound allenyl unit is an area of rapid growth. This fragment exhibits a variety of bonding types. Most common for binuclear complexes are the $\mu\text{-}\eta^1,\eta^2$ and $\mu\text{-}\eta^2,\eta^3$ interactions, whereas for trinuclear clusters $\mu\text{-}\eta^1,\eta^2,\eta^2$ dominates. The search for novel reactivity of heterometallic allenyl bridged complexes has proved less fruitful to date than for the homometallic counterparts, the chemistry of which appear to be dominated by regiospecific nucleophilic attack at C_β. This reactivity pattern has been shown to be governed by orbital and charge factors. Initial studies indicate a promising chemistry for this ligand as a source of unsaturated C_3 fragments in synthesis.

The higher nuclearity allenyl clusters provide an attractive potential model for the chemistry of surface-bound C_3 fragments. Such species could form from the combination of a surface-bound vinylidene and carbyne or an acetylide and methylene unit. These fragments have been identified in the heterogeneously catalyzed Fischer–Tropsch chemistry.

Although the area of allenyl coordination chemistry is developing rapidly, exploration of the chemistry and synthetic potential of these C_3 ligands is in its infancy. There will be a need for theoretical analyses of orbital, charge, and steric control of allenyl reactivity to complement synthetic work on mono- and polynuclear systems. Much can be expected from this area in the years ahead.

REFERENCES

(1) For recent review articles, see (a) Bruce, M. I. *Chem. Rev.* **1991**, *91*, 197; (b) Holton, J.; Lappert, M. F.; Pearce, R.; Yarrow, P. I. W. *Chem. Rev.* **1983**, *83*, 135; (c) Stone, F. G. A. *Adv. Organomet. Chem.* **1990**, *31*, 53; (d) Mayr, A.; Hoffmeister, H. *Adv. Organomet. Chem.* **1991**, *32*, 227; (e) Antonova, A.; Ioganson, A. A. *Russ. Chem. Rev.* **1989**, *58*, 693; (f) Bruce, M. I. *Pure Appl. Chem.* **1990**, *62*, 1021; (g) Schubert, U. (Ed.) "Advances in Metal Carbene Chemistry"; Kluwer Academic: Dordrecht, The Netherlands, 1989; (h) Kreigbel, F. R. (Ed.) "Transition Metal Carbyne Complexes"; Kluwer Academic: Boston, 1993; (i) Moiseev, I. *Russ. Chem. Rev.* **1989**, *58*, 682; (j) Stang, P. L. *Acc. Chem. Res.* **1982**, *15*, 348; (k) Dötz, K. H. *New J. Chem.* **1990**, *14*, 433; (l) Rubezhou, Z. *Russ. Chem. Rev.* **1991**, *60*, 169; (m) Feldman, J. R.; Schrock, R. R. *Prog. Inorg. Chem.* **1991**, *39*, 1; (n) Mayr, A.; Bastos, C. M. *Prog. Inorg. Chem.* **1992**, *40*, 1; (o) Hofmann, P.; Hammelei, M.; Unfried, G. *New J. Chem.* **1991**, *15*, 769.

(2) For a historical perspective to the emergence of organometallic chemistry and the important role of unsaturated C_1 and C_2 ligands, see, for example, Pruchnik, F. P. "Organometallic Chemistry of the Transition Elements"; Plenum: New York, 1990; Chapters 5 and 6; Elschenbroich, Ch.; Salzer, A. "Organometallics: A Concise Introduction"; 2nd ed.; VCH: Weinheim and New York, 1992; Chapters 14 and 15.

(3) Veillard, A. *Chem. Rev.* **1991**, *91*, 743; Tsipis, C. A. *Coord. Chem. Rev.* **1991**, *108*, 163; Mingos, D. M. P. In "Comprehensive Organometallic Chemistry"; Wilkinson, G.; Stone, F. G. A.; Abel, E. W., Eds.; Pergamon: New York, 1982; Vol. 3, Chapter 19. We also note the many important contributions of Hoffmann and co-workers to this area; see Hoffmann, R. *Science* **1981**, *211*, 995; Eisenstein, O.; Hoffmann, R. *J. Am. Chem. Soc.* **1981**, *103*, 4308; See also Kostic, N. M.; Fenske, R. F. *Organometallics* **1982**, *1*, 974; Albright, T. A.; Burdett, J. K.; Whangbo, M.-H. "Orbital Interactions in Chemistry"; Wiley: New York, 1985; Chapters 19 and 20.

(4a) Collman, J. P.; Hegedus, L. S.; Norton, J. R., Finke, R. G. "Principles and Applications of Organotransition Metal Chemistry"; University Science Books: Mill Valley, California, 1987; pp. 175–180 and Chapter 19, and references therein.

(4b) Wilkinson, G.; Stone, F. G. A.; Abel, E. W. (Eds.) "Comprehensive Organometallic Chemistry"; Pergamon: New York, 1982; Vol. 6, Sections 37.6, 38.7, and 39.9.

(5a) Welker, M. E. *Chem. Rev.* **1992**, *92*, 97.

(5b) Wojcicki, A.; Schuchart, C. E. *Coord. Chem. Rev.* **1990**, *105*, 35.

(6a) Lichtenberg, D. W.; Wojcicki, A. *J. Organomet. Chem.* **1975**, *94*, 311.

(6b) Foxman, B.; Marten, D.; Rosan, A.; Raghu, S.; Rosenblum, M. *J. Am. Chem. Soc.* **1977**, *99*, 2160.

(7a) Rosenblum, M. *Acc. Chem. Res.* **1974**, *7*, 122.

(7b) Bell, P. B.; Wojcicki, A. *Inorg. Chem.* **1981**, *20*, 1585.

(7c) Wojcicki, A. In "Fundamental Research in Organometallic Chemistry".

(7d) Tsutsui, M.; Ishii, T.; Huang, Y., Eds.; Van Nostrand-Reinhold: New York, 1982; pp. 569–597.

(8) Hopf, H. In "The Chemistry of Allenes"; Landor, S. R.; Ed.; Academic Press: New York, 1982; Vol. 12, Chapter 4.2.

(9a) Keng, R.-S.; Lin, Y.-C. *Organometallics* **1990**, *9*, 289.

(9b) Pu, J.; Peng, T.-S.; Arif, A. M.; Gladysz, J. A. *Organometallics* **1992**, *11*, 3232.

(9c) Chen, M.-C.; Keng, R.-S.; Lin, Y.-C.; Wang, Y.; Cheng, M.-C.; Lee, G.-H. *J. Chem. Soc., Chem. Commun.* **1990**, 1138.

(10) Schuchart, C. E.; Willis, R. R.; Wojcicki, A. *J. Organomet. Chem.* **1992**, *424*, 185.

(11a) Johnson, M. D.; Mayle, C. *J. Chem. Soc., Chem. Commun.* **1969**, 192.

(11b) Ariyaratne, J. K. P.; Green, M. L. H. *J. Organomet. Chem.* **1963**, *1*, 90.

(11c) Jolly, P. W.; Pettit, R. *J. Organomet. Chem.* **1968**, *12*, 491.

(12) Collman, J. P.; Cawse, J. N.; Kang, J. W. *Inorg. Chem.* **1969**, *8*, 2574.

(13) Casey , C. P.; Yi, C. S. *J. Am. Chem. Soc.* **1992**, *114*, 6597.

(14) Watson, P. L.; Bergman, R. G. *J. Am. Chem. Soc.* **1980**, *102*, 2698.

(15) Feher, F. J.; Green, M.; Rodrigues, R. A. *J. Chem. Soc., Chem. Commun.* **1987**, 1206.

(16a) Gamble, A. S.; Birdwhistle, K. R.; Templeton, J. L. *J. Am. Chem. Soc.* **1990**, *112*, 1818.

(16b) Collins, M. A.; Feng, S. G.; White, P. A.; Templeton, J. L. *J. Am. Chem. Soc.* **1992**, *114*, 3771.

(17) Heeres, H. J.; Teuben, J. H. *Organometallics* **1990**, *9*, 1508.

(18) Horton, A. D.; Orpen, A. G. *Organometallics* **1992**, *11*, 8.

(19) Blosser, P. W.; Gallucci, J. C.; Wojcicki, A. *J. Am. Chem. Soc.* **1993**, *15*, 2994.

(20a) Berke, H.; Härter, P.; Huttner, G.; Zsolnai, L. *Chem. Ber.* **1982**, *115*, 695.

(20b) Berke, H.; Gröbmann, U.; Huttner, G.; Zsolnai, L. *Chem. Ber.* **1984**, *117*, 3432.

(21a) Selegue, J. P. *Organometallics* **1982**, *1*, 217.

(21b) Selegue, J. P. *J. Am. Chem. Soc.* **1983**, *105*, 5921.

(21c) Le Bozec, H.; Ouzzine, K.; Dixneuf, P. H. *J. Chem. Soc., Chem. Commun.* **1989**, 219.

(21d) Wolinska, A.; Touchard, D.; Dixneuf, P. H.; Romero, A. *J. Organomet. Chem.* **1991**, *420*, 217.

(21e) Pilette, D.; Ouzzine, K.; Le-Bozec, H.; Dixneuf, P. H.; Rickard, C. E. F.; Roper, W. R. *Organometallics,* **1992**, *11*, 809.

(22) Pirio, N.; Touchard, D.; Dixneuf, P. H.; Fettouhi, M.; Ouahab, L. *Angew. Chem., Int. Ed. Engl.* **1992**, *31*, 651.

(23a) Seyferth, D.; Womack, G. B.; Dewan, J. C. *Organometallics* **1985**, *4*, 398.

(23b) Seyferth, D.; Womack, G. B.; Archer, C. M.; Dewan, J. C. *Organometallics* **1989**, *8*, 430.

(24) Seyferth, D.; Archer, C. M.; Dewan, J. C. *Organometallics* **1991**, *10*, 3759.

(25) Meyer, A.; McCabe, D. J.; Curtis, M. D. *Organometallics* **1987**, *6*, 1491.

(26a) Chen, J.-T.; Huang, T.-M.; Cheng, M.-C.; Lin, Y.-C.; Wang, Y. *Organometallics* **1992**, *11*, 1761.

(*26b*) Huang, T.-S.; Chen, J.-T.; Lee, G.-H.; Wang, Y. *J. Am. Chem. Soc.* **1993,** *115,* 1170.

(*26c*) Blosser, P. W.; Schimplf, D. G.; Gallucci, J. C.; Wojcicki, A. *Organometallics* **1993,** *12,* 1993.

(*26d*) Wouters, J. M. A.; Avis, M. W.; Elsevier, C. J.; Kyriakidis, C. E.; Stam, C. H. *Organometallics* **1990,** *9,* 2203.

(*26e*) Wouters, J. M. A.; Klein, R. A.; Elsevier, C. J.; Zoutberg, M.; Stam, C. H. *Organometallics* **1993,** *12,* 3864.

(*27a*) Aime, S.; Milone, L.; Osella, D.; Tiripicchio, A.; Lanfredi, A. M. M. *Inorg. Chem.* **1982,** *21,* 501.

(*27b*) Aime, S.; Osella, D.; Milone, L.; Tiripicchio, A. *Polyhedron* **1983,** *2,* 77.

(*28a*) Chi, Y.; Hwang, D.-K.; Chen, S.-F.; Liu, L.-K. *J. Chem. Soc., Chem. Commun.* **1989,** *1540.*

(*28b*) Chi, Y.; Chuang, S.-H.; Liu, L.-K.; Wen, Y.-S. *Organometallics* **1991,** *10,* 2485.

(*28c*) Chi, Y.; Wu, C.-H.; Peng, S.-M.; Lee, G.-H. *Organometallics* **1991,** 10, 1676.

(*28d*) Lin, R.-C.; Chi, Y.; Peng, S.-M.; Lee, G.-H. *Inorg. Chem.* **1992,** *31,* 3818.

(*29a*) Wojcicki, A. *J. Cluster Sci.* **1993,** *4,* 59.

(*29b*) Wojcicki, A. *New J. Chem.* **1994,** *18,* 61.

(*30*) Wido, T. M.; Young, G. H.; Wojcicki, A.; Calligaris, M.; Nardin, G. *Organometallics* **1988,** *7,* 452.

(*31a*) Young, G. H.; Wojcicki, A. *J. Am. Chem. Soc.* **1989,** *111,* 6890.

(*31b*) Young, G. H.; Raphael, M. V.; Wojcicki, A.; Calligaris, M.; Nardin, G.; Bresciani-Pahor, N. *Organometallics* **1991,** *10,* 1934.

(*32*) Schuchart, C. E.; Young, G. H.; Wojcicki, A.; Caligaris, M.; Nardin, G. *Organometallics* **1990,** *9,* 2417.

(*33*) Young, G. H.; Willis, R. R.; Wojcicki, A.; Calligaris, M.; Faleschini, P. *Organometallics* **1992,** *11,* 154.

(*34a*) Berke, H. *J. Organomet. Chem.* **1980,** *185,* 75.

(*34b*) Kolobova, N. E.; Ivanov, L. L.; Zhvanko, O. S.; Aleksandrov, G. G.; Struchkov, Yu. T. *J. Organomet. Chem.* **1982,** *228,* 265.

(*35*) Berke, H.; Härter, P.; Huttner, G.; Zsolnai, L. *Chem. Ber.* **1984,** *117,* 3423.

(*36*) Froom, S. F. T.; Green, M.; Mercer, R. J.; Nagle, K. R.; Orpen, A. G.; Schwiegk, S. *J. Chem. Soc., Chem. Commun.* **1986,** 1666.

(*37*) Binger, P.; Langhauser, F.; Gabor, B.; Mynott, R.; Herrmann, A. T.; Krüger, C. *J. Chem. Soc., Chem. Commun.* **1992,** 505.

(*38a*) Nucciarone, D.; Taylor, N. J.; Carty, A. J. *Organometallics* **1986,** *5,* 1179.

(*38b*) Carlton, N.; Corrigan; J. F.; Doherty, S.; Pixner, R.; Sun, Y.; Taylor, N. J.; Carty, A. J. *Organometallics,* in press.

(*39*) Breckenridge, S. M.; MacLaughlin, S. A.; Taylor, N. J.; Carty, A. J. *J. Chem. Soc., Chem. Commun.* **1991,** 1718.

(*40a*) MacLaughlin, S. A.; Johnson, J. P.; Taylor, N. J.; Carty, A. J.; Sappa, E. *Organometallics* **1983,** *2,* 352.

(*40b*) Nucciarone, D.; MacLaughlin, S. A.; Taylor, N. J.; Carty, A. J. *Organometallics* **1988,** *7,* 106.

(*41*) Suades, J.; Dahan, F.; Mathieu, R. *Organometallics* **1988,** *7,* 47.

(*42a*) Casey, C. P.; Meszaros, M. W.; Fagan, P. J.; Bly, R. K.; Marder, S. R.; Austin, E. A. *J. Am. Chem. Soc.* **1986,** *108,* 4043.

(*42b*) Casey, C. P.; Meszaros, M. W.; Fagan, P. J.; Bly, R. K.; Colborn, R. E. *J. Am. Chem. Soc.* **1986,** *108,* 4053.

(*42c*) Casey, C. P.; Woo, K. L.; Fagan, P. J.; Palermo, R. E.; Adams, B. R. *Organometallics* **1987,** *6,* 447.

(43) Howard, J. A. K.; Knox, S. A. R.; Terrill, N. J.; Yates, M. I. *J. Chem. Soc., Chem. Commun.* **1989**, 640.

(44) Casey, C. P.; Crocker, M.; Vosejpka, P. C.; Fagan, P. J.; Marder, S. R.; Gohdes, M. A. *Organometallics* **1988**, 7, 670.

(45) Casey, C. P.; Austin, E. A. *Organometallics* **1986**, 5, 584; Hoel, E. L.; Ansell, G. B.; Leta, S. *Organometallics* **1986**, 5, 585.

(46a) Lewandos, G. S.; Doherty, N. M.; Knox, S. A. R.; MacPherson, K. A.; Orpen, G. A. *Polyhedron* **1988**, 7, 837, and references therein.

(46b) Ben-Shoshan, R.; Pettit, R. *J. Chem. Soc., Chem. Commun.* **1968**, 247.

(46c) Hughes, R. P.; Powell, J. *J. Organomet. Chem.* **1973**, 54, 345.

(47) Doherty, N. M.; Elschenbroich, C.; Kneuper, H.-J.; Knox, S. A. R. *J. Chem. Soc., Chem. Commun.* **1985**, 170.

(48) Mercer, R. J.; Green, M.; Orpen, G. A. *J. Chem. Soc., Chem. Commun.* **1986**, 567.

(49) Aime, S.; Milone, L.; Osella, D.; Valle, M. *J. Chem. Res. (Miniprint)* **1978**, 0785(M).

(50) Ziller, J. W.; Bower, D. K.; Dalton, D. M.; Keister, J. B.; Churchill, M. R. *Organometallics* **1989**, 8, 492.

(51) Deeming, A. J.; Arce, A. J.; De Sanctis, Y.; Bates, P. A.; Hursthouse, M. B. *J. Chem. Soc., Dalton Trans.* **1987**, 2935.

(52) Dickson, R. S.; Jenkins, S. M.; Skelton, B. W.; White, A. H. *Polyhedron* **1988**, 7, 859.

(53a) Evans, M.; Hursthouse, M.; Randall, E. W.; Rosenberg, E.; Milone, L.; Valle, M. *J. Chem. Soc., Chem. Commun.* **1972**, 545.

(53b) Gambino, O.; Valle, M.; Aime, S.; Vaglio, G. A. *Inorg. Chim. Acta* **1974**, 8, 71.

(53c) Gervasio, G.; Osella, D.; Valle, M. *Inorg. Chem.* **1976**, 15, 1221.

(54) Bruce, M. I. In "Comprehensive Organometallic Chemistry"; Wilkinson, G.; Stone, F. G. A.; Abel, E., Eds; Pergamon: Oxford, 1982; Chapter 32.5, p. 843.

(55) Humphries, A. P.; Knox, S. A. R. *J. Chem. Soc., Dalton Trans.* **1975**, 1710.

(56a) Aime, S.; Osella, D.; Deeming, A. J.; Arce, A. J.; Hursthouse, M. B.; Dawes, H. M. *J. Chem. Soc., Dalton Trans.* **1986**, 1459.

(56b) Aime, S.; Osella, D.; Arce, A. J.; Deeming, A. J.; Hursthouse, M. B.; Galas, A. M. R. *J. Chem. Soc., Dalton Trans.* **1984**, 1981.

(56c) Aime, S.; Jannon, G.; Osella, D.; Deeming, A. J. *J. Organomet. Chem.* **1981**, 214, C15.

(57) Sappa, E.; Tiripicchio, A.; Braunstein, P. *Chem. Rev.* **1983**, 83, 203.

(58) Deeming, A. J.; Kabir, S. E.; Nuel, D.; Powell, N. I. *Organometallics* **1989**, 8, 717.

(59) Predieri, G.; Tiripicchio, A.; Tiripicchio-Camellini, M.; Costa, M.; Sappa, E. *Organometallics* **1990**, 9, 1729.

(60) Deeming, A. J.; Hasso, S.; Underhill, M. *J. Chem. Soc., Dalton Trans.* **1975**, 1614.

(61a) Day, M.; Hajela, S.; Hardcastle, K. I.; McPhillips, T.; Rosenberg, E.; Botta, M.; Gobetto, R.; Milone, L.; Osella, D.; Gellert, W. *Organometallics* **1990**, 9, 913.

(61b) Day, M.; Hajela, S.; Kabir, S. E.; Irving, M.; McPhillips, T.; Wolf, E.; Hardcastle, K. I.; Rosenberg, E.; Milone, L.; Gobetto, R.; Osella, D. *Organometallics* **1991**, 10, 2743.

(61c) Sutterfield, C. N.; Guttekin, S. *Ind. Eng. Chem. Proc. Des. Dev.* **1981**, 20, 62.

(62) Attali, S.; Dahan, F.; Mathieu, F. *Organometallics* **1986**, 5, 1376.

(63) Krivykh, U. V.; Taits, E. S.; Petrovskii, P. V.; Struchkov, Yu. T.; Yanovskii, A. I. *Mendeleev Commun.* **1991**, 103.

(64a) Bowden, F. L.; Giles R., *Coord. Chem. Rev.*, **1976**, 20, 8.

(64b) Lewis, L. N.; Huffman, J. C.; Caulton, K. G. *Inorg. Chem.* **1980**, 19, 1246.

(64c) Bailey, W. I., Jr.; Chisholm, M. H.; Cotton, F. A.; Murillo, C. A.; Rankel, L. A. *J. Am. Chem. Soc.* **1978**, 100, 802.

(65) Froom, S. F. T.; Green, M.; Nagle, K. R.; Williams, D. J. *J. Chem. Soc., Chem. Commun.* **1987**, 1305.

(66a) Cherkas, A. A.; Breckenridge, S. M.; Carty, A. J. *Polyhedron* **1992**, *11*, 1075.

(66b) Carty, A. J.; Cherkas, A. A.; Randall, L. H. *Polyhedron* **1988**, *7*, 1045.

(67a) Fantazier, R. M.; Poutsma, M. L. *J. Am. Chem. Soc.* **1968**, *90*, 5490.

(67b) Crandall, C. K.; Sojka, S. A. *J. Am. Chem. Soc.* **1972**, *95*, 5084.

(67c) Levy, G. C.; Lichter, R. L.; Nelson, G. L. "Carbon-13 Nuclear Magnetic Resonance Spectroscopy"; 2nd ed.; Wiley: New York, 1980; p. 88.

(68a) "Dynamic Nuclear Magnetic Resonance Spectroscopy"; Cotton, F. A.; Jackman, L. M., Eds.; Academic Press: New York, 1975; Chapter 2.

(68b) Aime, S.; Milone, L. *Prog. Nucl. Magn. Reson. Spectros.* **1977**, *11*, 183.

(68c) Band, E.; Muetterties, E. L. *Chem. Rev.* **1978**, *78*, 639.

(68d) Gallop, M. A.; Johnson, B. F. G.; Khattar, R.; Lewis, J.; Raithby, P. R. *J. Organomet. Chem.* **1990**, *386*, 121.

(68e) Rosenberg, E.; Bracker-Novak, J.; Aime, S.; Gobetto, R.; Osella, D. *J. Organomet. Chem.* **1989**, *365*, 163.

(69a) Martin, H. A.; Lemaire, P. J.; Jellinek, F. *J. Organomet. Chem.* **1968**, *14*, 149.

(69b) Henc, B.; Jolly, P. W.; Salz, R.; Stobbe, S.; Wilke, G.; Benn, R.; Mynott, R.; Seevogel, K.; Goddard, R.; Krüger, C. *J. Organomet. Chem.* **1980**, *191*, 449.

(70) Aime, S.; Gobetto, R.; Osella, D.; Milone, L.; Rosenberg, E. *Organometallics* **1982**, *1*, 640.

(71a) Shapley, J. R.; Richter, S. I.; Tachikawa, M.; Keister, J. B. *J. Organomet. Chem.* **1975**, *94*, C43.

(71b) Clauss, A. D.; Tachikawa, M.; Shapley, J. R.; Pierpont, C. G. *Inorg. Chem.* **1981**, *20*, 1528.

(71c) Cherkas, A. A.; Doherty, S.; Cleroux, M.; Hogarth, G.; Randall, L. H.; Breckenridge, S. M.; Taylor, N. J.; Carty, A. J. *Organometallics* **1992**, *11*, 1701.

(72) Crowte, R. J.; Evans, J. *J. Chem. Soc., Chem. Commun.* **1984**, 1332.

(73) Liu, J.; Boyar, E.; Deeming, A. J.; Donovan-Mtunzi, S. *J. Chem. Soc., Chem. Commun.* **1984**, 1182.

(74) Carty, A. J.; Taylor, N. J.; Smith, W. F.; Lappert, M. F.; Pye, P. L. *J. Chem. Soc., Chem. Commun.* **1978**, 1017.

(75) Bailey, W. I., Jr.; Chisholm, M. H.; Cotton, F. A.; Rankel, L. A. *J. Am. Chem. Soc.* **1978**, *100*, 5764.

(76) Deeming, A. J. *J. Organomet. Chem.* **1977**, *128*, 63.

(77a) Reseta, M. E.; Mishra, R. K.; Cawood, S. A.; Welker, M. E.; Rheingold, A. L. *Organometallics* **1991**, *10*, 2936.

(77b) Reseta, M. E.; Cawood, S. A.; Welker, M. E. *J. Am. Chem. Soc.* **1989**, *111*, 8268.

(78) Lichtenberg, D. W.; Wojcicki, A. *J. Am. Chem. Soc.* **1972**, *94*, 8271.

(79) Tseng, T.-H.; Wu, I.-Y.; Lin, Y.-C.; Chen, C.-T.; Chen, M.-C.; Tsai, Y.-J.; Chen, M.-C.; Wang, Y. *Organometallics* **1991**, *10*, 43.

(80) Castiglioni, M.; Milone, L.; Osella, D.; Vaglio, G. A.; Valle, M. *Inorg. Chem.* **1976**, *15*, 394.

(81a) Vaglio, G. A.; Valle, M. *Inorg. Chim. Acta* **1978**, *30*, 161.

(81b) Valle, M.; Osella, D.; Vaglio, G. A. *Inorg. Chim. Acta* **1976**, *20*, 213.

(81c) Lausarot, P. Michelin; Vaglio, G. A.; Valle, M. *Inorg. Chim. Acta* **1979**, *36*, 213.

(82) Castiglioni, M.; Giordano, R.; Sappa, E.; Tiripicchio, A.; Tiripicchio-Camellini, M. *J. Chem. Soc., Dalton Trans.* **1986**, 23.

(83) Breckenridge, S. M.; Taylor, N. J.; Carty, A. J. *Organometallics* **1991**, *10*, 837.

(84a) Cherkas, A. A.; Hadj-Bagheri, N.; Carty, A. J.; Sappa, E.; Tiripicchio, A.; Pellinghelli, M. A. *Organometallics* **1990**, *9*, 1887.

(*84b*) Cherkas, A. A.; Randall, L. H.; Taylor, N. J.; Mott, G. N.; Yule, J. E.; Guinamant, J. L.; Carty, A. J. *Organometallics* **1990**, *9*, 1677.

(*84c*) Cherkas, A. A.; Carty, A. J.; Sappa, E.; Pellinghelli, M. A.; Tiripicchio, A. *Inorg. Chem.* **1987**, *26*, 3201.

(*85a*) Randall, S. M.; Taylor, N. J.; Carty, A. J.; Ben Haddah, T.; Dixneuf, P. H. *J. Chem. Soc., Chem. Commun.* **1988**, 870.

(*85b*) Breckenridge, S. M.; Blenkiron, P.; Carty, A. J.; Sappa, E.; Pellinghelli, M. A.; Tiripicchio, A. Submitted for publication.

(*86*) Henrick, K.; McPartlin, M.; Deeming, A. J.; Hasso, S.; Manning, P. *J. Chem. Soc., Dalton Trans.* **1982**, 899.

(*87*) Nucciarone, D.; Taylor, N. J.; Carty, A. J. *Organometallics* **1988**, *7*, 127.

(*88*) Nucciarone, D.; Taylor, N. J.; Carty, A. J.; Tiripicchio, A.; Tiripicchio-Camellini, M.; Sappa, E. *Organometallics* **1988**, *7*, 118.

(*89*) Aime, S.; Botta, M.; Osella, D. In *"Organometallic Syntheses"*; King, R. B.; Eisch, J. J., Eds., Elsevier: New York, 1988; Vol. IV.

NOTE ADDED IN PROOF: The synthesis of [RhCl(=C=C=CRR′)(P^{i}Pr$_3$)$_2$] (*1*), the ylide complex [Mo$_2$(CO)$_4$Cp$_2$[μ-η^1,η^2-MeC=C=C(H)(PPh$_3$)}] (*2*), metal dimers of the type M=C=C=C=C=M′ (*3*), and the propargylating reagent [Pb(PPh$_3$){η^1-C(H)=C=CH$_2$}] (*4*) have recently been reported.

Reactions of platinum–allenyl derivatives (*5*), the transformations of η^3-propargyl and η^3-allenyl ligands into η^3-hydroxyallyls on mono- and binuclear complexes (*6*), the cyclization of diynes via metal–allenyl intermediates (*7*), and the oxidation of [Pt(Br)(PPh$_3$)$_2$(η^1-CH=C=CMe$_2$)] to [Pt(Br)(PPh$_3$)$_2${η^1-C≡CMe$_2$(OOH)}] (*8*) have also been described.

The formation of allenylidene derivatives from ethynyl–hexanol and alkenyl–vinylidene mononuclear complexes (*9*), the formation of mononuclear ruthenium allenyl complexes from terminal alkynes (*10*), the intermediacy of ruthenium–allenylidene complexes in forming propargylic alcohols (*11*), and in the cyclization of propargyl alcohols (*12*), and the use of mononuclear ruthenium compounds in allylic alkylation catalysis (*13*) have also been reported.

Finally, the transformations of vinylacetylide ligands into metal coordinated allenylidenes have been demonstrated (*14,15*).

References

(*1*) Werner, H.; Rappert, T.; Wiedemann, R.; Wolf, J.; Mahr, N. *Organometallics* **1994**, *13*, 2721. (2) El Amouri, H.; Gruselle, M.; Besace, Y.; Vaissermann, J.; Jaouen, G. *Organometallics* **1994**, *13*, 2244. (3) Weng, W.; Ramsden, J. A.; Arif, A. M.; Gladysz, J. A. *J. Am. Chem. Soc.* **1993**, *115*, 3284. (4) Seyferth, D.; Son, D. Y.; Shah, S. *Organometallics* **1994**, *13*, 2105. (5) Blosser, P. W.; Schimpff, D. G.; Galucci, J.-C.; Wojcicki, A. *Organometallics* **1993**, *12*, 1993. (6) Huang, T.-M.; Hsu, R.-H.; Yang, C.-S.; Chen, J.-T.; Lee, G.-H.; Wang, Y. *Organometallics* **1994**, *13*, 3657. (7) Matsuzaka, H.; Hirayama, Y.; Nishio, M.; Mizobe, Y.; Hidai, M. *Organometallics* **1993**, *12*, 30. (8) Wouters, J. M. A.; Vrieze, K.; Elsevier, C. J.; Zoutberg, M. C.; Goubitz, K. *Organometallics* **1994**, *13*, 1510. (9) Cadierno, V.; Gamasa, M. P.; Gimeno, J.; Lastra, E.; Borge, J.; Garcia-Granda, S. *Organometallics* **1994**, *13*, 745. (10) Touchard, D.; Morice, C.; Cadierno, V.; Paquette, P.; Toupet, L.; Dixneuf, P. H. *J. Chem. Soc., Chem. Commun.*, **1994**, 859. (11) Pirio, N.; Touchard, D.; Dixneuf, P. H. *J. Organomet. Chem.*, **1993**, *462*, C18. (12) Pilette, D.; Moreau, S.; Le Bozec, H.; Dixneuf, P. H.; Corrigan, J. F.; Carty, A. J. *J. Chem. Soc., Chem. Commun.*, **1994**, 409. (13) Zhang, S. W.; Mitsubo, T.; Kondo, T.; Watanabe, Y. *J. Organomet. Chem.*, **1993**, *450*, 197. (14) Cheng, P.-S.; Chi, Y.; Peng, S. M.; Lee, G.-H. *Organometallics* **1993**, *12*, 250. (15) Peng, J.-J.; Horng, K.-H.; Cheng, P.-S.; Chi, Y.; Peng, S. M.; Lee, G.-H. *Organometallics* **1994**, *13*, 2365.

ADVANCES IN ORGANOMETALLIC CHEMISTRY, VOL. 37

Ring-Opening Polymerization of Metallocenophanes: A New Route to Transition Metal-Based Polymers†

IAN MANNERS

Department of Chemistry
University of Toronto
Toronto, Ontario
Canada M5S 1A1

I

INTRODUCTION

Based on the well-studied characteristics of transition metal complexes, the incorporation of transition elements into a polymer main chain would be expected to allow access to processable speciality materials

†Dedicated to Derek S. Manners (1934–1994) and Ruth M. O'Hanlon (1935–1993).

with unusual and attractive electrical, optical, magnetic, preceramic, or catalytic properties (*1–4*). This is particularly the case where the metal atoms are present in the main chain in close spatial proximity to one another so as to promote cooperative interactions (*4*). However, virtually all of the polymers with skeletal transition metal atoms prepared to date have been synthesized by condensation (step-growth) routes, and these are usually either of low molecular weight, insoluble, or of poorly defined structure (*1,2*). Several, and in most cases recent, exceptions to this generalization exist (*5–7*), and two of the most well-established examples are rigid-rod, platinum polyynes such as **1** (*6*) and polymers such as **2** con-

1   **2**

taining ferrocene and siloxane units (*7*). Nevertheless, to facilitate the future development of transition metal-based polymer science it is clear that new synthetic routes to well-defined high molecular weight materials are critically needed.

Ring-opening polymerization (ROP) processes offer the attractive advantage that they generally proceed via a chain-growth mechanism and are therefore not subject to the stringent stoichiometry and conversion requirements which usually impede the preparation of high molecular weight polymers via condensation routes (*8*). For example, according to the theory of polycondensation reactions developed by Carothers, to obtain polymers with 200 repeat units even assuming the condensation proceeds completely (i.e., the extent of reaction is 100%) the reactant ratio for the difunctional monomers must be 0.99:1 or better (*9*). This is easily achieved with many difunctional organic monomers which are usually readily accessible and are often even available commercially in a high state of purity. However, synthesis of well-defined difunctional metal-containing monomers tends to be much more challenging. Moreover, even when compounds of this type are straightforward to prepare, as can be the case with dilithiated organometallic species, for example, they are so reactive toward air and moisture that completely pure samples of the monomer are extremely difficult to obtain. The use of metal-containing monomers of this type usually makes the assurance of exact stoichiometries during condensation polymerization reactions virtually impossible and therefore leads to the formation of low molecular weight oligomers rather than more desirable high molecular weight materials.

The ROP approach has now been developed to the stage where it represents a powerful route to a variety of organic polymer systems including polyamides, polyethers, polyolefins, polyacetylenes, and polycarbonates (*10*). In addition, the ROP of cyclic compounds with skeletons constructed from atoms of main group elements is of growing importance. For example, ROP provides an important route to polysiloxanes (silicones) (*11*), polyphosphazenes (*12*), and carbon–nitrogen–phosphorus and sulfur–nitrogen–phosphorus polymers (*13–15*), and the ROP of a variety of cyclic organosilicon compounds such as strained cyclotetrasilanes (*16*), silacyclobutanes and disilacyclobutanes (*17*), and silacyclopentenes (*18*) and related species (*19*) has also been well-established. However, reports of the use of ROP to prepare polymers with transition elements in the main chain are extremely rare and are limited to the polymerization of cyclometallaphosphazenes and cyclic trimeric metal nitrides (*20,21*).

This article focuses on recent work which has shown that [*x*]metallocenophanes (**3**), species which contain a metallocene group with a bridge containing *x* contiguous atoms linking the cyclopentadienyl ligands, can function as precursors to high molecular weight macromolecules **4** via ROP (*x* = *y*) or ROP-related (*x* > *y*) processes. The initial developments

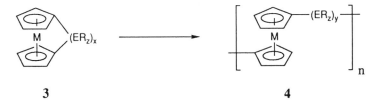

3   **4**

in this area were published in early 1992 (*22,23*). First, the sulfur atom abstraction-induced ROP of [3]trithiaferrocenophanes **3** (*x* = 3, E = S, *z* = 0) to yield polymers **4** (*y* = 2, E = S, *z* = 0) with backbones of ferrocene and S_2 units was described by Rauchfuss and co-workers (*22*). Although this novel process is not strictly a ROP and is probably better regarded as an atom abstraction-induced ring-opening polymerization (see Section VII), the polymers formed, polyferrocenylene persulfides, might be considered to arise because the hypothetical [2]dithiaferrocenophane intermediate **3** (*x* = 2, E = S, *z* = 0) is too strained to isolate. A few months after this discovery was reported, the first ROPs of strained [1]ferrocenophanes containing silicon in the bridge, **3** (*x* = 1, E = Si, *z* = 2), were published by our group (*23*). Subsequent work has shown that ROP of strained, ring-tilted metallocenophanes provides a general route to high molecular weight transition metal-based polymers with interesting properties. This article focuses on the progress to date in this area and covers the literature as far as the spring of 1994. After a brief overview of previously published work on metallocene-based polymers, research on the polymerization of [x]metallocenophanes (x = 1–3) is reviewed sequentially in order of increasing x.

II

FERROCENE-BASED POLYMERS

Because of the thermal stability and interesting physical (e.g., redox) properties associated with the ferrocene nucleus, attempts to introduce this organometallic unit into the side group structure of organic and inorganic polymers have been quite extensive and, in general, very successful. The synthesis of such materials generally requires only minor modifications of previously established synthetic methodologies. For example, poly(vinylferrocene) (6) can be prepared via the free radical addition poly-

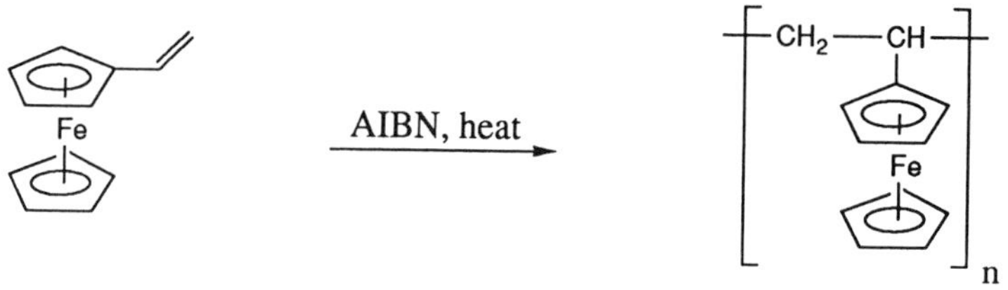

AIBN, heat

5   6

merization of vinylferrocene (5) (24), and polyphosphazenes, polysilanes, and polysiloxanes with ferrocenyl side groups have been synthesized via the incorporation of this organometallic moiety into the side group structure of the monomers used in the ROP or condensation routes to these materials (25–27). As a consequence of their accessibility and electroactive properties, polymers with ferrocenyl side groups have attracted significant interest as electrode mediators and as materials for the construction of electronic devices (28,29).

The same reasons for the interest in incorporating ferrocene units into the side group structure have also provided motivation for the synthesis of polymers in which ferrocenyl moieties are part of the main chain. An important additional reason for the synthesis of polymers with skeletal ferrocenyl units is provided by the observation that in molecular species in which two ferrocene units are linked together in close proximity the iron atoms can often interact to yield delocalized, mixed-valent species on one-electron oxidation. For example, the mixed-valence cations derived

7   **8**   **9**   **10**

X = Br or I

from biferrocenylene (**7**), the acetylene-bridged species **8**, and the biferrocenes such as **9** have been shown to be delocalized on the IR and/or Mössbauer time scale (*30–32*). These results suggest that macromolecules with skeletal ferrocene units might provide access to materials with interesting physical (e.g., electronic and magnetic) properties. However, in contrast to polymers with pendant ferrocenyl groups, prior to 1992 the pioneering work in this field yielded mainly low molecular weight ($M_n <<$ 10,000) and often poorly defined materials. For example, poly(ferrocenylenes) (**10**) with M_n above 5000 prepared via polycondensation processes such as the recombination of ferrocene radicals generated via the thermolysis of ferrocene in the presence of peroxides possess other fragments such as CH_2 and O in the main chain (*33*). Studies of the electrical properties of these polymers have indicated that on oxidation they become semiconducting with conductivities up to 10^{-6}–10^{-8} S/cm. More structurally well-defined poly(ferrocenylenes) (**10**, $M_n <$ 4000) have been prepared via the condensation reaction of 1,1-dilithioferrocene·tmeda with 1,1-diiodoferrocene (*34*), and, significantly, the reaction of 1,1′-dihaloferrocenes with magnesium has been shown to afford low molecular weight (M_n 4600 for soluble fractions) quasi-crystalline materials (*35*). In the case of the latter, oxidation with TCNQ affords doped polymers which are delocalized on the Mössbauer time scale ($\sim 10^{-7}$ second) at room temperature and which possess electrical conductivities of up to 10^{-2} S/cm. Bearing in mind that with low molecular weights the effects of end groups would be expected to significantly influence the electrical properties, these results suggest that high molecular weight polymers should be very interesting to study, particularly if such materials can be prepared in an highly ordered (i.e., crystalline) state.

Previous attempts to synthesize ferrocene-backbone macromolecules of type **4** ($y > 0$) in which the ferrocene units are separated via a spacer group have also focused on the use of condensation reactions. Polymers **4** ($y > 3$) in which the spacer is very long have been successfully prepared (*36*), but well-characterized materials of substantial molecular weight where the ferrocene units are held in close proximity to one another are extremely rare. Previous work on such polymers and other ferrocene-based materials with hydrocarbon or main group element spacers is discussed in the relevant sections below.

III

RING-OPENING POLYMERIZATION OF SILICON-BRIDGED [1]FERROCENOPHANES

A. *Condensation Routes to Main Chain Ferrocene–Organosilane Polymers*

Early attempts to prepare polymers of general structure 11 focused on reactions of dilithioferrocene·tmeda and appropriate dihaloorganosilanes. For example, in the late 1960s low molecular weight poly(ferrocenylsilanes) 11 ($x = 1$, R = Me or Ph) were prepared via condensation routes

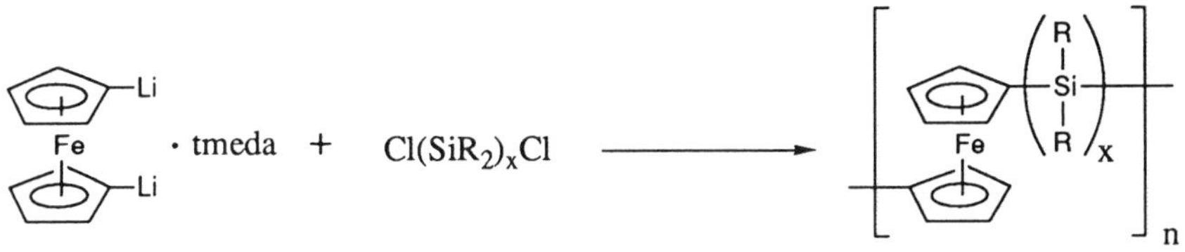

11

involving the reaction of dilithioferrocene·tmeda with R_2SiCl_2 in tetrahydrofuran (THF)/hexanes (*37*). The molecular weights of 1400–7000 reported for these materials are characteristic of condensation processes where exact reaction stoichiometries are virtually impossible because one reactant, in this case dilithioferrocene, cannot be readily prepared in pure form. This species, which is usually synthesized via the reaction of ferrocene with butyllithium·tmeda, is generally only about 90–95% pure (*38*). The main impurity, monolithioferrocene, functions as a chain capping agent because of its monofunctionality. Although the poly(ferrocenylsilanes) 11 ($x = 1$, R = Me or Ph) reported possessed only 5–20 repeat units, thermogravimetric analysis indicated that these materials exhibited reasonable thermal stability with no appreciable weight loss until 200–250°C (*37*).

Condensation approaches to ferrocene/organosilane polymers have been extended to the synthesis of low molecular weight materials 11 ($x = 6$, R = Me) with M_n 3500 from the reaction of dilithioferrocene·tmeda with the dichlorohexasilane $Cl(SiMe_2)_6Cl$ (*39*). These polymers were shown to possess electrical conductivities in the range of 10^{-5}–10^{-6} S/cm

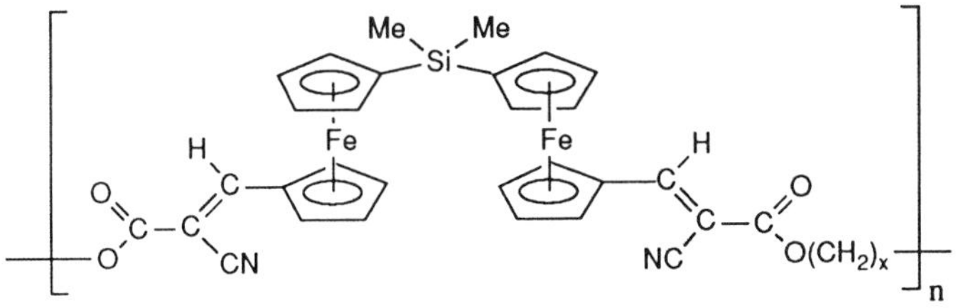

12

when doped with I_2 or AsF_5. Also, novel polymers **12** containing ferrocene units bonded to $SiMe_2$ groups together with long organic spacers have been reported, and these are of significantly higher molecular weight (M_w 19,000–55,000) (40).

B. Synthesis, Properties, and Ring-Opening Polymerization of Silicon-Bridged [1]Ferrocenophanes

Prior to 1992 no efforts to prepare polymers with main chains consisting of ferrocene and organosilane units via ring-opening methods had been reported. The first developments in this area involved attempts to polymerize silicon-bridged ferrocenophanes and related species (23,41). The first successful ROPs involved the use of [1]silaferrocenophanes as cyclic monomers, and the progress in this area to date together with relevant previous work is now reviewed.

1. Synthesis and Properties of [1]Silaferrocenophanes

The first reported [1]ferrocenophane containing a single silicon atom in the bridge, **13** (R = R' = Ph) was synthesized in 1975 via the reaction of

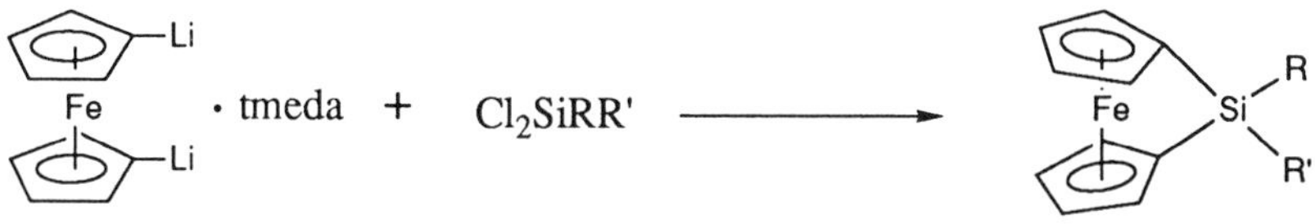

13

dilithioferrocene·tmeda with diphenyldichlorosilane in hexane (42). A subsequent X-ray crystal structure determination of **13** (R = R' = Ph) showed that the cyclopentadienyl ligands were significantly tilted with respect to one another by an angle of 19.1(10)°, which suggested that the compound was appreciably strained (43). This is in dramatic contrast to

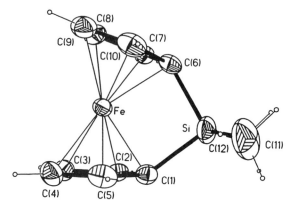

the situation in ferrocene where the cyclopentadienyl ligands are parallel (44).

During the period of 1975–1990 several species analogous to 13 (R = R' = Ph) were similarly prepared using the appropriate dichloroorganosilane (45–47). Several of these compounds have been characterized by X-ray diffraction, and the molecular structure of a representative example, 13 (R = R' = Me), is shown in Fig. 1 (48). The tilt angle in this species is 20.8(5)°, and other features of the structure which are typical for [1]sila-ferrocenophanes are the significant distortion from planarity at the cyclo-pentadienyl carbon atom bonded to silicon, the small C(Cp)–Si–C(Cp) angle, and the iron–silicon distance which is indicative of a weak dative interaction. The presence of a weak Fe–Si bond in these compounds is also supported by Mössbauer spectroscopy (49). Evidence for the presence of appreciable strain in [1]silaferrocenophanes is provided by NMR and UV/visible spectroscopic studies (46). For example, the ^{13}C NMR chemical shift for the ipso carbon atoms of the cyclopentadienyl groups bonded to silicon is approximately 30–40 ppm, which is dramatically different from the typical position of around 68–75 ppm for such reso-nances in unstrained systems. In addition, the UV/visible spectra of [1]silaferrocenophanes in the 200–800 nm range contain a band in the visible region at 470–478 nm (ε 235–280 M^{-1} cm^{-1}) which shows a batho-chromic shift and increase in intensity relative to the long wavelength band of ferrocene at 440 nm (ε 90 M^{-1} cm^{-1}) and that of bis(trimethyl-silylferrocene) at 448 nm (ε 130 M^{-1} cm^{-1}). Consequently, the color of these species is deep red rather than the pale orange color characteristic of most ferrocene compounds.

Previous work has also shown that the strain present in [1]silaferro-cenophanes profoundly influences the chemical reactivity observed for these molecules. Thus, stoichiometric ring-opening reactions have been found to occur with reagents containing hydroxyl groups (*45*). For example, reaction of **13** (R = R' = Me) with water or methanol affords the

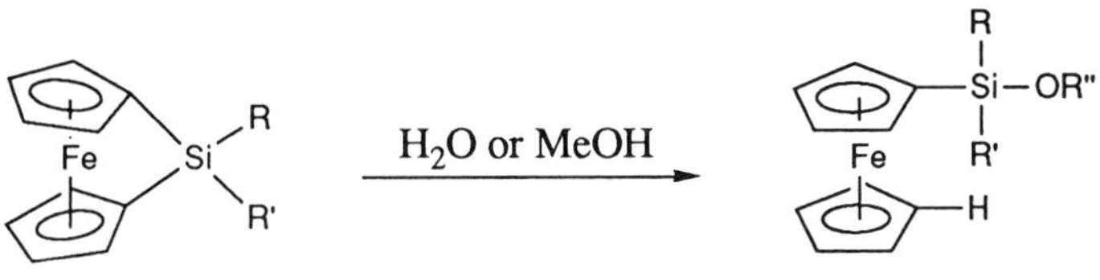

13   **14**

R" = H or Me

silanol **14** (R = R' = Me, R" = H) or the alkoxysilane derivative **14** (R = R' = Me, R" = Me), respectively (*45*). The analogous reaction of water with the more sterically encumbered diaryl [1]silaferrocenophane **13** (R = R' = Ph) is much slower, and no reaction was detected with methanol. The ring-opening reactions of [1]silaferrocenophanes such as **13** (R = R' = Me) with surface -OH functionalities have been exploited to derivatize silica, semiconductors, and metals (*45,50*).

2. Thermal and Anionic Ring-Opening Polymerization of [1]Silaferrocenophanes

In 1992, our group reported (*23*) the discovery that the [1]silaferro-cenophanes **13** (R = R' = Me or Ph) undergo spontaneous, exothermic, and quantitative thermally induced ROP reactions when heated above

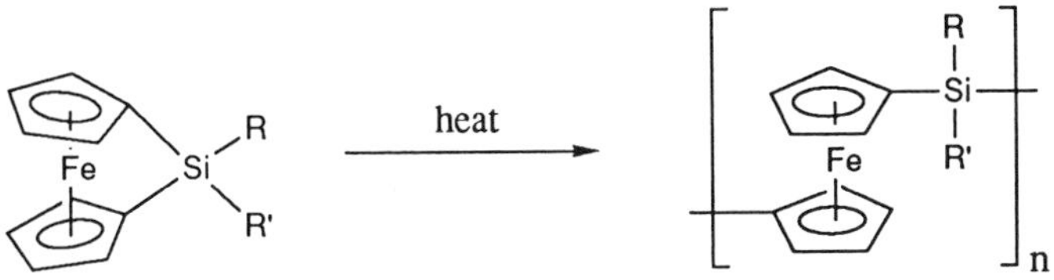

13   **15**

their melting points to yield poly(ferrocenylsilanes) **15** (R = R' = Me or Ph). The poly(ferrocenylsilane) **15** (R = R' = Me) with methyl substituents at silicon was found to be soluble in THF and was of high molecular weight (M_w 520,000, M_n 340,000) according to gel permeation chromatography (GPC) analysis using polystyrene standards. This corresponds to polymer chains with over 1200 repeat units.

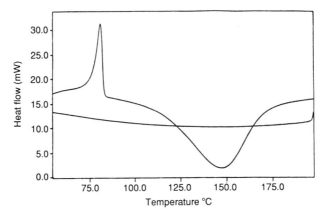

FIG. 2.   DSC thermogram for [1]silaferrocenophane **13** (R = R' = Me). Reprinted with permission from (*23*). Copyright (1992), American Chemical Society.

When the ROP of **13** (R = R' = Me) was followed by differential scanning calorimetry (DSC), a melting endotherm was detected at 78°C followed by a ROP exotherm at 120–170°C. Integration of the latter indicated that the strain energy was ca. 80 kJ/mol^{-1} (Fig. 2). A similar experiment for the more sterically encumbered monomer **13** (R = R' = Ph) yielded a strain energy of ca. 60 kJ/mol (*23*). This indicated that [1]ferrocenophanes possess appreciable strain energies in between those of cyclopentane (~42 kJ/mol) and cyclobutane (~118 kJ/mol) and which are far greater than that of the strained cyclotrisiloxane [Me$_2$SiO]$_3$(19 kJ/mol) (*51*).

This thermally induced ROP reaction has subsequently been found to be general for [1]silaferrocenophanes with a range of substituents at silicon and has been used to synthesize a variety of new symmetrically or unsymmetrically substituted poly(ferrocenylsilanes) (*48,51–54,59*). Moreover, because of the relatively large intrinsic strain present, even [1]silaferrocenophanes with bulky substituents such as norbornyl, ferrocenyl, or, as described above, phenyl groups still undergo ROP (*54,59*). This is unusual as the ROP of many organic and inorganic ring systems is normally only favorable when small side groups are present.

Additional work has shown that it is also possible to induce the ROP of [1]silaferrocenophanes in solution at ambient temperatures by using anionic initiators (*55*). For example, the reaction of **13** (R = R' = Me) with approximately 10 mol % ferrocenyllithium (FcLi) followed by hydrolytic workup yields a moderate molecular weight poly(ferrocenylsilane) **15** (R = R' = Me) with M_w 9500 and M_n 8000. With approximately equimolar

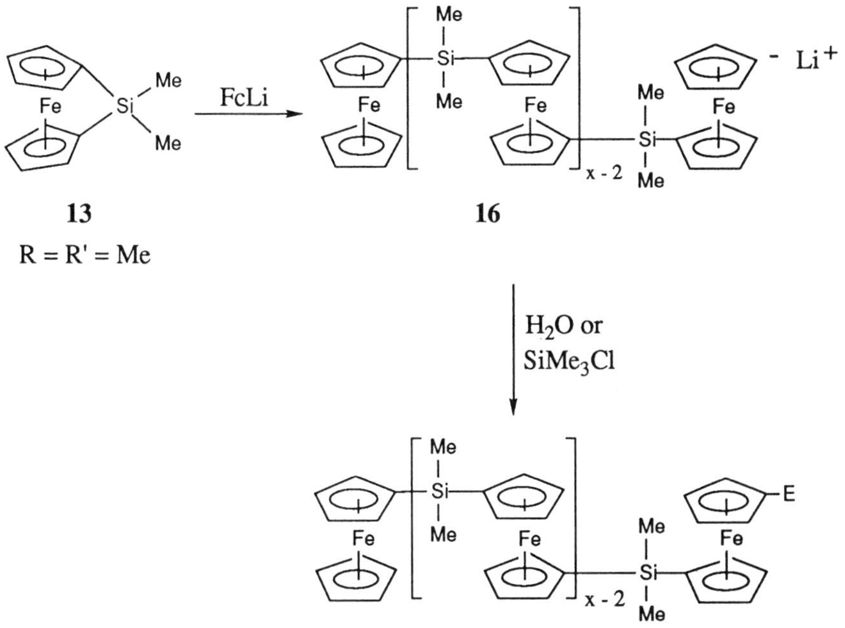

13
R = R' = Me

16

17 E = H or $SiMe_3$
x = 2 - 30

SCHEME 1

quantities of [1]silaferrocenophane monomer and initiator, oligo(ferrocenylsilanes) **17** are formed with between 2 and 8 ferrocenyl groups. These function as valuable models for the conformational and electrochemical properties of the corresponding high polymer **15** (R = R′ = Me) (see below). The probable mechanism for the anionic ring-opening reactions of **13** (R = R′ = Me) is shown in Scheme 1.

The poly(ferrocenylsilanes) synthesized to date by either thermal or anionic ROP are listed in Table I.

C. *Properties of Poly(ferrocenylsilanes)*

1. *General Characteristics*

Apart from the phenylated polymer **15** (R = R′ = Ph), most poly(ferrocenylsilanes) are readily soluble in organic solvents, and indeed some with long aliphatic chains attached to silicon such as **15** (R = R′ = *n*-Hex)

TABLE I

CHARACTERIZATION DATA FOR POLY(FERROCENYLSILANES) 15

R	R'	$M_w{}^a$	$M_n{}^a$	$\delta(^{29}Si)^b$ (ppm)	$T_g{}^c$ (°C)	$T_m{}^c$ (°C)
Me	Me	5.2×10^5	3.4×10^5	-6.4	33	122
Et	Et	7.4×10^5	4.8×10^5	-2.7	22	108
n-Bu	n-Bu	8.9×10^5	3.4×10^5	-4.9	3	116, 129
n-Hex	n-Hex	1.2×10^5	7.6×10^4	-2.3	-26	—
Ph	Ph	5.1×10^4	3.2×10^4	-12.9	—	—
Me	H	8.6×10^5	4.2×10^5	-20.0	9	87, 102
Me	TFP^d	2.7×10^6	8.1×10^5	-4.3	59	—
Me	Fc^d	1.6×10^5	7.1×10^4	-9.8	99	—
Me	n-C$_{18}$H$_{37}$	1.4×10^6	5.6×10^5	-5.3	1	16
Me	Vinyl	1.6×10^5	7.7×10^4	-12.9	28	—
Me	Ph	3.0×10^5	1.5×10^5	-10.9	54	—
Me	Nor^d	1.6×10^5	1.1×10^4	-3.4	81	—
Ph	Vinyl	1.1×10^5	7.6×10^4	-12.6	—	—

[a] Polymer molecular weights were determined by gel permeation chromatography in THF containing 0.1% by weight of [NBu$_4$]Br using polystyrene standards.
[b] ^{29}Si NMR data were recorded in C$_6$D$_6$.
[c] T_g and T_m values were determined by differential scanning calorimetry, the latter after annealing.
[d] TFP, Trifluoropropyl; Fc, ferrocenyl; Nor, norbornyl.

and 15 (R = Me, R' = n-C$_{18}$H$_{37}$) are even soluble in nonpolar hydrocarbons such as pentanes (52,59). The polymers 15 have been structurally characterized by a variety of spectroscopic and analytical techniques including ^{1}H, ^{13}C, and ^{29}Si NMR as well as elemental analysis. In contrast to the case of [1]silaferrocenophane monomers, both the ^{13}C NMR resonances for the ipso carbons of the cyclopentadienyl groups attached to silicon and the UV/visible spectra for poly(ferrocenylsilanes) are similar to those detected for ferrocene derivatives with unstrained structures (52). Usually the ^{29}Si NMR chemical shift values for poly(ferrocenylsilanes) are found around 2–3 ppm upfield of those of the respective monomers. The molecular weights of the poly(ferrocenylsilanes) have been studied by GPC and by low angle laser light scattering (LALLS). Typical values of M_w for the polymers by GPC (relative to polystyrene standards) are in the range of 3×10^5 to 3×10^6, and these data together with the corresponding values of M_n and the ^{29}Si NMR data are listed in Table I. The molecular weight values correspond to a number of repeat units (n) that vary from 100 to 10,000. An absolute molecular weight value of M_w = 2.29×10^5 for a sample of the poly(ferrocenylsilane) 15 (R = R' = n-Bu) has been determined by LALLS in THF (52). This value was significantly larger than the GPC estimate (M_w 1.7×10^5) for the same polymer sample determined in the same solvent using polystyrene standards for column calibration. This indicates that the hydrodynamic size of a coil of the poly(ferrocenylsilane) **15** (R = R' = *n*-Bu) is significantly smaller than that for a coil polystyrene of the same absolute molecular weight in THF under the same conditions (*52*). Similar results have been reported for the unsymmetrically substituted poly(ferrocenylsilane) **15** (R = Me, R' = Ph) (*59*).

2. Thermal Transition Behavior and Morphology

As would be expected, the physical properties of poly(ferrocenylsilanes) have been found to vary dramatically depending on the side groups attached to silicon. Most examples of these polymers are glassy, but some can be microcrystalline. Both types readily form solution-cast films (see Fig. 3). However, poly(ferrocenylsilanes) with long flexible side groups tend to be gummy materials (see Fig. 4). Bearing these observations in mind, the thermal transition behavior of symmetrically substituted poly (ferrocenylsilanes) has been studied by DSC and by dynamic mechanical analysis (DMA). In addition, the morphology of the polymers has been probed by wide angle X-ray scattering (WAXS) studies.

The glass transitions (T_g values) for poly(ferrocenylsilanes) (Table I) are, generally, significantly higher than those for the analogous polysi-

Fig. 3. Film of poly(ferrocenylsilane) **15** (R = R' = Me).

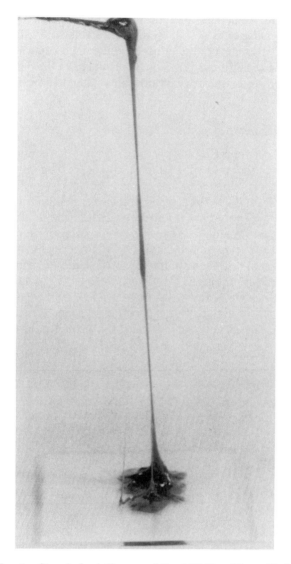

FIG. 4.   Strand of poly(ferrocenylsilane) **15** (R = R' = *n*-Hex).

lanes. For example, the T_g of the elastomeric polymer **15** (R = R' = *n*-Hex) is −26°C, whereas poly(di-*n*-hexylsilane) has a T_g of approximately −53°C. This indicates that the incorporation of a skeletal ferrocenyl moiety into a polysilane backbone decreases the conformational flexibility

(*52,53*). This effect can probably be attributed to the relative steric bulk and rigidity of the ferrocenyl unit.

Studies of the morphology of symmetrically substituted poly(ferrocenylsilanes) by DSC and DMA have indicated a significant dependence on sample history and perhaps molecular weight (*51–54,56*). When samples of the poly(ferrocenylsilanes) **15** (R = R' = Me, Et, or *n*-Bu) were annealed, melting transitions (T_m values) were detected by DSC and/or DMA (Table I). Weaker melting transitions were sometimes observed by DSC for unannealed samples. After the initial DSC scans melting transitions were not detected owing to the slow rate of recrystallization. This situation is illustrated for **15** (R = R' = Et) in Fig. 5. After annealing at 90°C for 1 hour a T_m was detected by DSC at 108°C. In subsequent scans no T_m was observed (*51*). Interestingly, when the poly(ferrocenylsilane) **15** (R = R' = *n*-Bu) was annealed for several hours at 90°C two melting

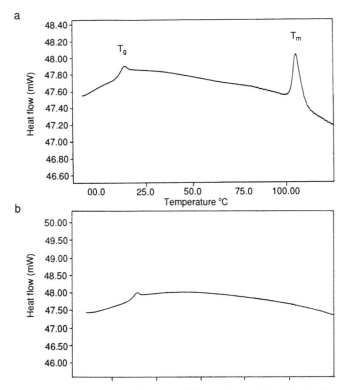

FIG. 5. DSC thermogram for poly(ferrocenylsilane) **15** (R = R' = Et). (a) After annealing at 90°C for 2 hours showing both a T_g and a T_m. (b) Subsequent scan showing only a T_g.

transitions were detected by DSC, with a new T_m at 116°C in addition to that previously detected at 129°C.

WAXS studies of both relatively amorphous (51,53) and crystalline (56) samples of the symmetrically substituted polymers **15** (R = R' = Me, Et) have been reported. The WAXS patterns for unnannealed samples of **15** (R = R' = Me, Et) show broad lines superimposed on amorphous halos, which suggest the presence of only short-range order (51,53). More crystalline samples of these polymers show several sharper lines consistent with the presence of significant long-range order (56). Interestingly, the WAXS pattern of **15** (R = R' = n-Bu) shows several fairly sharp peaks indicative of substantial order, whereas that of **15** (R = R' = n-Hex) shows only typical amorphous halos (51,53,56).

Work aimed at a detailed understanding of the conformations and morphology of poly(ferrocenylsilanes) clearly represents an interesting area for future research. Indeed, studies of a series of unsymmetrically substituted poly(ferrocenylsilanes) have shown that side-chain crystallization can occur if long n-alkyl substituents are present (59). Structural studies of short chain model oligo(ferrocenylsilanes) might be expected to facilitate progress in this area. A linear oligomer **17** (x = 5, E = H) with five ferrocenyl units formed by the anionic ring-opening oligomerization of **13** (R = R' = Me) has been shown to possess a trans planar, zig-zag conformation by single-crystal X-ray diffraction (see Fig. 6) (55). Thus, the conformation found for this species might also be preferred in the corresponding high polymer **15** (R = R' = Me).

3. Electrochemical Properties

One of the most well-studied characteristics of the ferrocene nucleus is its ability to undergo a reversible one-electron oxidation (57). Polymers such as poly(vinylferrocene) possess electroactive ferrocenyl moieties which are well-separated from one another by insulating organic units. The ferrocenyl units in these polymers are essentially noninteracting, and a single reversible oxidation wave is detected by cyclic voltammetry (58). The observation that the poly(ferrocenylsilane) **15** (R = R' = Me) exhib-

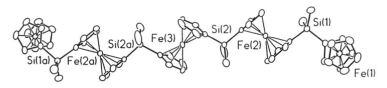

FIG. 6.   Molecular structure of **17** (x = 5, E = H). Reprinted with permission from (55). Copyright (1994), American Chemical Society.

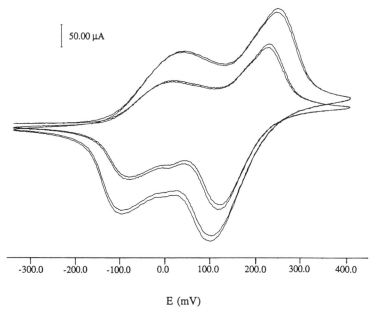

FIG. 7.   Cyclic voltammograms of poly(ferrocenylsilane) **15** (R = R' = Me) in CH₂Cl₂ at 500 and 1000 mV/second. Reprinted with permission from (*63*).

ited two reversible oxidation waves of equal intensity, as found for poly (ferrocenylene persulfides) (*22*) (see Section VII), therefore represented an interesting discovery which was interpreted in terms of initial oxidation at alternating iron sites (*23*).

Similar two-wave cyclic voltammetric behavior has been subsequently reported for all of the poly(ferrocenylsilanes) **15** (R = R' = Me, Et, *n*-Bu, *n*-Hex) under the same conditions (polymer 5 × 10^{-4} *M* in CH₂Cl₂, 0.1 *M* [Bu₄N][PF₆] as supporting electrolyte, scan rates of 250–1000 mV/second) and has been explained by the existence of interactions between the iron centers which, after oxidation of the first ferrocenyl moiety, make the subsequent removal of an electron from a neighboring iron center more difficult (*53*) (Fig. 7). This explanation is consistent with the results of extensive investigations on small molecule compounds, including α,ω-bis(ferrocenyloligosilanes), which contain ferrocenyl groups linked together by a bridging group (*30,58,60–62*). These studies show that the interaction between the ferrocenyl units, represented by the cyclic voltammetric wave separation Δ*E*, decreases as the length of the bridge between them increases.

Further examination of the electrochemical behavior of symmetrically substituted poly(ferrocenylsilanes) by cyclic voltammetry and chrono-coulometry has broadly supported these conclusions and has provided some additional insight (56). Studies of **15** (R = R' = Me, Et, n-Bu, n-Hex) by cyclic voltammetry (1×10^{-4} M CH_2Cl_2, scan rates of 10–100 mV/second, using 0.1 M [Et$_4$N][BF$_4$] as supporting electrolyte) have shown that although the polymers are soluble in this solvent they appear to deposit on the electrode surface as a monolayer. This alters the shape of the voltammogram by affecting the diffusion of supporting electrolyte ions to the electrode (Pt). Comparison of areas of the two oxidation waves and measurements of the amount of $FeCl_3$ required for complete chemical oxidation of the polymer in CH_2Cl_2 solution were consistent with the initial oxidation occurring at half the ferrocene centers.

The difference in oxidation potentials (ΔE) detected for the two waves found for the poly(ferrocenylsilanes) **15** (R = R' = Me, Et, n-Bu, n-Hex), which provides an indication of the degree of interaction between the iron sites, varies from 0.21 V (for **15** (R = R' = Me)) to 0.29 V (for **15** (R = R' = n-Bu or n-Hex)) (63). This indicates that the extent of the interaction between the ferrocenyl units in poly(ferrocenylsilanes) depends signifi-cantly on the nature of the substituents at silicon, which may be a result of electronic or conformational effects (63). Unsymmetrically substituted poly(ferrocenylsilanes) show similar electrochemical behavior (59). In ad-dition, polymer **15** (R = Me, R' = Fc) shows a complex cyclic voltammo-gram which indicates that interactions exist between the iron centers in the polymer backbone and the ferrocenyl side groups (59).

Studies of the electrochemical behavior of the oligo(ferrocenylsilanes) **17** (E = H x = 2–8) have been found to be completely consistent with initial oxidation at alternating iron sites in poly(ferrocenylsilane) high polymers (55). Thus, cyclic voltammetric studies of the oligomers with an odd number of repeat units **17** (x = 3, 5, or 7) show two waves which become closer to the 1:1 intensity ratio detected for the high polymer **15** (R = R' = Me) as the chain length increases. The even oligomer **17** (x = 2) also shows two waves of equal intensity, whereas the longer chain even oligomers **17** (x = 4, 6, or 8) show a third wave in between the other two which decreases in intensity as the chain length increases. The remaining two outer waves become of more equal intensity as the chain length increases, and so in the limit the cyclic voltammogram for the high poly-mer **15** (R = R' = Me) is also reproduced (55).

The oxidation of poly(ferrocenylsilanes) can also be achieved using either electrode-confined or free-standing films of these materials (51). The films are electrochromic and can be reversibly cycled between amber (reduced) and dark blue (oxidized) states. This suggests that these materials are worthy of further study with respect to possible applications in display devices.

4. *Optical and Electrical Properties*

As poly(ferrocenylsilanes) possess potentially conjugated σ, $p\pi$, and $d\pi$ units, and in view of the evidence for interactions between the skeletal ferrocenyl moieties provided by electrochemical studies, the possibility that appreciable delocalization of electrons can occur in these materials has been investigated by various techniques such as UV/visible spectroscopy (*52*), Mössbauer spectroscopy (*51,53*), and electrical conductivity studies (*53*). These are indicative of an essentially localized electronic structure in the materials investigated to date. For example, the solution UV/visible spectra of the poly(ferrocenylsilanes) **15** (R = R' = Me, Et, *n*-Bu, *n*-Hex) with alkyl side groups in the visible region were found to be similar to those for ferrocene and monomeric ferrocene derivatives with organosilicon groups attached to each cyclopentadienyl ring. The similarity of the λ_{max} for the visible absorption band in the poly(ferrocenylsilanes) with that for small molecule species such as ferrocene and silylated ferrocenes indicates that the electronic structure of the polymer main chain can, at least to a first approximation, be considered as localized (*52*). This situation contrasts with that for polysilanes where the λ_{max} of the HOMO–LUMO σ–σ^* transition shifts dramatically to longer wavelengths on moving from small molecule oligomers to high molecular weight polymers, which provides excellent evidence for the delocalization of σ electrons in these materials (*64,65*). Similar trends are apparent in the UV/visible spectra of π-conjugated polyenes where the π electrons are delocalized.

Further evidence for an essentially localized electronic structure is provided by the Mössbauer spectra of partially oxidized samples of **15** (R = R' = Me), which show the existence of discrete Fe^{II} and Fe^{III} environments, and the electronic conductivity of this material, which is approximately 10^{-13}–10^{-14} S/cm for the unoxidized polymer but rises into the weak semiconductor range (10^{-7}–10^{-8} S/cm) for I_2-doped samples at 25°C (*51,53*). These conductivity values for the doped polymers are consistent with an electron-hopping mechanism and are similar to those found for low molecular weight, amorphous poly(ferrocenylene) (*33*). In contrast, conductivities of up to 10^{-1} S/cm have been reported for AsF_5-doped polysilanes, and values as high as 10^5 S/cm have been reported for polyacetylene (*64,65*). Although the conductivities reported to date for poly(ferrocenylsilanes) are quite low, the delocalization detected in small molecule biferrocenes has been found to be very dependent on both the substitutents and the solid state environment (*66*). This suggests that the realization of materials with appreciable electrical conductivity might be possible, but considerably more work is clearly needed.

5. Formation of Iron Silicon Carbide Ceramics via Pyrolysis

Because of processing advantages, polymers containing inorganic elements are of considerable interest as pyrolytic precursors to ceramics (*3*). Polysilanes and polycarbosilanes have attracted particular attention in this respect as sources of silicon carbide materials (*64,65*). However, the use of polymers as precursors to transition metal-containing solids, which are known to possess a wide range of interesting electrical, magnetic, and optical properties, is virtually unexplored. Studies of the thermal stability of **15** (R = R' = Me, Ph) by thermogravimetric analysis (TGA) have

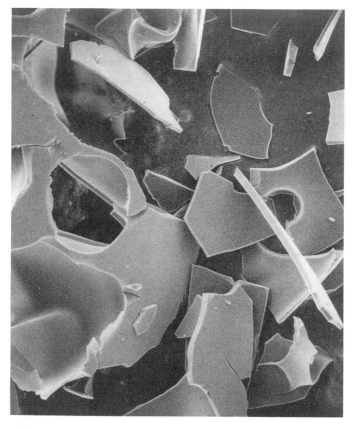

FIG. 8.   Scanning electron micrograph (at 100× magnification) of ceramic derived from **15** (R = R' = Ph). Reprinted with permission from (*67*) and The Royal Society of Chemistry.

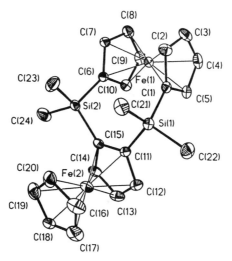

FIG. 9. Molecular structure of depolymerization product [Fe(η-C_5H_4)$_2$(μ-SiMe$_2$)$_2$ (ηC_5H_3)Fe(η-C_5H_5)] derived from **15** (R = R' = Me). Reprinted with permission from (*67*) and The Royal Society of Chemistry.

indicated that the polymers undergo significant weight loss at 350–500°C to yield ceramic residues in around 35–40% yield with no further weight loss up to 1000°C (*67*). The lustrous ceramic products (Fig. 8) formed when **15** (R = R' = Me) or **15** (R = R' = Ph) are heated at 500°C under a slow flow of dinitrogen are ferromagnetic and have been characterized as amorphous iron silicon carbide materials by techniques such as X-ray photoelectron spectroscopy (XPS) and energy-dispersive X-ray (EDX) microanalysis. In addition, orange-yellow depolymerization products have been isolated and in some cases characterized. For example, the pyrolysis of **15** (R = R' = Me) yields, in addition to the ceramic, mainly an unusual, unsymmetrical dimer which has been characterized crystallographically (Fig. 9) (*67*).

IV

RING-OPENING POLYMERIZATION OF GERMANIUM-BRIDGED [1]FERROCENOPHANES

The discovery of the facile thermal ROP of silicon-bridged [1]ferrocenophanes suggested that strained metallocenophanes with other elements in the bridge might also polymerize (*23*). The first [1]ferrocenophane containing a single germanium atom in the bridge, **18** (R =

R′ = Ph), was prepared in 1980 from the reaction of $Fe(\eta\text{-}C_5H_4Li)_2\cdot tmeda$ with Ph_2GeCl_2 in hexanes (46). An X-ray crystal diffraction study of **18** (R = R′ = Ph) revealed that this species was very slightly less ring-tilted [tilt angle 16.6(15)°] than the analogous compound **13** (R = R′ = Ph) with silicon in the bridge [tilt angle 19.2(10)°] (68). The only other reported germanium-containing [1]ferrocenophane, the spirocyclic species **19**, was isolated in low yield from the reaction of 2 equiv of $Fe(\eta\text{-}C_5H_4Li)_2\cdot tmeda$ with $GeCl_4$ and was characterized spectroscopically and by elemental analysis (69).

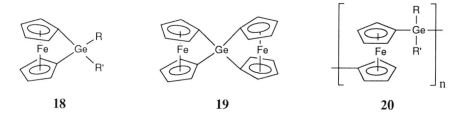

18   **19**   **20**

A variety of [1]germaferrocenophanes **18** have now been prepared via the salt elimination reactions of $Fe(\eta\text{-}C_5H_4Li)_2\cdot tmeda$ with the appropriate dihalogermanium species (51). The ^{13}C NMR chemical shifts for the resonances assigned to the ipso carbon atoms of the cyclopentadienyl ligands bonded to germanium show a shift to high field compared to those for analogous unbridged species in a similar way to [1]silaferrocenophanes. This feature is characteristic of strained [1]ferrocenophanes with main group elements in the bridge (46). The methylated [1]germaferrocenophane **18** (R = R′ = Me) was shown to undergo rapid thermal ROP at relatively low temperatures (90°C), with an onset polymerization temperature around 30°C lower than for the silicon analog (70). The resulting yellow, film-forming poly(ferrocenylgermane) **20** (R = R′ = Me) was of very high molecular weight with $M_w = 2.3 \times 10^5$, $M_n = 5.2 \times 10^4$ and $M_w = 2.0 \times 10^6$, $M_n = 8.5 \times 10^5$ determined by GPC for two separate polymerization experiments (70). The ROP route has now been extended to provide approaches to a range of poly(ferrocenylgermanes) **20** (R = R′ = Et, n-Bu, or Ph) (Table II) (51).

To date, poly(ferrocenylgermanes) have been much less studied than the silicon analogs, and they represent an interesting area for future work. Nevertheless, their thermal transition behavior has been explored (Table II) (51). Studies of the electrochemistry of the poly(ferrocenylgermane) **20** (R = R′ = Me) have shown that this material exhibits two reversible oxidation waves in CH_2Cl_2, which indicates that the iron atoms interact with one another in a similar way to poly(ferrocenylsilanes) (63).

TABLE II

MOLECULAR WEIGHT AND GLASS TRANSITION DATA FOR
POLY(FERROCENYLGERMANES) **20**

R	R'	$M_w{}^a$	$M_n{}^a$	$T_g{}^b$ (°C)
Me	Me	2.0×10^6	8.5×10^5	28
Et	Et	7.4×10^5	4.8×10^5	12
n-Bu	n-Bu	8.9×10^5	3.4×10^5	-7
Ph	Ph	1.0×10^6	8.2×10^5	114

^{a} Polymer molecular weights were determined by gel permeation chromatography in THF containing 0.1% by weight of [NBu$_4$]Br using polystyrene standards.
^{b} T_g values were determined by differential scanning calorimetry.

V

RING-OPENING POLYMERIZATION OF PHOSPHORUS-BRIDGED [1]FERROCENOPHANES

A. Condensation Routes to Poly(ferrocenylphosphines)

Condensation type routes have been used to prepare low molecular weight, relatively poorly defined poly(ferrocenylphosphines) **21** and poly-(ferrocenylphosphine oxides) **22** (*71,72*). For example, the products (M_n

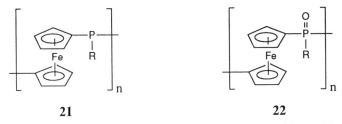

21   **22**

<6000) formally assigned the structure **22** from the reaction of ferrocene with PhPCl$_2$ in the presence of ZnCl$_2$ at 80–170°C followed by oxidation with H$_2$O$_2$ actually possess 1,2-, 1,3-, and 1,1'-ferrocenylene linkages (*71,72*).

Much better defined, moderate molecular weight (M_w 8900) poly(ferro-cenylphosphines) **24** were prepared by Seyferth and co-workers via the polycondensation reaction of 1,1'-dilithioferrocene·tmeda with PhPCl$_2$ in dimethoxyethane at 25°C followed by conversion of the proposed Cp–Li and P–Cl end groups of the condensation polymer **23** to Cp–H and P–OH or P–Ph, respectively (*73*). Furthermore, under certain carefully controlled conditions using a slow addition of the dilithioferrocene·tmeda to PhPCl$_2$ in Et$_2$O at 25°C or DME at −40°C, surprisingly high molecular weight products (M_w 131,000–161,000) were formed (73). The formation of high molecular weight condensation polymers would not be expected using dilithioferrocene·tmeda as normally prepared from ferrocene and n-butyllithium in the presence of tmeda as this is seldom of higher than 90–95% purity (38). Although it is possible that the 1,1'-dilithioferrocene·tmeda used for these polymerization experiments was unusually pure, the observation that [1]silaferrocenophanes will polymerize in the presence of anionic initiators in solution (55) suggests, with hindsight, the alternative possibility that under these conditions the poly(ferrocenylphosphines) may arise from the anionically induced ROP of the in situ generated [1]phosphaferrocenophane 25 (R = Ph). Indeed, as discussed in the next section, anionic ring-opening oligomerization (but not polymerization) has been previously reported for this species, and several [1]phosphaferrocenophanes have now been shown to undergo thermally induced ROP.

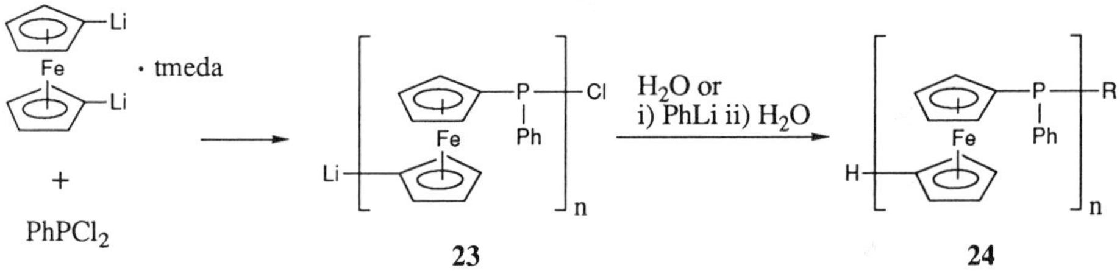

23   24

R = OH or Ph

B. Synthesis, Properties, and Ring-Opening Polymerization of Phosphorus-Bridged [1]Ferrocenophanes

A variety of [1]phosphaferrocenophanes 25 with different substituents at phosphorus have been reported since 1980, and these have been prepared via the reaction of dilithioferrocene·tmeda with the appropriate organodichlorophosphine RPCl$_2$ in a nonpolar solvent (68,73,74). X-Ray structural studies of several of these compounds have shown that the tilt angle in these is even greater than that found for silicon-bridged [1]ferrocenophanes. For example, the phenylated species 25 (R = Ph) possesses a tilt angle of around 26.9° (68). However, compared to Si- and Ge-bridged analogs, the angle between the plane of the cyclopentadienyl ligand and the Cp–phosphorus bond is smaller (74). The iron–phosphorus distance in [1]phosphaferrocenophanes is once again indicative of a significant dative interaction between iron and the bridging atom. The [1]phosphaferrocenophanes studied to date also possess high-field ipso ^{13}C NMR resonances and UV/visible spectra indicative of the presence of strained structures (46).

Facile, stoichiometric ring-opening reactions with nucleophilic reagents have been detected for [1]phosphaferrocenophanes, and such reactions have been used to prepare a range of interesting metallocenophanes with transition elements in the bridge (74,75). Furthermore, the reaction of **25** (R = Ph) with an equimolar quantity or a slight deficiency of lithio(diphenylphosphino)ferrocene (which is derived from 1:1 reaction of the same [1]phosphaferrocenophane with PhLi) leads to anionic ring-opening oligomerization to afford **26** (x = 2–5). However, attempts to carry out ROP by using low concentrations of initiator were reported to be unsuccessful and yielded only unreacted **25** (R = Ph) together with the same low molecular weight oligo(ferrocenylphosphine) products **26** (x = 2–5) (73). The failure to observe the formation of polymeric products via ROP under the conditions studied was attributed to a reduction of the reactivity of the growing oligomeric anion because of either steric effects or insolubility (73).

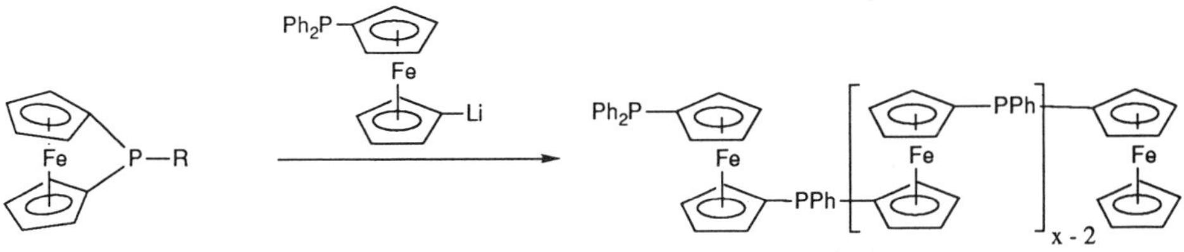

Compound **25** (R = Ph) has been found to polymerize thermally at elevated temperatures to yield the poly(ferrocenylphosphine) **27** (R = Ph), which is spectroscopically identical to the material previously prepared by condensation routes (51). Similar ROP reactions have been detected for a variety of analogous species (51,63).

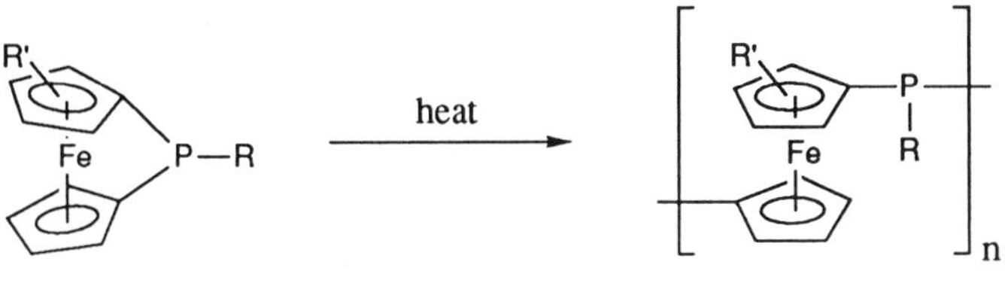

C. *Properties of Poly(ferrocenylphosphines)*

To date, the properties of poly(ferrocenylphosphines) have not been investigated in detail, and much further work remains to be done in this area. They have been structurally characterized by ^{31}P NMR and IR spectroscopy. Interestingly, the molecular weights have been measured in solution using low angle laser light scattering as attempts to use GPC were unsuccessful (73). The polymers 24 (R = Ph) formed by condensation routes are reported to be tan-colored, air-stable, and thermally stable up to 350°C (73). In addition, materials derived from the reaction of the poly(ferrocenylphosphine) 24 (R = Ph) with $Co_2(CO)_8$ have been investigated as hydroformylation catalysts (76). Poly(ferrocenylphosphines) 27 (R = Ph, R′ = n-Bu) formed by ROP have been shown by cyclic voltammetry to possess similar cooperative interactions between the iron centers as other polymers derived from the ROP of [1]ferrocenophanes (63).

VI

RING-OPENING POLYMERIZATION OF HYDROCARBON-BRIDGED [2]METALLOCENOPHANES

A wide variety of [2]ferrocenophanes with different elements in the bridge structure are known (77). In general, however, because of the longer bridge, these species would be expected to be less strained than the [1]metallocenophane counterparts. Indeed, attempts to extend the ROP methodology to [2]ferrocenophanes which possess two silicon atoms in the bridge using either thermal or catalytic initiation have been unsuccessful (48). The reduced propensity for such species to polymerize has been attributed to the lower degree of ring strain present, which is reflected by the very small cyclopentadienyl ring tilt angle of only about 4° (see Fig. 10). However, as described below, by incorporating smaller atoms in the bridge structure it has been possible to produce strained [2]metallocenophanes which do undergo ROP.

A. *Ring-Opening Polymerization of Hydrocarbon-Bridged* *[2]Ferrocenophanes*

Condensation routes to polymers with ferrocene groups separated by short hydrocarbon spacers have mainly resulted in the preparation of low molecular weight and poorly defined materials (2). For example, poly(ferrocenylmethylene), formally 28, has been previously prepared via the

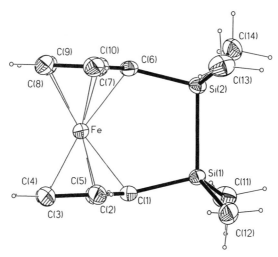

FIG. 10.   Molecular structure of disilane-bridged [2]ferrocenophane Fe(η-C_5H_4)$_2$(SiMe$_2$)$_2$.
Reprinted with permission from (48). Copyright (1993), American Chemical Society.

ZnCl$_2$/HCl-catalyzed polymerization of (dimethylamino)methylferrocene.
However, the resulting material was of low molecular weight (M_n 2400),
and a mixture of 1,2-, 1,3-, and 1,1'-disubstituted ferrocene units was
found to be present in the main chain (78). Recently, however, the synthesis of well-defined polymers of structure 29 (M_w 18,000–139,000) containing face to face ferrocene units with naphthalene bridging groups was
achieved (79).

28

29

Hydrocarbon-bridged [2]ferrocenophanes **30** are quite well known and
are significantly more strained than their disilane-bridged analogs because
of the smaller size of carbon relative to silicon. For example, tilt angles up
to around 21° are known for these species. The first ROP of a [2]metallo-

cenophane was reported by our group in 1993 (*80*). Thus, the hydrocarbon-bridged species **30** (R = H or Me) will undergo ROP, providing access to the first poly(ferrocenylethylenes) **31**. These polymers, which
possess backbones consisting of alternating ferrocene groups and
aliphatic C_2 units, were found to be insoluble if R = H but are readily
soluble in solvents such as THF if R is an organic group such as methyl.
The molecular weight distribution for **31** (R = Me) was found to be bimodal, with an oligomeric fraction (M_w 4800) and a high molecular weight
fraction (M_w 81,000) (*80*).

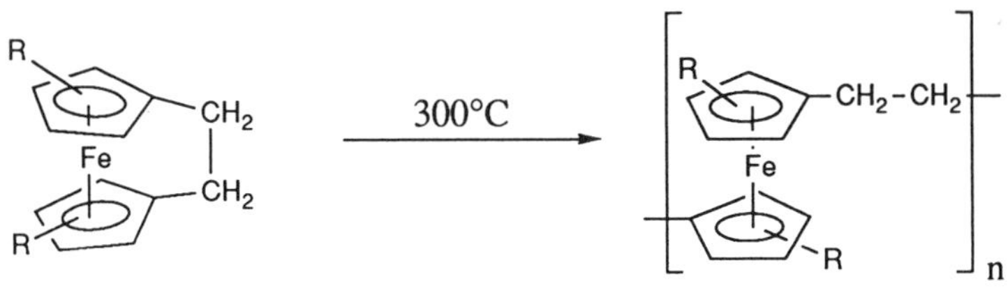

30   **31**

Poly(ferrocenylethylenes) **31** possess ferrocene units which are further
separated from one another than in polymers derived from [1]ferro-
cenophanes, where electrochemical evidence is indicative of the presence
of substantial cooperative interactions between the iron centers. In contrast, studies of the electrochemistry of **31** (R = Me) showed the presence
of only a single reversible oxidation wave (Fig. 11), indicating that the
ferrocene groups interact to much less significant extent (*63*).

B. Synthesis and Ring-Opening Polymerization of Hydrocarbon-Bridged [2]Ruthenocenophanes

To prepare [2]metallocenophanes that are even more strained than
[2]ferrocenophanes **30**, species with a larger ruthenium atom in the place
of iron have been synthesized (*81*). Such [2]ruthocenophanes would be
expected to possess much greater ring-tilt angles; moreover, because
ruthenocene is known to possess significantly different electrical properties compared to ferrocene, modified polymer properties would be anticipated (*57*).

The first examples of [2]ruthenocenophanes **32** were prepared via the
reaction of the salt $Li_2[(C_5H_3R)_2CH_2CH_2]$ with *cis*-$RuCl_2(DMSO)_4$, and **32**
(R = H) possesses a tilt angle of 29.6(5)°, the largest known to date for a
neutral iron group metallocenophane (Fig. 12) (*81*). The hydrocarbon-
bridged [2]ruthenocenophanes **32** undergo ROP more readily than the iron

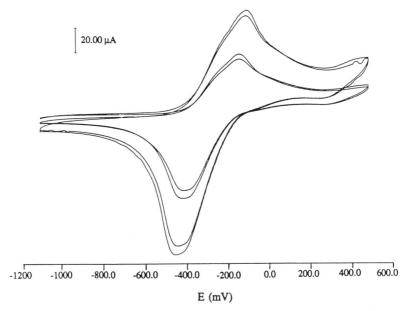

FIG. 11.   Cyclic voltammograms of poly(ferrocenylethylene) **31** (R = Me) in CH$_2$Cl$_2$ at 500 and 1000 mV/second. Reprinted with permission from (*63*).

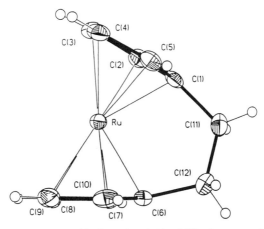

FIG. 12.   Molecular structure of hydrocarbon-bridged [2]ruthenocenophane **32** (R = H). Reprinted with permission from (*81*).

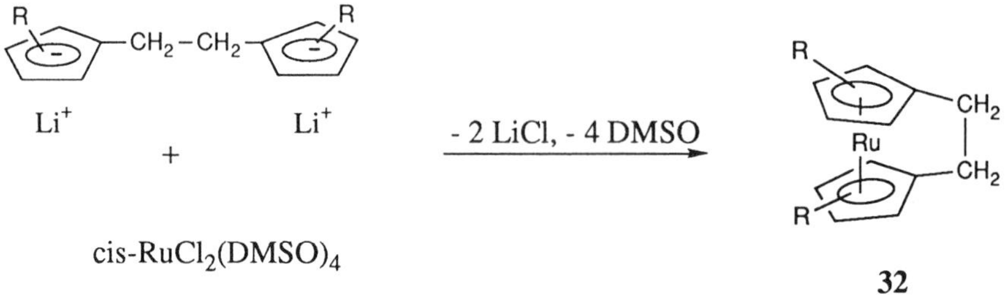

32

R = H or Me

analogs to yield poly(ruthenocenylethylenes) **33**. As with the iron analogs, the polymer **33** (R = H) is insoluble, whereas the methylated material **33** (R = Me) is soluble and possesses a bimodal molecular weight distribution. In contrast to poly(ferrocenylethylenes), however, poly(ruthenocenylethylenes) undergo irreversible electrochemical oxidation (*81*).

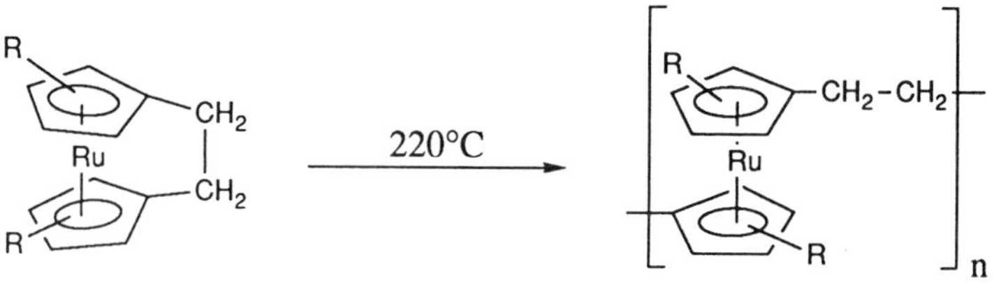

32   **33**

R = H or Me

VII

ATOM ABSTRACTION-INDUCED POLYMERIZATION OF [3]TRITHIAMETALLOCENOPHANES

Metallocenophanes which possess three contiguous bridging atoms would be expected to be less ring-tilted and strained than the [1]- and [2]metallocenophane analogs. Indeed, attempts to induce the thermal or anionic ROP of [3]ferrocenophanes with a siloxane bridge have been unsuccessful (*82*). Nevertheless, in 1992 it was reported by Rauchfuss and co-workers that [3]trithiaferrocenophanes (*83–85*), which are essentially unstrained as indicated by their molecular structures that possess untilted cyclopentadienyl rings, function as precursors to polyferrocenylene persulfides via a novel atom abstraction-induced ROP route (*22*). Thus, reaction of the [3]trithiaferrocenophanes **34** (R = H or Bu) with PBu$_3$ led to the formation of the phosphine sulfide S=PBu$_3$ and the polymer **35**.

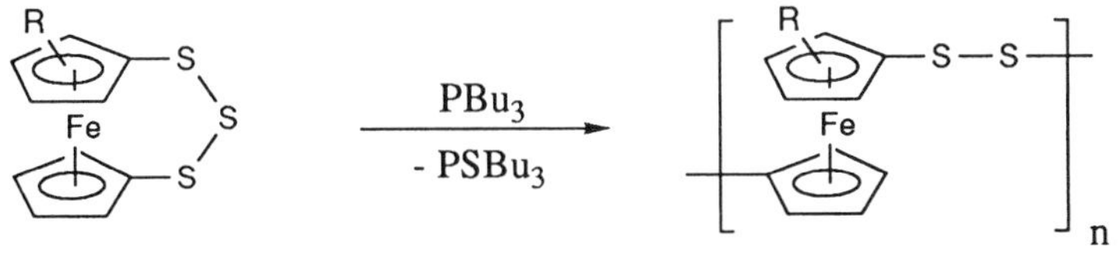

R = H or nBu

The presence of a butyl substituent is necessary for the polymer to be soluble. The molecular weight (M_w) of **35** (R = n-Bu) was determined to be 40,000 by gel permeation chromatography. The [3]ferrocenophanes **34** (R = H and n-Bu) can be copolymerized to give soluble approximately 1:1 copolymers with M_w 25,000. Remarkably, the polymers **35** can be reversibly reductively cleaved with Li[BEt$_3$H] (to give **36**) and then regenerated on oxidation with I$_2$.

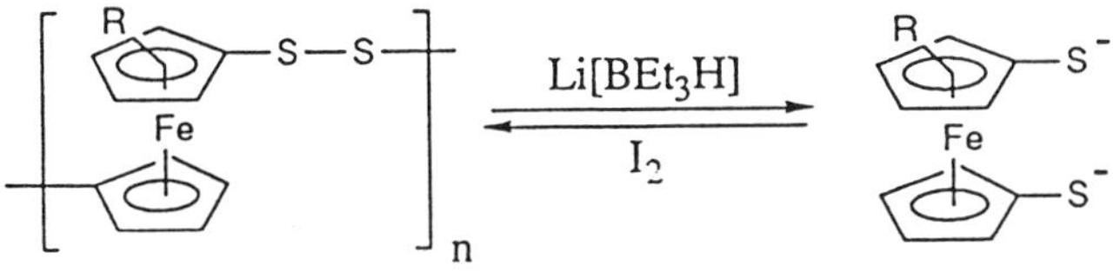

Electrochemistry of the polymer **35** (R = n-Bu) (1 × 10^{-3} M in CH$_2$Cl$_2$ using 0.1 M [NBu$_4$][PF$_6$] as supporting electrolyte) showed the presence of two chemically reversible oxidation waves separated by 0.32 V, attributed to an initial oxidation occurring at alternating iron sites along the polymer chain. This wave separation (ΔE) is greater than that observed for most poly(ferrocenylsilanes) and other polymers prepared from the ROP of [1]ferrocenophanes and this suggests that the interactions between the iron centers in **35** (R = n-Bu) are even larger. Further investigations of the properties of these polymers should be of considerable interest.

The atom abstraction route using PBu$_3$ as desulfurization agent has been extended to the preparation of very high molecular weight (M_w 50,000–1,000,000) network polymers (Fig. 13) by using [3]ferrocenophanes with two trisulfido bridges as monomers (*86*).

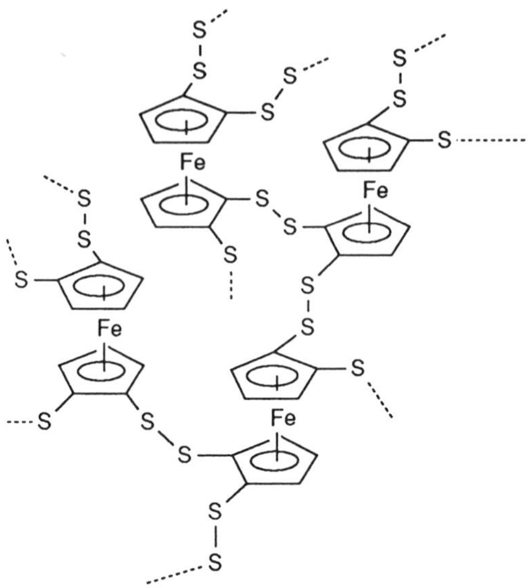

FIG. 13.   Random network poly(ferrocenylenepersulfide).

VIII

SUMMARY, CONCLUSIONS, AND FUTURE WORK

The results described in this article indicate that the ROP of metallocenophanes represents a new and versatile route to a variety of high
molecular weight metallocene-based polymer systems. It is very likely
that the ROP route could be extended still further to other metallocenophanes, including those which contain transition elements other than
iron and ruthenium. For example, [1]ferrocenophanes with elements in
the bridge such as arsenic, titanium, and zirconium are known and also
possess strained, ring-tilted structures, but their polymerization behavior
has not yet been studied (74, 87). In addition, ring-tilted and therefore
presumably strained bis(arene) complexes have also been reported (88),
and these may also undergo ROP. Polymerization of such species would
provide further examples of transition metal-based polymers with properties which would be expected to differ from those of the materials prepared to date.

The discovery of the ROP of metallocenophanes also provides impetus
for the preparation of new strained organometallic species. For example,
[1]ruthenocenophanes, which would be expected to be even more ring-

tilted and strained than their iron counterparts owing to the larger size of ruthenium compared to iron, are still, at present, unknown and represent an obvious synthetic target. In addition, despite several attempts, [1]stannaferrocenophanes have not yet been prepared from the reaction of dilithioferrocene·tmeda with the organohalostannanes R_2SnCl_2. The isolated products are either cyclic dimers or low molecular weight polymers (46,63,74,89,90). In addition, although disilane-bridged [2]ferrocenophanes do not appear to undergo ROP, even less strained species with heavier group 14 elements in the bridge may polymerize as the strength of the bonds formed by these elements to carbon is less, making cleavage of the Cp–bridge bond more facile. Underlying all of this work is the need for a deeper understanding of ring strain in organometallic chemistry. Compared to organic chemistry where ring strain is a well-studied phenomenon, very little detailed work has been reported concerning strained organotransition metal compounds.

One of the key areas for future research concerns the mechanisms for the ROP reactions. Based on the observed ROP of silicon-bridged [1]ferrocenophanes in solution in the presence of anionic initiators (Scheme 1), it is possible that the thermally induced ROP reactions of [1]ferrocenophanes are similarly initiated by trace amounts of nucleophilic impurities. This is supported by the observation that higher molecular weight polymers are formed from monomer samples which have been more carefully purified. However, alternative mechanisms such as ring fusion via σ-bond metathesis are also possible (89). According to this mechanism, the Si–C(Cp) bonds of the [1]ferrocenophane add to those of another molecule to yield a cyclic dimer via a four-center transition state. This process is illustrated for the dimerization of [1]silaferrocenophanes in Scheme 2. The cyclic dimer formed could then undergo the same reaction with monomer to yield a cyclic trimer, and these ring-fusion processes could continue until polymer is generated. The driving force for the reactions would be the relief of strain from the [1]ferrocenophane monomer, and if the mechanism is correct the polymers formed would be macrocyclic. There is some evidence to support this as the detection of polymer end

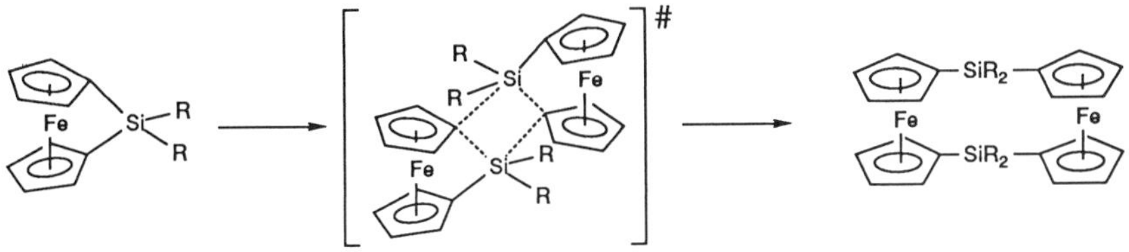

SCHEME 2

groups has not yet been reported, but this may be a consequence of the very high molecular weights of the polymers prepared to date. One piece of additional experimental evidence which is consistent with this mechanism is the detection of cyclic oligomers which are formed in trace amounts during the thermally induced ROP reactions (80).

Another possible polymerization mechanism involves cyclopentadienyl ring exchange which would once again be driven by the relief of ring strain. Slow ligand exchange reactions of this type have been detected when unstrained ferrocene derivatives are heated at elevated temperatures. An equimolar mixture of 1,1'-dimethylferrocene and ferrocene is heated at 250°C for 14 days, the product is the unsymmetrical monomethylated species **37** (92). Such ring-exchange reactions might be expected for metallocenophanes where the bonds between the cyclopentadienyl ligand and the bridging atoms are relatively strong and resistant to cleavage as in the case of hydrocarbon-bridged species. Indeed, hydrocarbon-bridged [2]metallocenophanes polymerize to yield polymers with a bimodal molecular weight distribution, suggestive of the operation of two different polymerization mechanisms (80,81). The mechanism for the atom abstraction-induced ROP of [3]trithiaferrocenophanes is almost certainly entirely different from those for the strained, ring-tilted species. Ideas for meaningful mechanistic experiments abound for all of the polymerizable metallocenophanes, and interesting results can be anticipated in the future.

37

Finally, the polymers produced from the ROP reactions, which are very rare examples of high molecular weight, well-defined polymers with skeletal transition metal atoms held in close proximity to one another, are of considerable interest with respect to their properties and possible applications (93). The most promising materials derived from [1]- or [2]metallocenophanes appear to be the poly(ferrocenylsilanes) which are easy to prepare from the readily available starting materials, ferrocene and dichloroorganosilanes, of which many are provided by the silicone industry. Although the main features of these polymers appear to have been elucidated, much detailed work remains to be done to fully understand the characteristics of these unusual materials. It is expected that many interesting developments will result in the near future from studies of the materials science of both poly(ferrocenylsilanes) and the other new poly-mers, such as poly(ferrocenylene persulfides) now available by ROP or ROP-related routes.

ACKNOWLEDGMENTS

Our research on the use of ROP as a route to transition metal-based polymers has been supported by the Ontario Center of Materials Research (OCMR), the ACS Petroleum Research Fund (PRF), the Natural Sciences and Engineering Research Council of Canada (NSERC), and the Institute of Chemical Science and Technology (ICST). Research on the detailed properties of the polymers is being carried out in collaboration with the Polymer Materials Science Research Group of Professor G. Julius Vancso at the University of Toronto. I would also like to acknowledge and thank my very talented and enthusiastic co-workers who have carried out this research and whose names are found in the references. I am also grateful to the Alfred P. Sloan Foundation for a fellowship (1994–1996).

REFERENCES

(1) Sheats, J. E.; Carraher, C. E.; Pittman, C. U. "Metal Containing Polymer Systems"; Plenum; New York, 1985; Pittman, C. U., Jr.; Carraher, C. E.; Reynolds, J. R. Organo-metallic polymers. *Encycl. Polym. Sci. Eng.* **1987**, *10*, 541.

(2) Neuse, E. W.; Rosenberg, H. *J. Macromol. Sci. Rev. Macromol. Chem.* **1970**, *C4(1)*, 1.

(3) Peuckert, M.; Vaahs, T.; Brück, M. *Adv. Mater.* **1990**, *2*, 398.

(4) Kahn, O.; Pei, Y.; Journaux, Y. In "Inorganic Materials"; Bruce, D. W.; O'Hare, D., Eds.; Wiley: New York, 1992; Chapter 2, p. 59; Miller, J. S. (Ed.) "Extended Linear Chain Compounds"; Plenum: New York, 1982; Vols. 1–3).

(5) For recent work involving well-characterized materials see, for example, Fyfe, H. B.; Mlekuz, M.; Zargarian, D.; Taylor, N. J.; Marder, T. B. *J. Chem., Soc. Chem. Com-mun.* **1991**, 188; Davies, S. J.; Johnson, B. F. G.; Khan, M. S.; Lewis, J. *J. Chem. Soc., Chem. Commun.* **1991**, 187; Lewis, J.; Khan, M. S.; Kakkar, A. K.; Johnson, B. F. G.; Marder, T. B.; Fyfe, H. B.; Wittmann, F.; Friend, R. H.; Dray, A. E. *J. Organomet. Chem.* **1992**, *425*, 165; Sturge, K. C.; Hunter, A. D.; McDonald, R.; Santarsiero, B. D. *Organometallics* **1992**, *11*, 3056; Tenhaeff, S. C.; Tyler, D. R. *Organometallics* **1992**, *11*, 1466; Bayer, R.; Pöhlmann, T.; Nuyken, O. *Makromol. Chem. Rapid Commun.* **1993**, *14*, 359.

(6) Hagihara, N.; Sonogashira, K.; Takahashi, S. *Adv. Polym. Sci.* **1981**, *41*, 149; Abe, A.; Kimura, N.; Tabata, S. *Macromolecules* **1991**, *24*, 6238.

(7) Patterson, W. J.; McManus, S.; Pittman, C. H. *J. Polym. Sci., Chem. Ed.* **1974**, *12*, 837.

(8) Young, R. J.; Lovell, P. A. "Introduction to Polymers"; 2nd ed.; Chapman & Hall: London, 1991.

(9) Young, R. J.; Lovell, P. A. "Introduction to Polymers"; 2nd ed.; Chapman & Hall: London, 1991; pp. 23–25.

(10) Ivin, K. J.; Saegusa, T. (Eds.) "Ring-Opening Polymerization"; Elsevier: New York, 1984; McGrath, J. E. (Ed.) "Ring Opening Polymerization"; ACS Symp. Ser. 286; American Chemical Society: Washington, D.C., 1985; Höcker, H.; Keul, H. *Adv. Mater.* **1994**, *6*, 21. Gibson, V. C. *Adv. Mater.* **1994**, *6*, 37.

(11) Zeigler, J. M.; Fearon, F. W. G. (Eds.) "Silicon-Based Polymer Science"; Advances in Chemistry Vol. 224; American Chemical Society: Washington, D.C., 1990.

(*12*) Allcock, H. R. *Chem. Eng. News* **1985**, *63(11)*, 22; Allcock, H. R. *J. Inorg. Organomet. Polym.* **1992**, *2(2)*, 197.

(*13*) Manners, I.; Renner, G.; Allcock, H. R.; Nuyken, O. *J. Am. Chem. Soc.* **1989**, *111*, 5478; Allcock, H. R.; Coley, S. M.; Manners, I.; Renner, G.; Nuyken, O. *Macromolecules* **1991**, *24*, 2024.

(*14*) Dodge, J. A.; Manners, I.; Renner, G.; Allcock, H. R.; Nuyken, O. *J. Am. Chem. Soc.* **1990**, *112*, 1268; Allcock, H. R.; Dodge, J. A.; Manners, I.; *Macromolecules* **1993**, *26*, 11.

(*15*) Liang, M.; Manners, I. *J. Am. Chem. Soc.* **1991**, *113*, 4044; Ni, Y.; Stammer, A.; Liang, M.; Massey, J.; Vancso, G. J.; Manners, I. *Macromolecules* **1992**, *25*, 7119; Manners, I. *Polym. News* **1993**, *18*, 133; Manners, I. *Coord. Chem. Rev.* **1994**, in press.

(*16*) Cypryk, M.; Gupta, Y.; Matyjaszewski, K. *J. Am. Chem. Soc.* **1991**, *113*, 1046.

(*17*) See Cundy, C. S.; Eaborn, C.; Lappert, M. F. *J. Organomet. Chem.* **1972**, *44*, 291; Wu, H. J.; Interrante, L. V. *Chem. Mater.* **1989**, *1*, 564, and references cited therein.

(*18*) Sargeant, S. J.; Zhou, S. Q.; Manuel, G.; Weber, W. P. *Macromolecules* **1992**, *25*, 2832.

(*19*) West, R.; Hayase, S.; Iwahara, T. *J. Inorg. Organomet. Polym.* **1991**, *1*, 545; Suzuki, M.; Obayashi, T.; Saegusa, T. *J. Chem. Soc., Chem. Commun.* **1993**, 717; Anhaus, J. T.; Clegg, W.; Collingwood, S. P.; Gibson, V. C. *J. Chem. Soc., Chem. Commun.* **1991**, 1720.

(*20*) Roesky, H. W.; Lücke, M. *Angew. Chem., Int. Ed. Engl.* **1989**, *28*, 493.

(*21*) Roesky, H. W.; Lücke, M. *J. Chem. Soc., Chem. Commun.* **1989**, 748.

(*22*) Brandt, P. F.; Rauchfuss, T. B. *J. Am. Chem. Soc.* **1992**, *114*, 1926.

(*23*) Foucher, D. A.; Tang, B. Z.; Manners, I. *J. Am. Chem. Soc.* **1992**, *114*, 6246.

(*24*) Pittman, C. U.; Lai, J. C.; Vanderpool, D. P.; Good, M.; Prado, R. *Macromolecules* **1970**, *3*, 746.

(*25*) Allcock, H. R.; Dodge, J. A.; Manners, I.; Riding, G. H. *J. Am. Chem. Soc.* **1991**, *113*, 9596.

(*26*) Pannell, K. H.; Rozell, J. M.; Zeigler, J. M. *Macromolecules* **1988**, *21*, 276.

(*27*) Inagaki, T.; Lee, H. S.; Skotheim, T. A.; Okamoto, Y. *J. Chem. Soc., Chem. Commun.* **1989**, 1181.

(*28*) Hale, P. D.; Inagaki, T.; Karan, H. I.; Okamoto, Y.; Skotheim, T. A. *J. Am. Chem. Soc.* **1989**, *111*, 3482.

(*29*) Kittlesen, G. P.; White, H. S.; Wrighton, M. S.; *J. Am. Chem. Soc.* **1985**, *107*, 7373.

(*30*) Mueller-Westerhoff, U. T. *Angew. Chem., Int. Ed. Engl.* **1986**, *25*, 702, and references cited therein.

(*31*) Morrison, W. H.; Hendrickson, D. N. *Inorg. Chem.* **1975**, *14*, 2331, and references cited therein.

(*32*) Kramer, J. A.; Hendrickson, D. N. *Inorg. Chem.* **1980**, *19*, 3330.

(*33*) Pittman, C. U.; Jr.; Sasaki, Y.; Mukherjee, T. K. *Chem. Lett.* **1975**, 383; Bilow, N.; Landis, A.; Rosenberg, H. *J. Polym. Sci. Chem. Ed.* **1969**, *7*, 2719.

(*34*) Neuse, E. W.; and Bednarik, L. *Macromolecules* **1979**, *12*, 187.

(*35*) Yamamoto, T.; Sanechika, K.; Yamamoto, A.; Katada, M.; Motoyama, I.; Sano, H. *Inorg. Chim. Acta* **1983**, *73*, 75, and references cited therein.

(*36*) Gonsalves, K.; Zhanru, L.; Rausch, M. D. *J. Am. Chem. Soc.* **1984**, *106*, 3862.

(*37*) Rosenberg, H. U.S. Patent 3,426,053 **1969**; Neuse, E. W.; Rosenburg, H. *J. Macromol. Sci. Rev., Macromol. Chem.* **1970**, *C4(1)*, 110.

(*38*) Bishop, J. J.; Davidson, A.; Katcher, M. L.; Lichtenberg, D. W.; Merrill, R. E.; Smart, J. C. *J. Organomet. Chem.* **1971**, *27*, 241; Butler, I. R.; Cullen, W. R. *Organometallics* **1986**, *5*, 2537 (see p. 2540).

(*39*) Tanaka, M.; Hayashi, T. *Bull. Chem. Soc. Jpn.* **1993**, *66*, 334.

(40) Wright, M. E.; Sigman, M. S. *Macromolecules* **1992**, *25*, 6055.

(41) Finckh, W.; Tang, B. Z.; Lough, A.; Manners, I. *Organometallics* **1992**, *11*, 2904.

(42) Osborne, A. G.; Whiteley, R. H. *J. Organomet. Chem.* **1975**, *101*, C 27.

(43) Stoeckli-Evans, H.; Osborne, A. G.; Whiteley, R. H. *Helv. Chim. Acta* **1976**, *59*, 2402.

(44) Dunitz, J. D.; Orgel, L. E.; Rich, A. *Acta Crystallogr.* **1956**, *9*, 373.

(45) Fischer, A. B.; Kinney, J. B.; Staley, R. H.; Wrighton, M. S. *J. Am. Chem. Soc.* **1979**, *101*, 6501.

(46) Osborne, A. G.; Whiteley, R. H.; Meads, R. E. *J. Organomet. Chem.* **1980**, *193*, 345.

(47) Butler, I. R.; Cullen, W. R.; Rettig, S. J. *Can J. Chem.* **1987**, *65*, 1452.

(48) Finckh, W.; Ziembinski, R.; Tang, B. Z.; Foucher, D. A.; Zamble, D. B.; Lough, A.; Manners, I. *Organometallics* **1993**, *12*, 823.

(49) Silver, J. *J. Chem. Soc., Dalton Trans.* **1990**, 3513.

(50) Fischer, A. B.; Bruce, J. A.; McKay, D. R.; Maciel, G. E.; Wrighton, M. S. *Inorg. Chem.* **1982**, *21*, 1766.

(51) Foucher, D. A.; Ziembinski, R.; Rulkens, R.; Nelson, J.; Manners, I. In "Inorganic and Organometallic Polymers"; Allcock, H. R.; Wynne, K.; Wisian-Neilson, P.; Eds. ACS Symp. Ser. 572; American Chemical Society: Washington, D.C., 1994.

(52) Foucher, D. A.; Ziembinski, R.; Tang, B. Z.; Macdonald, P. M.; Massey, J.; Jaeger, R.; Vancso, G. J.; Manners, I. *Macromolecules* **1993**, *26*, 2878.

(53) Manners, I. *J. Inorg. Organomet. Polym.* **1993**, *3*, 185.

(54) Manners, I. *Adv. Mater.* **1994**, *6*, 68.

(55) Rulkens, R.; Lough, A. J.; Manners, I. *J. Am. Chem. Soc.* **1994**, *116*, 797.

(56) Nguyen, M. T.; Diaz, A. F.; Dement'ev, V. V.; Pannell, K. H. *Chem. Mater.* **1993**, *5*, 1389.

(57) Connelly, N. G.; Geiger, W. E. *Adv. Organomet. Chem.* **1984**, *23*, 1.

(58) Flanagan, J. B.; Margel, S.; Bard, A. J.; Anson, F. C. *J. Am. Chem. Soc.* **1978**, *100*, 4248.

(59) Foucher, D.; Ziembinski, R.; Petersen, R.; Pudelski, J.; Edwards, M.; Ni, Y.; Massey, J.; Jaeger, C. R.; Vansco, G. J.; Manners, I. *Macromolecules* **1994**, *27*, 3992.

(60) Dong, T. Y.; Hwang, M. Y.; Wen, Y. S.; Hwang, W. S. *J. Organomet. Chem.* **1990**, *391*, 377, and references cited therein.

(61) Bocarsly, A. B.; Walton, E. G.; Bradley, M. G.; Wrighton, M. S. *J. Electroanal. Chem. Interfacial Electrochem.* **1979**, *100*, 283.

(62) Dement'ev, V. V.; Cervantes-Lee, F.; Parkanyi, L.; Sharma, H.; Pannell, K. H.; Nguyen, M. T.; Diaz, A. F. *Organometallics* **1993**, *12*, 1983.

(63) Foucher, D. A.; Honeyman, C.; Nelson, J. M.; Tang, B. Z.; Manners, I. *Angew Chem., Int. Ed. Engl.* **1993**, *32*, 1709.

(64) West, R. *J. Organomet. Chem.* **1986**, *300*, 327.

(65) Miller, R. D.; Michl, J. *Chem. Rev.* **1989**, *89*, 1359.

(66) Webb, R. J.; Hagen, P. M.; Wittebort, R. J.; Sorai, M., Hendrickson, D. N. *Inorg. Chem.* **1992**, *31*, 1791.

(67) Tang, B. Z.; Petersen, R.; Foucher, D. A.; Lough, A.; Coombs, N.; Sodhi, R.; Manners, I. *J. Chem. Soc., Chem. Commun.* **1993**, *6*, 523.

(68) Stoeckli-Evans, H.; Osborne, A. G.; Whiteley, R. H. *J. Organomet. Chem.* **1980**, *194*, 91.

(69) Blake, A. J.; Mayers, F. R.; Osborne, A. G.; Rosseinsky, D. R. *J. Chem. Soc. Dalton Trans.* **1982**, 2379.

(70) Foucher, D. A.; Manners, I. *Makromol. Chem., Rapid Commun.* **1993**, *14*, 63.

(71) Neuse, E. W.; Chris, G. J. *J. Macromol. Sci. Chem.*, **1967**, *3*, 371; see Neuse, E. W.; Rosenberg, H. *J. Macromol. Sci. Rev. Macromol. Chem.* **1970**, *C4(1)*, 122.

(72) Pittman, C. U. *J. Polym. Sci., Polym. Chem. Ed.* **1967**, *5*, 2927.

(73) Withers, H. P.; Seyferth, D.; Fellmann, J. D.; Garrou, P. E.; Martin, S. *Organometallics* **1982,** *1,* 1283.

(74) Seyferth, D.; Withers, H. P. *Organometallics* **1982,** *1,* 1275; Butler, I. R.; Cullen, W. R.; Einstein, F. W. B.; Rettig, S. J.; Willis, A. J. *Organometallics* **1983,** *2,* 128.

(75) Butler, I. R.; Cullen, W. R.; Rettig, S. J. *Organometallics* **1987,** *6,* 872, and references cited therein.

(76) Fellmann, J. D.; Garrou, P. E.; Withers, H. P.; Seyferth, D.; Traficante, D. D. *Organometallics* **1983,** *2,* 818.

(77) See, for example, Burke Laing, M.; Trueblood, K. N. *Acta Crystallogr.* **1965,** *19,* 373; Lentzner, H. L.; Watts, W. E. *Tetrahedron* **1971,** *27,* 4343; Yasufuku, K.; Aoki, K.; Yamzaki, H. *Inorg. Chem.* **1977,** *16,* 624.

(78) Neuse, E. W.; Khan, F. B. D. *Macromolecules* **1986,** *19,* 269.

(79) Nugent, H. M.; Rosenblum, M.; Klemarczyk, P. *J. Am. Chem. Soc.* **1993,** *115,* 3848.

(80) Nelson, J. M.; Rengel, H.; Manners, I. *J. Am. Chem. Soc.* **1993,** *115,* 7035.

(81) Nelson, J. M.; Lough, A.; Manners, I. *Angew Chem., Int. Ed. Engl.* **1994,** *33,* 989.

(82) Angelakos, C.; Zamble, D. B.; Foucher, D. A.; Lough, A. J.; Manners, I. *Inorg. Chem.* **1994,** *33,* 1709.

(83) Davison, A.; Smart, J. C. *J. Organomet. Chem.* **1979,** *174,* 321.

(84) Osborne, A. G.; Hollands, R. E.; Howard, J. A. K.; Bryan, R. F. *J. Organomet. Chem.* **1981,** *205,* 395.

(85) Leitner, P.; Herberhold, M. *J. Organomet. Chem.* **1991,** *411,* 233.

(86) Galloway, C. P.; Rauchfuss, T. B. *Angew Chem., Int. Ed. Engl.* **1993,** *32,* 1319.

(87) See, for example, Broussier, R.; Da Rold, A.; Gautheron, B.; Dromzee, Y.; Jeannin, Y. *Inorg. Chem.* **1990,** *29,* 1817.

(88) Elschenbroich, C.; Hurley, J.; Metz, B.; Massa, W.; Baum, G. *Organometallics* **1990,** *9,* 889.

(89) Clearfield, A.; Simmons, C. J.; Withers, H. P.; Seyferth, D. *Inorg. Chim. Acta* **1983,** *75,* 139.

(90) The low molecular weights of the poly(ferrocenylstannanes) (M_n <5000) are consistent with formation via a condensation route. However, based on the results described in this review it is also possible that they arise from the ROP of an *in situ* generated [1]stannaferrocenophane (see Section V,A).

(91) For a discussion of a σ-bond metathesis mechanism for the dehydrogenative coupling route to polysilanes, see Tilley, T. D. *Acc. Chem. Res.* **1993,** *26,* 22.

(92) Allcock, H. R.; McDonnell, G. S.; Riding, G. H.; Manners, I. *Chem. Mater.* **1990,** *2,* 425.

(93) Dagani, R. *Chem. Eng. News,* **1993,** *Aug. 2,* 22.

ADVANCES IN ORGANOMETALLIC CHEMISTRY, VOL. 37

Alkyl(pentacarbonyl) Compounds of the Manganese Group Revisited

JO-ANN M. ANDERSEN and JOHN R. MOSS

Department of Chemistry
University of Cape Town
Rondebosch 7700
Cape Town, South Africa

I

INTRODUCTION

Transition metal complexes with alkyl ligands lie at the heart of organo-metallic chemistry (1). Even though such compounds have been known for many years, their study is an exciting and ever-growing field. One reason is that the novel properties of these compounds (both physical and chemical) can lead to useful applications in several areas including the catalysis industry, metal vapor deposition (MOCVD), isotope separations, and reagents for organic synthesis. The importance of transition metal alkyls in the catalysis industry is partly due to the proposed inter-mediacy of these species in almost every catalytic reaction, either homo-geneous or heterogeneous, that involves hydrocarbons and transition metals (2). Examples of such reactions include alkene hydrogenation, isomerization or polymerization, the Fischer–Topsch synthesis (3), the

Monsanto acetic acid synthesis (4), and the hydroformylation or OXO reaction (5). A key mechanistic step in catalytic carbonylation reactions is the migration of an alkyl group onto an adjacent carbonyl ligand. This reaction involves the formation of a new carbon–carbon bond and has been termed a "carbonyl insertion" reaction since a CO ligand has been formally inserted into the transition metal–carbon σ-bond. Because of the industrial and commercial importance of these catalytic reactions, the search for stoichiometric systems in which this step can be observed directly has been, and still is, one of great endeavor.

Classic examples of complexes showing the carbonyl insertion reaction are the complexes $[Mn(R)(CO)_5]$ (R = alkyl). Alkylpentacarbonyl complexes of the manganese group are also important from both fundamental and theoretical standpoints since they contain the alkyl group and CO in activated forms by virtue of coordination of the metal.

Methyl manganesepentacarbonyl, $[Mn(CH_3)(CO)_5]$, was the first transition metal carbonyl alkyl compound to be prepared (6). Transition metal compounds with alkyl ligands had, however, been prepared previously, and one of the first was trimethyl platinum iodide in 1907 (7), although this was regarded as an exception. Prior to the early 1950s, transition metal alkyls were assumed to be impossible to prepare (8), and the "magic formula" of Mulliken (9) was invoked to account for this (10). Fortunately, the field of organometallic chemistry has progressed enormously since that time, and alkyl compounds of almost all the transition metals in the periodic table are now known.

Although $[Mn(CH_3)(CO)_5]$ remains to date the most well-studied transition metal alkyl carbonyl compound in terms of both structure and reactivity, until recently very little was known about higher alkyl derivatives of the manganese group. Such studies should prove valuable since in many catalytic reactions (e.g., the Fischer–Tropsch synthesis and Ziegler–Natta polymerization) C_3 and higher hydrocarbon fragments are involved. Ethyl and n-propyl manganesepentacarbonyl have been prepared (11) but were found to be extremely unstable, decomposing even under vacuum in the dark at $-10°C$ (12). The list is even shorter for reported n-alkyl derivatives of rheniumpentacarbonyl, where only the methyl (13) and ethyl (14) compounds are known. For technetium, the list is shorter still! No n-alkyl derivatives of technetiumpentacarbonyl are known at present, although a theoretical study on bond strengths in $[Tc(CH_3)(CO)_5]$ has been performed (15). As a result of the instability of $[Mn(C_2H_5)(CO)_5]$, it may have been assumed that derivatives with longer alkyl chains would be even more unstable. There was certainly a widely held belief that $[Mn(C_2H_5)(CO)_5]$ decomposed by the β-hydride elimination reaction, and if this were the case then the longer chain alkyl compounds of manganesepentacarbonyl would also have been expected to be unstable. This perception may have deterred further investigations, and attempts to synthesize the longer chain alkyl manganesepentacarbonyl complexes were apparently not made. It has been found (16–18), however, that these assumptions were not correct, and an extensive series of manganese- and rheniumpentacarbonyl alkyl compounds $[M(R)(CO)_5]$ ($R = CH_3$ to $n\text{-}C_{18}H_{37}$; $M = Mn$ or Re) has now been synthesized and fully characterized.

The purpose of this article is to provide an overall picture of the synthesis, structure, and reactivity of mononuclear n-alkyl compounds of the manganese group, $[M(R)(CO)_5]$ ($M = Mn$, Tc, Re; $R = n$-alkyl). This includes the two new homologous series $[M(R)(CO)_5]$ where $M = Mn$, Re and $R = CH_3$ to $n\text{-}C_{18}H_{37}$. Several binuclear hydrocarbyl-bridged compounds, $[(CO)_5M(CH_2)_nM(CO)_5]$ are known (16,19–26), as well as several phenyl derivatives ($R = C_6H_5$ or C_6H_4X) (27–33), but these compounds are not included in this review. It should also be borne in mind, however, that because $[Mn(CH_3)(CO)_5]$ is probably the most well-studied transition metal alkyl carbonyl compound, a comprehensive review of every physical measurement and reaction reported for that compound would prove too lengthy for the present article. Thus, we have had to be selective, but we hope that we have included all of the most important reports on the subject.

II

SYNTHESIS OF ALKYL(PENTACARBONYL) COMPOUNDS OF THE MANGANESE GROUP

A. Synthesis of Manganesepentacarbonyl Alkyls

Methyl manganesepentacarbonyl, $[Mn(CH_3)(CO)_5]$, reported in 1957 (6), was the first manganesepentacarbonyl alkyl compound and the first transition metal carbonylalkyl compound to be prepared. It was synthesized by the direct reaction of the manganesepentacarbonyl anion with methyl iodide, as shown in Eq. (1).

$$Na[Mn(CO)_5] + CH_3I \rightarrow [Mn(CH_3)(CO)_5] + NaI \qquad (1)$$

This route was also used by Hieber and Wagner in 1958 in attempts to prepare the ethyl and n-propyl analogs (34). These authors reported that the reaction of $Na[Mn(CO)_5]$ with ethyl iodide or n-propyl iodide gave the ethyl or n-propyl derivatives, respectively, as shown in Eqs. (2) and (3).

$$Na[Mn(CO)_5] + C_2H_5I \rightarrow [Mn(C_2H_5)(CO)_5] + NaI \qquad (2)$$

$$Na[Mn(CO)_5] + n\text{-}C_3H_7I \rightarrow [Mn(n\text{-}C_3H_7)(CO)_5] + NaI \qquad (3)$$

They maintained that both compounds existed as colorless, needlelike crystals, with the ethyl derivative melting at 58°C and the n-propyl derivative melting at 40°C. However, in response to a communication by Coffield, 2 years later Hieber $et\ al.$ proposed that the "ethyl manganesepentacarbonyl" they had reported in 1958 was in fact propionyl manganesepentacarbonyl, $[Mn(COC_2H_5)(CO)_5]$ (27). They went on to say they had in fact now managed to synthesize the ethyl derivative, which existed as a yellow oil. However, they gave no experimental details, nor did they mention the "propyl manganesepentacarbonyl" they had reported in 1958. This compound was later found to be the n-butanoyl derivative, $[Mn(COC_3H_7)(CO)_5]$ (16). Ethyl and n-propyl manganesepentacarbonyl, and the corresponding acyl derivatives, were also prepared by Calderazzo and Cotton in 1962 (11). However, no experimental details for the preparation of the alkyl compounds were given. The acyl derivatives were prepared by reaction of the alkyl compounds with carbon monoxide. Ethyl manganesepentacarbonyl was then subsequently synthesized by Green and Nagy (12), who found the compound to be unstable, decomposing in the dark under vacuum at $-10°C$. The n-propyl derivative was also found to be unstable (16).

The method first used to prepare $[Mn(CH_3)(CO)_5]$ is still a fairly common route to carbonyl-containing alkyl compounds of manganese. This route involves the preparation of $Na[Mn(CO)_5]$ $in\ situ$ (most commonly by reductive cleavage of $[Mn_2(CO)_{10}]$ over a sodium/mercury amalgam) which then acts as a nucleophile to displace a halide ion from an organic halide. Another synthetic route involves the treatment of the pentacarbonyl halides, $[Mn(X)(CO)_5]$, with an organic derivative of magnesium or lithium; thus, the reaction of $[Mn(Br)(CO)_5]$ with phenyllithium gives $[Mn(Ph)(CO)_5]$ (35), and treatment of $[Mn(Br)(CO)_5]$ with benzylmagnesium chloride gives the benzyl derivative (34). However, this route suffers from the disadvantages of low product yields and the formation of $[Mn_2(CO)_{10}]$ as a by-product. Nucleophilic attack by the carbanions can also occur at the carbon of a carbonyl ligand to give carbene species, and this may in part be responsible for the low yields by this route.

An alternative route used in the attempted preparation of $[Mn(C_2H_5)(CO)_5]$ was the reaction of $Na[Mn(CO)_5]$ with diethyl sulfate; however, a pure product was not obtained (36).

Although relatively few alkyl halides have been shown to react with $Na[Mn(CO)_5]$ to form $[Mn(R)(CO)_5]$ derivatives (37), owing in part to the low nucleophilicity of the manganesepentacarbonyl anion (38), virtually all acyl chlorides will react with Na[Mn(CO)$_5$] to form the acyl compounds, [Mn(COR)(CO)$_5$], as represented in Eq. (4). Thus, numerous acyl derivatives have been prepared, including the compounds where COR is COPh (6), COCH$_2$CH$_3$ (39), CO(i-Bu) (39), CO(CH$_2$)$_4$CH$_3$ (22), COCH$_2$CH$_2$Ph (40), and COCH$_2$Ph (39).

$$Na[Mn(CO)_5] + RCOCl \rightarrow [Mn(COR)(CO)_5] + NaCl \qquad (4)$$

$$[Mn(COR)(CO)_5] \rightarrow [Mn(R)(CO)_5] + CO \qquad (5)$$

The facile decarbonylation [Eq. (5)] of the acyl derivatives permits the preparation of many alkyl and aryl derivatives of manganese which are not obtainable by the direct reaction of Na[Mn(CO)$_5$] with alkyl halides. This decarbonylation process can be thermally or photolytically effected. Chemical decarbonylation is also possible with reagents such as trimethylamine N-oxide (41).

Thus, via thermal decarbonylation, the phenyl and phenylethyl derivatives, [Mn(Ph)(CO)$_5$] and [Mn(CH$_2$CH$_2$Ph)(CO)$_5$], were prepared from the benzoyl (39) and the phenylpropionyl (40) derivatives, respectively. In many cases this carbonylation reaction is reversible; for example, methyl manganesepentacarbonyl reacts with carbon monoxide at room temperature to give acetyl manganesepentacarbonyl in good yield (39). This facile carbonylation process for some manganesepentacarbonyl alkyl species may in fact be the main reason for their instability, as is found for ethyl and n-propyl manganesepentacarbonyl (37). The fact that ethyl and n-propyl manganesepentacarbonyl undergo a very rapid carbonylation process has been demonstrated by Calderazzo and Cotton (11) and Andersen and Moss (18,42).

There appears to be some discrepancy in the literature as to the reasons for the instability of [Mn(C$_2$H$_5$)(CO)$_5$] and, presumably, [Mn(n-C$_3$H$_7$)(CO)$_5$]. Several reports attribute the instability to a very facile β-hydride transfer/alkene elimination process (1,34,43,44). Thus, the observation that [Mn(CH$_2$SiMe$_3$)(CO)$_5$] is more stable than [Mn(C$_2$H$_5$)(CO)$_5$] was attributed to the lack of β-hydrogens in the former complex (45). However, a β-elimination reaction from [Mn(C$_2$H$_5$)(CO$_5$] would be expected to result in the formation of ethylene and [Mn(H)(CO)$_5$] (which could decompose to [Mn$_2$(CO)$_{10}$]). However, neither ethylene nor [Mn(H)(CO)$_5$] have been observed in the decomposition of [Mn(C$_2$H$_5$)(CO)$_5$]. Propionyl manganesepentacarbonyl and [Mn$_2$(CO)$_{10}$] (which could derive from any source of [Mn(R)(CO)$_5$]) are the observed decomposition products (12,16,37,46).

In addition to this fact, substituted alkyl manganesepentacarbonyl complexes which contain β-hydrogen atoms, for example,

$[Mn(CH_2CH_2Ph)(CO)_5]$ (*40*) and $[Mn\{CH_2CH_2CH{=}C(Ph)(CH_3)\}(CO)_5]$ (*47*), are known to be relatively stable. Green and Nagy (*12*) obtained a red oil as the second decomposition product of $[Mn(C_2H_5)(CO)_5]$, but this does not appear to have been positively identified. This decomposition product may be identical in nature to the red polynuclear manganesecarbonyl compound obtained by Freudenberger and Orchin from the reaction of $[Mn(CH_2Ph)(CO)_5]$ with H_2 (*48*). Gismondi and Rausch thermally degraded $[Mn(C_2H_5)(CO)_5]$ (at room temperature) and sublimed crystals of $[Mn(COC_2H_5)(CO)_5]$ from the residue (*46*). No mention of any other decomposition products in the residue was made.

Bearing in mind the high carbonylation rate measured for $[Mn(C_2H_5)(CO)_5]$ (*11,18*), it seems to be reasonable to conclude that ethyl manganesepentacarbonyl probably decomposes by facile carbonylation (to propionyl manganesepentacarbonyl) rather than by β-hydride elimination. A mechanism as outlined in Scheme 1 could be in operation for the thermal decomposition of $[Mn(C_2H_5)(CO)_5]$. Thus, the carbonylation process for $[Mn(C_2H_5)(CO)_5]$ is so rapid that some of the carbon monoxide produced in steps (ii) and (iii) is used in step (iv) to carbonylate the solvated acyl intermediate. The decomposition of *n*-propyl manganesepentacarbonyl may be expected to follow a similar pathway.

Because of the unstable nature of ethyl and *n*-propyl manganesepentacarbonyl, very little work has been reported on them, and until relatively recently no work at all had been carried out on higher *n*-alkyl derivatives. Work of this nature should prove to be valuable since in many catalytic reactions it is C_3 and higher alkyl fragments which are involved. The

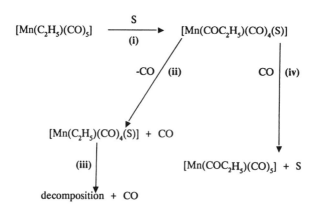

S = molecule of solvent, *e.g.* THF

SCHEME 1.

homologous series $[Mn(R)(CO)_5]$ has been completed, where R = CH_3 to $n\text{-}C_{18}H_{37}$ (16–18). This probably represents the most extensive series of simple metal carbonyl alkyl compounds to have been prepared so far. The above compounds were synthesized either (a) by thermal decarbonylation of the acyl compounds $[Mn(COR)(CO)_5]$ [as shown in Eq. (5)], for R = C_2H_5 to $n\text{-}C_9H_{19}$, $n\text{-}C_{11}H_{23}$, $n\text{-}C_{13}H_{27}$, $n\text{-}C_{15}H_{31}$, $n\text{-}C_{17}H_{35}$ [which were prepared by reaction of the appropriate acyl chlorides with $Na[Mn(CO)_5]$, as shown in Eq. (4)], or (b) by the direct reaction of $Na[Mn(CO)_5]$ with an alkyl bromide, as shown in Eq. (6) (16–18). In these studies, methyl iodide was used to prepare the methyl derivative.

$$Na[Mn(CO)_5] + RBr \xrightarrow[\text{3–5 days}]{\text{THF, 25°C}} [Mn(R)(CO)_5] + NaBr \qquad (6)$$

$$R = n\text{-}C_{10}H_{21},\ n\text{-}C_{12}H_{25},\ n\text{-}C_{14}H_{29},\ n\text{-}C_{16}H_{35},\ n\text{-}C_{18}H_{37}$$

With the exception of the methyl, ethyl, and n-propyl derivatives, all of the above alkyl compounds were new and have been fully characterized (4,16). Rather suprprisingly, the higher n-alkyl derivatives ($>C_5$) are all relatively stable, thermally and to air. No significant variation in the IR and NMR (^{1}H and ^{13}C) data was observed on changing the length of the alkyl chain, and the NMR data indicate that the effect of the metal is only felt in the first three carbon atoms of the alkyl chain (16,18), although the chemical shift in the ^{13}C NMR spectrum of the α-carbon of the alkyl chain does vary significantly with alkyl chain length.

B. Synthesis of Rheniumpentacarbonyl Alkyls

The first report of a rheniumpentacarbonyl alkyl derivative was that of Hieber and Braun in 1959 (13), who described the synthesis of methyl rheniumpentacarbonyl by the reaction of the rheniumpentacarbonyl anion with methyl iodide [Eq. (7)]. The same group then followed this with a report, in 1960, of the synthesis of $[Re(C_2H_5)(CO)_5]$, $[Re(CH_2C_6H_5)(CO)_5]$, $[Re(C_6H_5)(CO)_5]$, $[Re(COCH_3)(CO)_5]$, $[Re(COCH_2C_6H_5)(CO)_5]$, and $[Re(COC_6H_5)(CO)_5]$ (27). The same synthetic route as used for the methyl compound was applied, namely, reaction of $Na[Re(CO)_5]$ (generated in situ by the reductive cleavage of $[Re_2(CO)_{10}]$ over a Na/Hg amalgam) with the appropriate organic halide [Eqs. (8) and (9)]. The phenyl derivative, however, was formed via thermal decarbonylation of the benzoyl compound [Eq. (10)].

$$Na[Re(CO)_5] + CH_3I \rightarrow [Re(CH_3)(CO)_5] + NaI \qquad (7)$$

$$Na[Re(CO)_5] + RCOCl \rightarrow [Re(COR)(CO)_5] + NaCl \qquad (8)$$
$$R = CH_3,\ CH_2C_6H_5,\ C_6H_5$$

$$Na[Re(CO)_5] + RCl \rightarrow [Re(R)(CO)_5] + NaCl \qquad (9)$$
$$R = C_2H_5, CH_2C_6H_5$$

$$[Re(COC_6H_5)(CO)_5] \xrightarrow{120°C} [Re(C_6H_5)(CO)_5] + CO \qquad (10)$$

The ethyl derivative, however, was not isolated in a pure form; it was always contaminated by traces of solvent. Then, in 1963, Davison *et al.* (*14*) reported the synthesis of the ethyl and propionyl derivatives. The ethyl compound was isolated in a pure form from the treatment of $Na[Re(CO)_5]$ with ethyl iodide. $[Re(C_2H_5)(CO)_5]$ was also found to be much less susceptible to carbonylation than $[Mn(C_2H_5)(CO)_5]$ and was therefore considerably more stable: the reaction between $[Re(C_2H_5)(CO)_5]$ and CO (100 atm) to form $[Re(COC_2H_5)(CO)_5]$ was still not complete after 1 hour at 100°C (*14*).

A brief mention must be made of the high temperatures required to decarbonylate rheniumpentacarbonyl acyl species to the corresponding alkyl derivatives (usually over 100°C). Because rhenium–carbon bonds are considerably stronger than manganese–carbon bonds (*49*), temperatures 30–70°C higher than those required for decarbonylation of manganesepentacarbonyl acyl species are needed for the decarbonylation of the analogous rhenium compounds. The relatively high strength of the rhenium–carbon alkyl bond also results in rheniumpentacarbonyl alkyl compounds being considerably more stable than the analogous manganese compounds.

In 1985, Warner and Norton (*50*) reported an alternative synthesis of $[Re(C_2H_5)(CO)_5]$; they reacted $Na[Re(CO)_5]$ with excess ethyl tosylate to give the desired product [Eq. (11)]. This report also included the synthesis of isobutyl rheniumpentacarbonyl from the treatment of $Na[Re(CO)_5]$ with isobutyl tosylate [Eq. (11)].

$$Na[Re(CO)_5] + CH_3C_6H_4SO_2R \rightarrow [Re(R)(CO)_5] + CH_3C_6H_4SO_2Na \qquad (11)$$
$$R = C_2H_5, (CH_3)_2CHCH_2$$

Until recently, no higher *n*-alkyl derivatives ($>C_2$) of rheniumpentacarbonyl had been reported, even though they would be expected to be more stable than the manganesepentacarbonyl alkyl compounds. However, an extensive homologous series of rheniumpentacarbonyl alkyl compounds, $[Re(R)(CO)_5]$ ($R = CH_3$ to *n*-$C_{18}H_{37}$), has now been synthesized (*16,18*). All the compounds were found to be relatively stable, both thermally and to air. An extensive series of acyl compounds $[Re(COR)(CO)_5]$ ($R = CH_3$ to *n*-C_9H_{19}, *n*-$C_{11}H_{23}$, *n*-$C_{13}H_{27}$, *n*-$C_{15}H_{31}$, *n*-$C_{17}H_{35}$), prepared via the reaction of $Na[Re(CO)_5]$ with the appropriate acyl chloride [Eq. (12)], has also been described (*16,18*). The alkyl compounds were prepared either by thermal decarbonylation of the appropriate acyl precursor [Eq. (13)] or by direct reaction of the sodium salt of the rheniumpentacarbonyl anion with the appropriate alkyl halide [Eq. (14)].

$$Na[Re(CO)_5] + RCOCl \rightarrow [Re(COR)(CO)_5] + NaCl \qquad (12)$$

$$[Re(COR)(CO)_5] \xrightarrow{110°C} [Re(R)(CO)_5] + CO \qquad (13)$$
$$R = C_2H_5 \text{ to } n\text{-}C_9H_{19}, n\text{-}C_{11}H_{23}, n\text{-}C_{13}H_{27}, n\text{-}C_{15}H_{31}, n\text{-}C_{17}H_{35}$$

$$Na[Re(CO)_5] + RX \rightarrow [Re(R)(CO)_5] + NaX \qquad (14)$$
$$R = CH_3, n\text{-}C_{10}H_{21}, n\text{-}C_{12}H_{25}, n\text{-}C_{14}H_{29}, n\text{-}C_{16}H_{33}, n\text{-}C_{18}H_{37}; X = Br, I$$

III

STRUCTURAL AND SPECTROSCOPIC STUDIES

The molecular structures and bonding of $[M(R)(CO)_5]$ species (M = Mn, Tc, Re; R = alkyl) have received considerable attention over the years, mainly because they contain a simple σ bond between a transition metal and an alkyl group. The structures of the methyl compounds $[Mn(CH_3)(CO)_5]$ and $[Re(CH_3)(CO)_5]$ have been investigated by a wide variety of techniques, including X-ray, electron, and incoherent inelastic neutron diffraction, as well as spectroscopic techniques, namely, vibrational, NMR, and mass spectroscopy. Several theoretical studies have also been reported.

A. Diffraction Studies

The gas-phase electron diffraction study of $[Mn(CH_3)(CO)_5]$ showed the compound to have a C_{4V} symmetry with a manganese–methyl carbon–bond length of 2.185 Å (see Fig. 1) (51); hydrogen positions were not located. A separate gas-phase electron diffraction study of $[Re(CH_3)(CO)_5]$ gave a rhenium–methyl carbon bond length of 2.308 Å, rhenium–carbonyl bond lengths in the range 2.00 to 2.01 Å, and carbonyl carbon–oxygen bond lengths of 1.13 Å (52).

A problem arose when the X-ray crystal structure of $[Mn(CH_3)(CO)_5]$ was attempted (53). Results showed that the solid phase was almost entirely orientationally disordered; each of the six coordination sites around the manganese atom was found to be occupied by, on average, about five-sixths of a carbonyl ligand and one-sixth of a methyl ligand. The superimposed methyl and carbonyl carbon atoms could not be resolved, which precluded determination of any significant bonding parameters. However, the X-ray diffraction data did indicate that $[Mn(CH_3)(CO)_5]$ crystallizes in

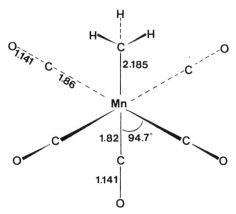

FIG. 1.   Structure of methyl manganesepentacarbonyl, with bond lengths in Å (*53*). A Mn–CO (axial) bond length of 1.82 Å was assumed.

the orthorhombic space group *Pcmn* with four molecules per unit cell. The disorder was confirmed by Raman measurements, which showed that some orientational disorder persisted even at 10 K (*53*).

B. *Spectroscopic Studies*

1. *Photoelectron Spectroscopy*

Several photoelectron spectroscopy (PES) studies on [M(R)(CO)$_5$] (M = Mn, Re) have been reported (*54–60*). However, it is beyond the scope of this article to cover each of these reports in detail; hence, we mention only a few of the most relevant and recent results.

Avanzino and Jolly, in 1976 (*58*), studied the oxygen 1*s* X-ray PES spectrum of gaseous [Mn(COCH$_3$)(CO)$_5$] and identified the two different oxygens (acetyl and terminal). Their results suggested that the positive charge on the acetyl carbon atom was less than that on any of the terminal carbonyl carbon atoms. Avanzino *et al.* (*59*) subsequently used X-ray PES to study the effect of R in [Mn(R)(CO)$_5$] on the carbonyl stretching force constants. From this study, they concluded that changes in the degree of back-bonding to carbonyl groups are determined principally by changes in the σ-donor nature of R. Ricco *et al.* (*60*) conducted a study on the X-ray PES of [Mn(R)(CO)$_5$] (R= CH$_3$, *n*-C$_3$H$_7$, σ-C$_3$H$_5$) and concluded that the methyl, propyl, and σ-allyl groups are negatively charged.

2. Vibrational Spectroscopy

The vibrational studies conducted by Andrews *et al.* (*53*) for [Mn(CH$_3$)(CO)$_5$] show that the manganese–methyl carbon stretching frequency occurs at a much lower value than most of the manganese–carbonyl carbon stretching frequencies. Together with the lower CH$_3$ versus CO oscillator mass, this was taken as indicative of a weaker force constant for the Mn–CH$_3$ versus Mn–CO stretch. Huang *et al.* (*61*) have conducted a vibrational study on the effects of high external pressures on the structures and phase behaviors of [M(CH$_3$)(CO)$_5$] (M = Mn, Re). The vibrational data indicated that both [Mn(CH$_3$)(CO)$_5$] and [Re(CH$_3$)(CO)$_5$] undergo a phase transition (most probably second order); the manganese species at around 9 bar and the rhenium species at around 22 bar.

A monosubstituted metal carbonyl of the type [M(R)(CO)$_5$] (M = Mn, Tc, Re) possesses C_{4v} symmetry, for which formal symmetry rules predict the following vibrations for the carbonyl groups (*62*): E, allowed, x,y polarized; $2A_1$, allowed, z polarized; B_1, forbidden. One can expect a strong band, a band which is less strong by a factor of four, and a third, much weaker band (*62*). This is in fact the pattern observed for most [M(R)(CO)$_5$] (M = Mn, Re) species. When M is Mn, these bands occur typically at around 2115 cm^{-1} (weak, A_1), 2012 cm^{-1} (strong, E), and 1990 cm^{-1} (medium, A_1) (in hydrocarbon solvents) (*28*). The corresponding values for rhenium are 2125, 2010, and 1979 cm^{-1} (*63*). Studies (*16–18,42*) on the homologous series [M(R)(CO)$_5$] (M = Mn, Re; R = CH$_3$ to n-C$_{18}$H$_{37}$) showed that there was no significant variation in ν(CO) on changing the length of the alkyl chain.

The B_1 transition can sometimes be observed for alkyl manganesepentacarbonyl species. Noack *et al.* (*28*) observed the B_1 band to be absent in a heptane solution of [Mn(CH$_3$)(CO)$_5$], but it appeared in diethoxydiethyl ether (at 2044 cm^{-1}); thus, the solvent may interact with these complexes, altering the symmetry sufficiently to bring about the appearance of the otherwise inactive B_1 band. The frequencies of the terminal carbonyl vibrations are relatively insensitive to the nature of R. Long and co-workers (*64,65*) have recorded infrared spectra in the CH and CD stretching regions for various methyl manganese- and rheniumpentacarbonyl isotopomers (methyl = CH$_3$, CD$_3$, CHD$_2$, CH$_2$D). The spectra of these species were qualitatively interpreted in terms of a C–H stretching force constant which varies with the internal rotation angle of an essentially freely rotating methyl group. An A_1 force field calculation for all vibrations showed that all the metal–carbonyl carbon bonds increase in strength from manganese to rhenium, whereas the methyl carbon–hydrogen bonds are weakened. The axial and equatorial metal–carbonyl bonds in [Re(CH₃)(CO)₅] are equal in strength, indicating a negligible trans effect on the part of the methyl ligand.

The above two studies concluded that internal rotation of the methyl group in [Mn(CH₃)(CO)₅] is essentially free, with the C–H stretching frequency showing a slight dependence on orientation. However, these conclusions were based on measurement of infrared bands with no resolution of rotational structure, and this limited the information available. High-resolution infrared studies by Gang *et al.* *(66)* confirmed that any restriction of rotation of the methyl group was below the limit of detection by this method.

3. NMR Spectroscopy

Very few studies on alkyl and acyl derivatives of manganese- and rheniumpentacarbonyl report the use of NMR spectroscopy as a tool to probe structure. The majority of studies have used NMR data for kinetic measurements (67) or merely to obtain characterization data to aid in identification. The first detailed NMR study was that of Davison *et al.* *(14)*, who obtained ¹H NMR spectra of a series of [M(R)(CO)₅] compounds (M = Mn, Re; R = H, CH₃, C₂H₅, COC₂H₅). The spectra of the methyl and ethyl derivatives showed that, as with non-transition metal alkyl compounds *(14)*, the screening of the alkyl protons could not be adequately described on the basis of inductive and paramagnetic effects. The terminal carbonyl groups could conceivably contribute to this lack of correlation by producing long-range screening contributions at the alkyl proton positions. In general, proton resonances for alkyl groups σ-bonded to the manganesepentacarbonyl moiety range from −0.89 to 7.23 ppm (R = CH₃ to CF₂H) and those for rhenium, −0.77 to 1.77 ppm (R = CH₃ to CH₂CH₃).

Calderazzo *et al.* *(68)* have measured ⁵⁵Mn NMR spectra for various [Mn(R)(CO)₅] compounds and found the chemical shift (in ppm) to increase as the electron-donating ability of R increased. Webb and Graham *(69)* have reported ¹³C NMR data for some [Re(R)(CO)₅] compounds which showed that the carbonyl carbon atoms trans to R are more shielded than those cis to R. When compared to ¹³C NMR data known for [Mn(R)(CO)₅] species *(70)*, it appears that the generally observed increase in shielding of carbonyl carbon atoms on descending a periodic group (for transition metals) is in evidence for manganese and rhenium.

Studies *(16–18,42)* on the homologous series [M(R)(CO)₅] (M = Mn, Re; R = CH₃ to *n*-C₁₈H₃₇) showed that there is no significant variation in chemical shift of any of the peaks in the ¹H NMR spectra when changing the length of the alkyl chain. Thus, integration is the only method of distinguishing between these compounds using ^{1}H NMR measurements.

4. *Mass Spectrometry*

Few mass spectral data are available for [Mn(R)(CO)$_5$] compounds and almost none for the analogous [Re(R)(CO)$_5$] species. Available information has mainly been reported as characterization data and has not been analyzed in any detail. The only detailed study known is that of Mays and Simpson *(71)*, who reported complete positive ion mass spectra of four complexes: [Mn(CH$_3$)(CO)$_5$], [Mn(C$_6$H$_5$)(CO)$_5$], [Mn(CF$_3$)(CO)$_5$], and [Mn(SO$_2$CH$_3$)(CO)$_5$]. The metal-containing ions were classified according to the mode of derivation from the molecular ion as follows: (i) loss of carbonyl groups, (ii) loss of R, (iii) loss of ligand fragments from the atom attached to manganese, and (iv) transfer of groups or atoms from R to manganese. Each process may involve more than one mechanism, for example, for [Mn(CH$_3$)(CO)$_5$] (Scheme 2).

The only mass spectra reported for alkyl rheniumpentacarbonyl species are those for the binuclear compounds [(CO)$_5$ReCH$_2$CH(CH$_3$)Re-(CO)$_5$] *(20)* and (CO)$_5$Re(CH$_2$)$_n$Re(CO)$_5$] (*n* = 3,4) *(23)*. For the latter two compounds, molecular ions of low intensity were observed. [(CO)$_5$Re(CH$_2$)$_4$Re(CO)$_5$] followed a fragmentation pattern similar to that of the analogous manganese compound *(23)*, namely (i) loss of CO, (ii) loss of the hydrocarbon bridge to form [Re$_2$(CO)$_{10}$], and (iii) loss of [Re(CO)$_5$]. However, for [(CO)$_5$Re(CH$_2$)$_3$Re(CO)$_5$], the major decomposition pathway involved initial elimination of C$_3$H$_6$.

Studies *(16–18,42)* on the homologous series [M(R)(CO)$_5$] (M = Mn, Re; R = CH$_3$ to *n*-C$_{18}$H$_{37}$) showed similar patterns, that is, sequential loss

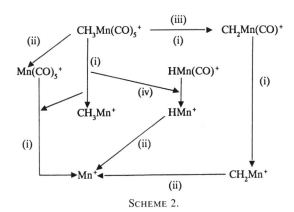

SCHEME 2.

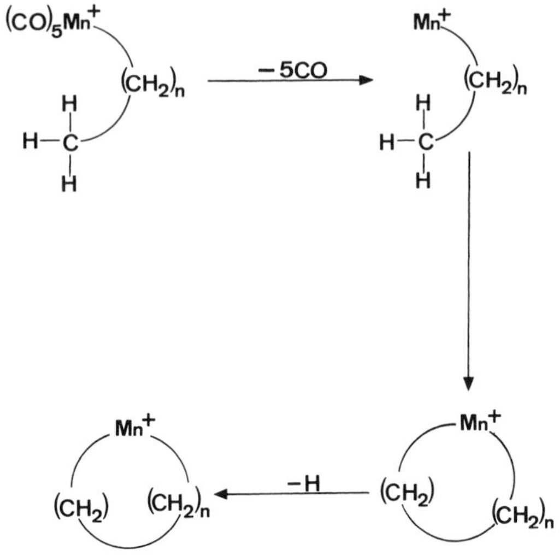

SCHEME 3.

of carbonyl ligands followed by sequential loss of methylene fragments. A high-resolution mass spectral study on the compound
$[Mn\{(CH_2)_{17}CH_3\}(CO)_5]$ also showed evidence for remote functionaliza-
tion (18,72) whereby a C–H bond of the methyl group of the alkyl chain
can be activated by a coordinatively unsaturated metal atom to give a
metallocyclic species. Further loss of hydride could occur by β-hydride
elimination without cleavage of the remaining alkyl residue, as shown in
Scheme 3.

C. Theoretical Studies

A theoretical study (molecular orbital calculations based on density
functional theory, using a modified Hartree–Fock–Slater method) on the
bond lengths and bond strengths in $[M(CH_3)(CO)_5]$ (M = Mn, Tc, Re) has
been reported by Ziegler et al. (15). The results are shown in Table I. The
results for the manganese and rhenium species are in good agreement with
reported experimental values (49,51,52). The increase in bond dissociation energies as the triad is descended can be accounted for by the increased steric crowding from Mn to Tc to Re. This report is, to the best of
our knowledge, the only report of an alkyl compound of technetiumpentacarbonyl in the literature. It is, however, a theoretical study; no alkyl
compounds of technetiumpentacarbonyl have as yet been synthesized.

TABLE I

CALCULATED BOND DISTANCES AND ENERGIES FOR
$[M(CH_3)(CO)_5]$

M	Bond length (Å)	Bond energy (M–CH₃) (kJ/mol)
Mn	2.16	153
Tc	2.26	178
Re	2.27	200

IV

REACTIONS OF ALKYL(PENTACARBONYL) COMPOUNDS OF THE
MANGANESE GROUP

A. Reactions of Manganesepentacarbonyl Alkyls

Methyl manganesepentacarbonyl, $[Mn(CH_3)(CO)_5]$, is one of the most widely studied compounds in organo-transition metal chemistry. The reactions that have been reported for this and related compounds are extensive; however, they can be divided into six main categories, namely, (1) reaction with transition metal hydrides, for example, $[CpW(H)(CO)_3]$; (2) addition of Lewis acids, for example $AlCl_3$; (3) orthometallation reactions; (4) reaction with cationic species, for example, H^+; (5) reaction with anionic nucleophiles, for example, $[CpFe(CO)_2]^-$; (6) reaction with neutral nucleophiles, for example, PPh_3. Each of these categories will be discussed separately, although it should be borne in mind that this account is not fully comprehensive since it would be beyond the scope of this article to provide a complete review of every reaction ever reported for $[Mn(R)(CO)_5]$ compounds. Consequently, only several of the more important and representative reactions from each category have been selected for discussion. Furthermore, reaction of $[Mn(R)(CO)_5]$ with neutral nucleophiles is covered as a separate subsection (Section III,B) because of the extensive body of information available.

1. Reaction of $[Mn(R)(CO)_5]$ with Transition Metal Hydrides

The reaction of $[Mn(R)(CO)_5]$ with transition metal hydrides usually results in the formation of di- or polynuclear complexes. For example, the reaction of $[Mn(CH_3)(CO)_5]$ with various transition metal hydrides has been reported (50) [Eq. (15) and (16)]. Thus, in coordinating solvents the metal-containing products are solvated dinuclear species, and in noncoordinating solvents they are polynuclear hydrides formed by coordination of a second equivalent of hydride (the vacant coordination site is created on manganese).

$$[Mn(CH_3)(CO)_5] + M\!-\!H \xrightarrow{CH_3CN} [MMn(CH_3)(CO)_4] + CH_3CHO \qquad (15)$$
$$M\!-\!H = [Re(H)(CO)_5], [CpW(H)(CO)_3]\ (Cp = \eta^5 - C_5H_5)$$

$$[Mn(CH)_3(CO)_5] + 2[Re(H)(CO)_5] \xrightarrow{C_6H_6} [Mn(H)Re_2(CO)_{14}] + CH_3CHO \qquad (16)$$

The reaction of $[Mn(R)(CO)_5]$ with $[Mn(H)(CO)_5]$ in noncoordinating solvents results in the formation of dimanganese η^1-aldehyde complexes, $[Mn_2(\eta^1\text{-}RCHO)(CO)_9]$ (47). The reductive elimination of $[Mn(R)(CO)_5]$ by $[Mn(H)(CO)_5]$ can yield either RH or RCHO compounds, depending on the reaction conditions $(73,74)$ [Eqs. (17) and (18)].

$$[Mn(R)(CO)_5] + [Mn(H)(CO)_5] \xrightarrow{C_6H_6} [Mn_2(CO)_{10}] + RH \qquad (17)$$

$$[Mn(R)(CO)_5] + [Mn(H)(CO)_5] \xrightarrow[C_6H_6,\ CO]{CH_3CN\ or} [Mn_2(CO)_{10}] + RCHO \qquad (18)$$
$$R = p\text{-}CH_3OC_6H_4CH_2,\ CH_2C_6H_5$$

2. Reaction of $[Mn(R)(CO)_5]$ with Lewis Acids

One of the earliest reports concerning reaction of $[Mn(R)(CO)_5]$ with Lewis acids came from Shriver and co-workers (75), who reacted $[Mn(R)(CO)_5]$ with $AlBr_3$ (Scheme 4). The purpose of the Lewis acid is to

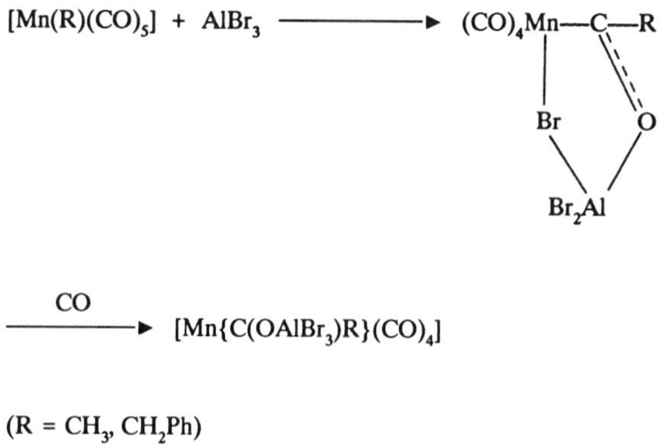

$$[Mn(R)(CO)_5] + AlBr_3 \longrightarrow (CO)_4Mn\!-\!\overset{}{C}\!-\!R$$

$$\xrightarrow{CO} [Mn\{C(OAlBr_3)R\}(CO)_4]$$

$$(R = CH_3,\ CH_2Ph)$$

Scheme 4.

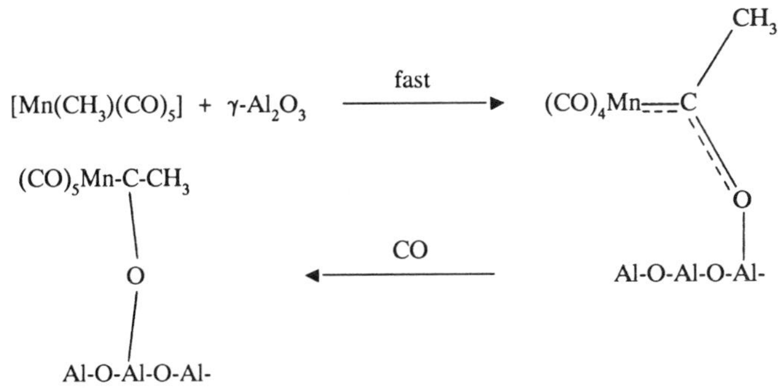

$$[Mn(CH_3)(CO)_5] + \gamma\text{-}Al_2O_3 \xrightarrow{\text{fast}} (CO)_4Mn{=}\!{=}\!{=}C\overset{CH_3}{\underset{O}{\diagdown}}$$

$$
\begin{array}{c}
(CO)_5Mn\text{-}C\text{-}CH_3 \\
| \\
| \\
O \\
/ \\
Al\text{-}O\text{-}Al\text{-}O\text{-}Al\text{-}
\end{array}
\xleftarrow{\quad CO \quad}
\begin{array}{c}
\\
| \\
Al\text{-}O\text{-}Al\text{-}O\text{-}Al\text{-}
\end{array}
$$

SCHEME 5.

facilitate alkyl migration by providing an electron-rich atom to fill the vacant coordination site. The Lewis acid, therefore, does not merely increase the rate of reaction with CO; it actually induces a CO insertion, providing an alternative reaction route by way of two intermediate steps. Similar results were obtained using $AlCl_3$ and BF_3 in place of $AlBr_3$.

Richmond et al. (76) have reported the Lewis acid-assisted carbonylation of $[Mn(CH_3)(CO)_5]$ with $AlBr_3$, $AlCl_3$, $AlCl_2Et$, and $AlClEt_2$ and have obtained kinetic data. The reactions involving $AlCl_2Et$ and $AlClEt_2$ gave the following results: k_{OBS} for $AlCl_2Et$, 10 mol/second; k_{OBS} for $AlClEt_2$, 0.37 mol/second. Nolan et al. (77) measured the enthalpy of reaction for the formation of $[Mn\{C(OAlBr_3)(CH_3)\}(CO)_4]$ from $AlBr_6$ and $[Mn(CH_3)(CO)_5]$ and reported a ΔH value of -167.4 kJ/mol. Correa et al. (78) reported acceleration of the carbonylation reaction by γ-alumina, also via formation of a cyclic adduct (Scheme 5).

Two theoretical studies on this topic have been reported (79,80) which indicate that when the Lewis acid coordinates to the carbonyl oxygen, it activates that group toward alkyl migration by withdrawal of electron density. This lowers the energies of both bonding and antibonding orbitals.

3. Orthometallation Reactions

Orthometallation reactions are reactions between an organic compound containing a substituted benzyl or phenyl group (which will then act as a donor ligand) and an organometallic compound. Electron-withdrawing substituents on the aromatic ring tend to activate it. The reaction is thought to occur via formation of an initial metallated complex of the organic compound through carbonyl displacement. Further dissociation of CO gives a sixteen-electron intermediate that can undergo oxidative

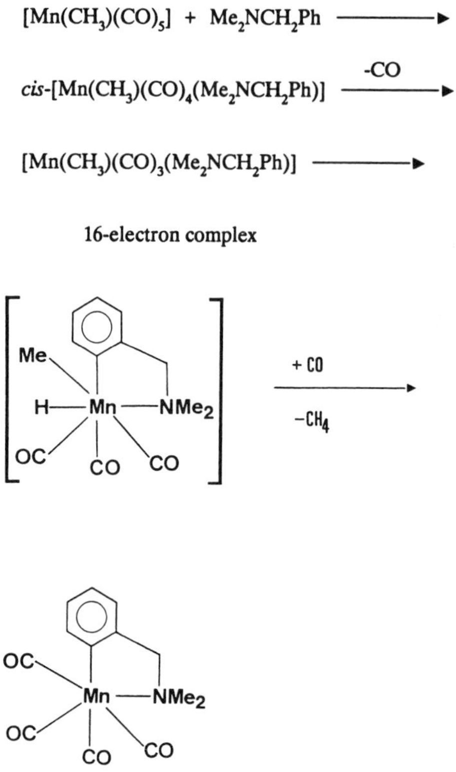

$$[Mn(CH_3)(CO)_5] + Me_2NCH_2Ph \longrightarrow$$

$$\textit{cis-}[Mn(CH_3)(CO)_4(Me_2NCH_2Ph)] \xrightarrow{\quad -CO \quad}$$

$$[Mn(CH_3)(CO)_3(Me_2NCH_2Ph)] \longrightarrow$$

16-electron complex

Scheme 6.

addition to an ortho C–H group of the aryl ring. Reductive elimination of methane then gives the final product. For example, the reactions shown in Scheme 6 have been reported by Bennett *et al.* (*81*).

Most orthometallation studies have involved reactions between [Mn(R)(CO)$_5$] compounds and either nitrogen or phosphorus donor ligands such as PhN=CHPh or P(OPh)$_3$. Relatively few authors have investigated orthometallation with oxygen donor ligands, although McKinney *et al.* (*82*) have described the metallation of acetophenone with [Mn(CH$_3$)(CO)$_5$] [Eq. (19)].

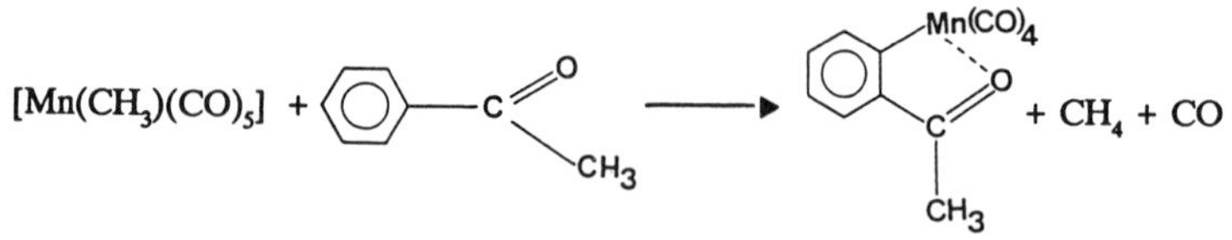

Another report (*83*) of an orthometallation reaction involves the reaction of 1,3-diacetylbenzene with [Mn(CH$_2$Ph)(CO)$_5$] to give the mono- and

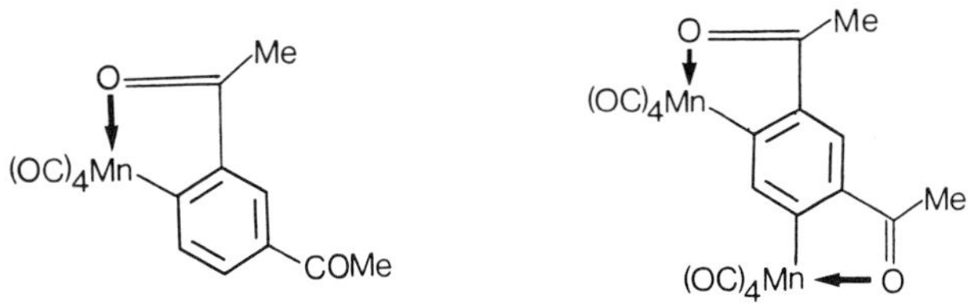

FIG. 2.   Products of the reaction of 1,3-diacetylbenzene with $[Mn(CH_2Ph)(CO)_5]$.

dicyclamanganated complexes shown in Fig. 2. The structure of the dimanganated complex was confirmed by X-ray crystal structure determination. Analogous orthometallated compounds were formed with 1,4-diacetylbenzene. An extensive review of orthometallation reactions using $[Mn(R)(CO)_5]$ compounds prior to 1982 is given by Treichel *et al.* (*43*, pages 78–85).

4. *Reaction of [Mn(R)(CO)₅] with Cationic Species*

Reaction of $[Mn(R)(CO)_5]$ with cationic species often results in cleavage of the manganese–alkyl carbon bond. Thus, the reaction of $[Mn(CH_3)(CO)_5]$ with sulfuric or hydrochloric acid produces methane (*84*) [Eq. (20)]. The reaction is believed to occur via oxidative addition of the acid followed by reductive elimination of CH_4.

$$[Mn(CH_3)(CO)_5] + HA \rightarrow [Mn(A)(CO)_5] + CH_4 \qquad (20)$$
$$HA = H_2SO_4, HCl$$

The gas-phase reaction of $[Mn(CH_3)(CO)_5]$ with various proton donors (BH^+) has been reported (*85*). The products of the reaction were found to depend on the nature of B.

$$[Mn(CH_3)(CO)_5] + HA + CO \rightarrow [Mn(COCH_3)(CO)_5] + HA \qquad (21)$$

A study by Butts *et al.* (*86*) was carried out on the reaction of $[Mn(CH_3)(CO)_5]$ with various acids in the presence of CO [Eq. (21)]. The acid was found to increase the rate of alkyl migration. Interestingly, in most cases, the acid does not cleave the $Mn–COCH_3$ or $Mn–CH_3$ bonds. The order of increase in the alkyl migration rate was $HA = CF_3CO_2H >$ $CCl_2HCO_2H > CClH_2CO_2H$. When HA is HBr, manganese–carbon bond cleavage occurred to give $[Mn(Br)(CO)_5]$ and CH_4. The binuclear complex $[(CO)_5Mn(CH_2)_4Mn(CO)_5]$ was found to react with HCl to give the expected cleavage products (*23*) [Eq. (22)].

$$[(CO)_5Mn(CH_2)_4Mn(CO)_5] + 2HCl \rightarrow 2[Mn(Cl)(CO)_5] + n\text{-}C_4H_{10} \qquad (22)$$

A relatively recent study by Motz *et al.* (*40*) involved the reactions of a variety of $[Mn(R)(CO)_5]$ compounds with acids (CF_3SO_3H or HBF_4) to give the expected cleavage products [Eq. (23)]. The rate of manganese–alkyl carbon bond cleavage was found to increase in the order R = H, CH_3, Ph, p-$CH_3C_6H_4$, p-$CF_3C_6H_4$ > CH_2Ph > p-$ClC_6H_4CH_2$ = p-$CH_3OC_6H_4CH_2$ ≫ $PhCH_2CH_2$. All reactions were slower with HBF_4 than with CF_3SO_3H.

$$[Mn(R)(CO)_5] + HA \rightarrow [Mn(A)(CO)_5] + RH \qquad (23)$$

The reaction of ethyl manganesepentacarbonyl with Ph_3CBF_4 results in β-hydride abstraction, giving a complex with coordinated ethylene as a ligand (12) [Eq. (24)].

$$[Mn(CH_2CH_3)(CO)_5] + Ph_3CBF_4 \rightarrow [Mn(CH_2{=}CH_2)(CO)_5]BF_4 \qquad (24)$$

5. Reaction of [Mn(R)(CO)$_5$] with Anionic Nucleophiles

The reaction of $[Mn(R)(CO)_5]$ with anionic nucleophiles most commonly proceeds via alkyl migration followed by nucleophilic attack at the now coordinatively unsaturated manganese atom. For example, the reaction of $[Mn(CH_3)(CO)_5]$ with lithium iodide is known (87) [Eq. (25)]. The same types of products result from the reaction of $[Mn(CH_3)(CO)_5]$ with OCH_3^-, SCN^-, and CN^-.

$$[Mn(CH_3)(CO)_5] + LiI \rightarrow Li[Mn(COCH_3)(I)(CO)_4] \qquad (25)$$

The reaction of $[Mn(CH_3)(CO)_5]$ with X^- (X = H or OCH_3) was investigated, with a view to protonating the expected acyl anionic species. However, the isolated product was a trinuclear hydride (88) [Eq. (26)].

$$[Mn(CH_3)(CO)_5] + X^- \rightarrow [Mn_3H_3(CO)_{12}] \qquad (26)$$

The reaction of $[Mn(CH_3)(CO)_5]$ with X^- (X = Cl, Br, or I) leads to the formation of halogenoacyl anions which can be protonated to give hydroxycarbenes (89) [Eq. (27)]. These were some of the first hydroxycarbene complexes to be isolated and characterized.

$$[Mn(CH_3)(CO)_5] + X^- \rightarrow [Mn(COCH_3)(X)(CO)_4]^- \xrightarrow{H_3PO_4}$$
$$cis\text{-}[Mn\{C(OH)(CH_3)\}(X)(CO)_4] \qquad (27)$$

The reaction of $[Mn(CH_3)(CO)_5]$ with $Na[M(CO)_5]$ (M = Mn, Re) followed by CH_3OSO_2F gives bimetallic carbene complexes $[(CO)_5MMn\{C(OCH_3)(CH_3)\}(CO)_4]$ (90). The reaction of $[Mn(CH_3)(CO)_5]$ with OH^- followed by allyl bromide led to the formation of $[Mn(\eta^3\text{-}C_3H_5)(CO)_4]$ and reductive elimination of methane (91) (Scheme 7). The reaction of $[Mn(CH_3)(CO)_5]$ with germyl lithium species gave anionic compounds [Eq. (28)] which were either isolated as tetraethylammonium salts or al-

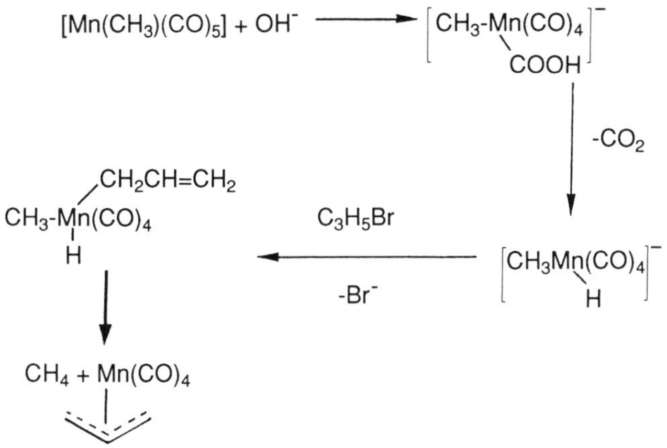

$$[Mn(CH_3)(CO)_5] + OH^- \longrightarrow \left[\begin{array}{c} CH_3\text{-}Mn(CO)_4 \\ \diagdown \\ COOH \end{array} \right]^-$$

$$\downarrow -CO_2$$

$$\begin{array}{ccc} & CH_2CH{=}CH_2 & \\ & \diagup & \\ CH_3\text{-}Mn(CO)_4 & & C_3H_5Br \\ | & & \\ H & \xleftarrow{\hspace{1cm}} & \left[\begin{array}{c} CH_3\underset{\diagdown}{Mn}(CO)_4 \\ H \end{array} \right]^- \\ & -Br^- & \end{array}$$

$$\downarrow$$

$$CH_4 + Mn(CO)_4$$

SCHEME 7.

kylated to give carbene complexes (92). The reaction of $[Mn(CH_3)(CO)_5]$ with Ph_3CBF_4 gives $[Mn(FBF_3)(CO)_5]$ which can subsequently be converted to the binuclear complex $[(CO)_5MnCH_2CH_2Mn(CO)_5]$ (20).

$$[Mn(CH_3)(CO)_5] + R_3GeLi \rightarrow Li[Mn(COCH_3)(GeR_3)(CO)_4] \qquad (28)$$
$$R_3 = Ph_3 \text{ or } MePh\text{-}1\text{-naphthyl}$$

A study by Wang and Atwood (93) involved the reaction of $[Mn(R)(CO)_5]$ (R = CH_3, CH_2Ph, Ph) with $Na[CpFe(CO)_2]$. The reaction resulted in transfer of R to $Na[CpFe(CO)_2]$ to give $[CpFe(R)(CO)_2]$ and generation of $[Mn(CO)_5]^-$. The reactions were found to be first order in $Na[CpFe(CO)_2]$ and in $[Mn(R)(CO)_5]$. A dependence of the rate on R was observed: $CH_2Ph > CH_3 > Ph$; this is consistent with a nucleophilic attack mechanism.

B. Reaction of Manganesepentacarbonyl Alkyls with Neutral Nucleophiles

Reaction of $[Mn(R)(CO)_5]$ with neutral nucleophiles is by far the most widely studied type of reaction for $[Mn(R)(CO)_5]$ compounds. The reaction usually involves addition of the neutral neucleophile, L, and is accompanied by CO insertion/alkyl migration to form an acyl species [Eq. (29)]. L is usually a tertiary phosphine (PR_3), an alkylated amine (RNH_2), or free carbon monoxide. Besides being a carbon–carbon bond forming reaction of fundamental importance, alkyl migration reactions of transition metal alkyl species have direct relevance to catalysis, especially for the OXO or hydroformylation process (2), the Monsanto acetic acid synthesis (2), and the synthesis of ethylene glycol (94).

$$[Mn(R)(CO)_5] + L \rightarrow [Mn(COR)(CO)_4(L)] \qquad (29)$$

1. Kinetics and Mechanism of Alkyl Migration

a. *Mechanistic Studies.* Over the years, various mechanisms have been proposed for the alkyl migration reaction. For example, Calderazzo and Cotton proposed *(95)* that an activated complex was formed by direct combination of L with $[Mn(R)(CO)_5]$ without any solvent participation. However, it is now generally accepted that the reaction essentially proceeds by migration of the alkyl group to an adjacent (i.e., cis) carbonyl group *(43,96)* to form the coordinatively unsaturated species $[Mn(COR)(CO)_4]$ (see Scheme 8). Coordination of the incoming ligand then follows to form $[Mn(COR)(CO)_4(L)]$ products. That the reaction proceeds via alkyl migration rather than by direct CO insertion into the metal–alkyl carbon σ bond has been demonstrated by labeling studies *(97,98)*.

The general kinetic scheme shown in Scheme 8 has been developed for alkyl migration in $[Mn(R)(CO)_5]$ compounds. Thus, two pathways to the final product are available: (i) a second-order pathway $(k[Mn(R)(CO)_5][L])$ and (ii) a two-step sequence via the coordinatively unsaturated intermediate $[Mn(COR)(CO)_4]$. In polar solvents, this intermediate may exist as a solvated hexacoordinate species, $[Mn(COR)(CO)_4(S)]$ (S = molecule of solvent). In the absence of a suitably polar solvent, the intermediate is thought to adopt a square-based pyramidal conformation *(99,100)* with the acetyl ligand occupying a basal position, although a π-acyl derivative has also been proposed *(101,102)*.

Alkyl migration in $[Mn(R)(CO)_5]$ species is assumed to follow a concerted reaction pathway, namely, concomitant bond breaking and bond formation *(3)*. Calderazzo and Cotton investigated the reaction of CO with $[Mn(CH_3)(CO)_5]$ and obtained an activation energy of 61.9 kJ/mol for the alkyl migration process *(95)*. This value is well below the reported bond dissociation energy of 184 kJ/mol for the manganese–methyl carbon bond

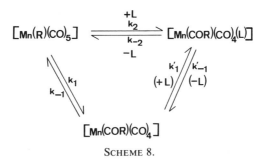

SCHEME 8.

in $[Mn(CH_3)(CO)_5]$ (*49*), supporting a concerted mechanism, that is, a three-centered transition state.

The acyl compounds which are formed from the alkyl migration reaction initially adopt a cis geometry, but where the incoming ligand has a large steric profile they often isomerize to the trans form. This shows that the initial reaction is kinetically rather than thermodynamically controlled. A study reported in 1991, however, notes an example where the incoming ligand enters trans to the acetyl group in a related $[Mn(R)-(CO)_3(L_2)]$ compound (*103*). The isomerization process has been shown to occur via a dissociative intermolecular process, namely, dissociation of L (*104*).

b. *Effect of Solvent.* In solvents of relatively high polarity, it has been intimated that a solvent molecule may coordinate to the coordinatively unsaturated intermediate, $[Mn(COR)(CO)_4]$, to form a solvated hexacoordinate species which stabilizes the reactive intermediate (*43,96,105*). A large, negative entropy of activation supports this suggestion (*105*). It has been proposed that the cleavage of the manganese–alkyl bond is assisted by the solvent, which then enters the coordination sphere of the metal. In a number of instances, the reaction rate has been found to increase with an increase in polarity of the solvent; the rate acceleration depends on the electron-donating ability of the solvent rather than its dielectric constant (*106*).

Several studies on the effect of the solvent on alkyl migration have been reported by Cotton and co-workers (*96,107–109*), who found that electron-donating solvents increase the rate of alkyl migration in substituted benzyl manganesepentacarbonyl compounds, $[Mn(CH_2C_6H_{5-n}X_n)(CO)_5]$. In some instances, the solvent-coordinated species was stable enough to be detected (*96*). In addition to influencing the reactivity, the coordinating ability of the solvent can affect the mechanism of alkyl migration. If a poorly coordinating solvent is used in conjunction with a strong nucleophile, L, the k_1 path (Scheme 8) can be suppressed and the k_2 path may dominate.

c. *Effect of Entering Ligand.* The effect of the entering ligand, L, is very closely related to the nature of the solvent. If the solvent lacks coordinating ability, then alkyl migration proceeds via attack of L on the alkyl compound. In a coordinating solvent, however, there will be competition between the reaction of the intermediate, $[Mn(COR)(CO)_4(S)]$, with L to give the acyl product, $[Mn(COR)(CO)_4 (L)]$, and its reversion to the alkyl, $[Mn(R)(CO)_5]$. If L is a strong enough nucleophile to capture the intermediate (solvated or not) then the k_1 path will prevail, alone or in conjunction with the k_2 path. Very few quantitative data are available on the effect of L, although one study was reported by Cotton *et al.* (*107*), who reacted [Mn(CH$_2$Ph)(CO)$_5$] with a series of tertiary phosphines and found that the rate of alkyl migration increases with a decrease in the Tolman electronic parameter, ν, of the phosphine (*110*), that is, with an increase in electron donation.

d. *Effects of Alkyl Group.* Although both the solvent and incoming nucleophile can affect the rate of alkyl migration considerably, the nature of the alkyl group, R, has a far greater influence. Both steric and elecronic factors are important in determining the magnitude of this effect. Early studies indicated the following trend in reactivity with different R groups: R = *n*-Pr > Et > Ph > CH$_3$ > CH$_2$F >> CF$_3$ (*11,111*). A somewhat more recent study has substantiated this trend and added to the list of alkyl groups: R = *n*-Pr > Et > CH$_2$C$_6$H$_{11}$ > CH$_2$CH$_2$OCH$_3$ > CH$_2$C$_6$H$_5$ > CH$_2$CO$_2$H (*112*). This very facile carbonylation process for the ethyl and *n*-propyl derivatives may in fact be the reason for their high instability.

There appears to be some discrepancy in the literature as to the reasons for the instability of [Mn(C$_2$H$_5$)(CO)$_5$]. Several reports attribute the instability to a facile β-hydride elimination process (*1,34,43,44*). In fact, Collman *et al.* (*1*) make the following statement: "... the β-hydride elimination reaction dominates the chemistry of alkyl ligands." However, a β-hydride elimination reaction would result in the formation of [Mn(H)(CO)$_5$] (which could decompose to [Mn$_2$(CO)$_{10}$]) and ethylene, yet it is propionyl manganesepentacarbonyl, not ethylene, which is one of the observed decomposition products of [Mn(C$_2$H$_5$)(CO)$_5$]. Gismondi and Rausch thermally degraded [Mn(C$_2$H$_5$)(CO)$_5$] (at room termperature) and sublimed crystals of [Mn(COC$_2$H$_5$)(CO)$_5$] from the residue (*46*). Bearing in mind the high carbonylation rate for [Mn(C$_2$H$_5$)(CO)$_5$], it seems reasonable to conclude that ethyl manganesepentacarbonyl probably decomposes by facile carbonylation (to propionyl manganesepentacarbonyl) rather than by β-hydride elimination. We believe that the decomposition of ethyl (and *n*-propyl) manganesepentacarbonyls is more complicated than was originally suggested, and these decompositions are currently under investigation (*113*).

Cotton and co-workers (*107,114*) investigated the reactions of a series of substituted benzyl manganesepentacarbonyl compounds, [Mn(CH$_2$C$_6$H$_4$X)(CO)$_5$] with PR$_3$ [Eq. (30)] in order to probe the steric and electronic effects of X on the alkyl migration reaction. The rate data were analyzed in terms of the reaction pathway proposed in Scheme 8. In the majority of cases there was no detectable contribution from the k_2 pathway. The rate constant, k_{OBS} (effectively k_1), was found to increase significantly with increasing electron-donating ability of the substituent. The most reactive substituent among those studied, p-OCH$_3$, reacted around five times faster than the least reactive, p-CH$_3$. The results were analyzed in terms of Hammet σ substituent parameters, giving a rather modest value (-0.97) of ρ, the reaction parameter.

$$[Mn(CH_2C_6H_4X)(CO)_5] + PR_3 \xrightarrow[30°C]{CH_3CN} cis\text{-}[Mn(COCH_2C_6H_4X)(CO)_4(PR_3)] \quad (30)$$

The following reactivity order was observed for the methyl-substituted compounds: p-CH$_3$ > m-CH$_3$ > o-CH$_3$, which is at variance with that predicted on electronic grounds. Although the electronic effect of the methyl group is small, the group is both inductively and by resonance electron-donating, and thus its influence should be greater from the ortho position than from the para position. However, as the steric influence at the ortho position is increased, rate enhancement is observed; for example, k_{OBS} for X = o-iPr is approximately the same as k_{OBS} for X = p-iPr, that is, the "ortho-inhibition" observed for the methyl substituent has been overcome. The low reactivity of the o-CH$_3$ compound may reflect competition between two opposing steric effects.

In addition to the enhancement arising from nonbonded interactions in the starting material, which will weaken the metal–carbon (benzyl) bond in the ground state, inhibition can also arise from interactions in the transition state associated with migration of the benzyl ligand to an adjacent carbonyl group. This effect is demonstrated by the large decreases in k_{OBS} for the large alkyl group di-ortho-substituted [CpMo(CH$_2$C$_6$H$_3$X$_2$)-(CO)$_3$] compounds (*115*) compared to the mono-substituted counterparts. For the manganese complexes studied by Cotton and co-workers, the limiting size necessary for the onset of the latter effect as a major influence on reactivity was not reached or observed (*114*). For the concept of two competing influences to be valid, the profile of each effect with respect to the substituent would have to be very different, and the profile would also have to vary from metal to metal. Cotton proposed (*108,114*) that the overall effect was to create a "steric window" within which the reactivity is enhanced. Comparison of the relative reactivities of the manganese (*96,107,108,114*) and molybdenum (*115,116*) systems studied by Cotton suggests that the "window" for the manganese compounds encompasses larger alkyl groups and that, overall, the manganese compounds are sterically less congested. Thus, the rate of alkyl migration increases as the electron-donating ability of the alkyl group in [Mn(R)(CO)$_5$] increases. It has also been shown that [Mn(CH$_2$I)(CO)$_5$], [Mn(SiMe$_3$)(CO)$_5$], and [Mn(SiPh$_3$)(CO)$_5$] do not undergo the CO insertion/

TABLE II

ΔH Values for Reaction of
CO with [Mn(R)(CO)$_5$]

R	ΔH (kJ/mol)
CH$_3$	-54 ± 8
C$_6$H$_5$	-63 ± 8
CF$_3$	-12 ± 7

alkyl migration reaction (*117–119*). Thus the rate of alkyl migration is slowed down, or even stopped, by electron-withdrawing groups.

To measure the migratory aptitudes of different alkyl groups into a carbonyl ligand, Connor *et al.* obtained values of ΔH for the reaction of CO with [Mn(R)(CO)$_5$] at 298 K (*120*) as shown in Table II. These results demonstrate conclusively that alkyl migration in [Mn(CH$_3$)(CO)$_5$] is much more facile, and more exothermic, than in [Mn(CF$_3$)(CO)$_5$]. Molecular orbital calculations indicate that the decrease in the rate of alkyl migration from R = CH$_3$ to R = CF$_3$ is not in fact connected with any strengthening of the manganese–alkyl carbon bond, but rather with a change in charge distribution on the alkyl carbon atom (*121*). This will then reduce the likelihood of a nucleophilic attack.

2. Reaction of [Mn(R)(CO)$_5$] with Tertiary Phosphines and Related Ligands

Reactions of [Mn(R)(CO)$_5$] with tertiary phosphines, PR$_3$, are extensive and range from the very early studies of Mawby *et al.* (*122*), Noack *et al.* (*104*), and Bannister *et al.* (*123*) to the far more recent studies of Cotton *et al.* (*96,107,114*) and Andersen and Moss (*16–18*). In the first study reported on this type of reaction, Mawby *et al.* (*122*) reacted [Mn(CH$_3$)(CO)$_5$] with PPh$_3$ and P(OPh)$_3$ in THF. The products of the reaction were the tertiary phosphine-substituted acyl complexes, *cis*-[Mn-(COCH$_3$)(CO)$_4$(PR$_3$)]. Similar results were subsequently obtained by other workers in the field (*67,123–126*). The later results also demonstrated the presence of the trans isomer of the phosphine-substituted acyl products. Noack *et al.* (*104*) reported a study of the isomerization process. Bannister *et al.* conducted a rather extensive study on the reaction of [Mn(R)(CO)$_5$] (R = CH$_3$, C$_6$H$_5$) with a range of nucleophiles (*123*). The results for [Mn(CH$_3$)(CO)$_5$] are shown in Table III. Thus, the reaction of [Mn(CH$_3$)(CO)$_5$] (and [Mn(Ph)(CO)$_5$]) with phosphites gave disubstituted products. Similar results were obtained in subsequent studies on this type of reaction (*21,41,67,99*).

TABLE III

REACTION OF $[Mn(CH_3)(CO)_5]$ WITH LIGANDS

Ligand, L	Product
PPh_3	$cis/trans-[Mn(COCH_3)(CO)_4(L)]$
$AsPh_3$	$cis/trans-[Mn(COCH_3)(CO)_4(L)]$
$SbPh_3$	$cis-[Mn(COCH_3)(CO)_4(L)]$
$P(OPh)_2Me$	$cis-[Mn(CH_3)(CO)_3(L)_2]$
$P(OPh)_3$	$trans-[Mn(COCH_3)(CO)_3(L)_2]$

When $[Mn(R)(CO)_5]$ compounds are reacted with chelating diter-tiary phosphines(diphos), mono-substituted products are formed initially which then, in a second step, undergo an intramolecular substitution reaction with elimination of CO to give a chelate complex (127). Thus, $[Mn(CH_3)(CO)_5]$ will react with $Ph_2P(CH_2)_nPPh_2$ (n = 1 to 3) (in a 2:1 molar ratio) to give fac-$[Mn(COCH_3)(CO)_3(diphos)]$ (123,127–129). A higher mole ratio of $[Mn(CH_3)(CO)_5]$ to $Ph_2PCH_2CH_2PPh_2$ results in the formation of $[\{Mn(COCH_3)(CO)_4\}_2\{Ph_2PCH_2CH_2PPh_2\}]$ (123,128). An exception to the above general process is the reaction of $[Mn(CH_3)(CO)_5]$ with 1,3-bis(diphenylphosphino)propane in the presence of triflic acid, which gave fac-[1,3-bis(diphenylphosphino)propane(methylhydroxycar-bene) tricarbonyl manganese(I)]triflate (130):

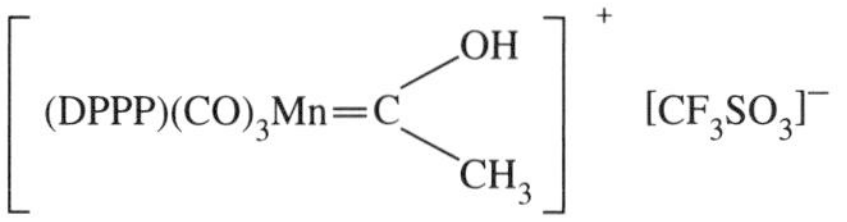

$$\left[(DPPP)(CO)_3Mn=C \diagup_{CH_3}^{OH} \right]^+ [CF_3SO_3]^-$$

The reaction of $[Mn(CH_3)(CO)_5]$ with chelating tridentate phosphorus ligands (denoted P-P-P) gives $[Mn(COCH_3)(CO)_2(P-P-P)]$ species (131,132). Equation (31) represents an example of this type of reaction (130).

$$[Mn(CH_3)(CO)_5] + (Me_2PCH_2CH_2)_2PPh + \rightarrow [Mn(COCH_3)(CO)_2\{(Me_2PCH_2CH_2)_2PPh\}]$$
(31)

Three of the more recent studies on reactions of $[Mn(R)(CO)_5]$ compounds with tertiary phosphines are as follows. Mapolie and Moss (23) reacted the binuclear complexes $[(CO)_5Mn(CH_2)_nMn(CO)_5]$ with PR_3 to give cis,cis-$[(PR_3)(CO)_4MnCO(CH_2)_nCOMn(CO)_4(PR_3)]$ (n = 4–6; PR_3 = PPh_3, PPh_2Me, $PPhMe_2$, PMe_3). Cotton et al. (107), as discussed in the previous section, studied reactions of $[Mn(CH_2C_6H_4X)(CO)_5]$ with PR_3. Andersen and Moss (16–18) investigated the reaction of an extensive

TABLE IV

KINETIC DATA FOR REACTION OF [Mn(R)(CO)$_5$]
WITH PPh$_3$ AT 32°C IN HEXANE[a]

R	k_{OBS} ($\times 10^4$ second^{-1})	log k_{OBS}	$t_{1/2}$ (minutes)
CH$_3$	0.63	−4.20	182.8
C$_2$H$_5$	3.26	−3.49	35.5
n-C$_3$H$_7$	6.37	−3.20	18.1
n-C$_4$H$_9$	4.14	−3.38	27.0
n-C$_5$H$_{11}$	2.83	−3.55	40.8
n-C$_6$H$_{13}$	2.21	−3.66	52.3
n-C$_7$H$_{15}$	2.02	−3.69	52.5
n-C$_8$H$_{17}$	2.01	−3.70	57.5
n-C$_9$H$_{19}$	1.99	−3.70	58.0
n-C$_{11}$H$_{23}$	1.96	−3.71	58.9
n-C$_{13}$H$_{27}$	1.91	−3.72	60.5
n-C$_{15}$H$_{31}$	1.89	−3.72	61.1
n-C$_{17}$H$_{35}$	1.85	−3.73	62.4
n-C$_{18}$H$_{37}$	1.84	−3.74	62.8

[a] From Andersen and Moss (18).

ogous series of alkyl manganesepentacarbonyl compounds, [Mn(R)(CO)$_5$] (R = CH$_3$ to n-C$_{18}$H$_{37}$), with PPh$_3$ [Eq. (32)]. The results were analyzed in terms of the k_1 pathway (Scheme 8) and are shown in Table IV and Fig. 3.

$$[Mn(R)(CO)_5] + PPh_3 \xrightarrow[32°C]{hexane} cis\text{-}[Mn(COR)(CO)_4(PPh_3)] \qquad (32)$$
$$R = CH_3 \text{ to } n\text{-}C_{18}H_{37}$$

The trend represented in Fig. 3 was accounted for by a combination of steric and electronic effects. When R is CH$_3$ to n-C$_3$H$_7$ electronic effects predominate, with the rate of alkyl migration increasing with an increase in electron-donating ability of the alkyl group. When R is n-C$_4$H$_9$ to n-C$_7$H$_{15}$, the elecronic effect of the alkyl group becomes more or less constant and steric effects dominate, with increasing steric bulk of the alkyl group decreasing the rate of alkyl migration. When R is n-C$_8$H$_{17}$ to n-C$_{18}$H$_{37}$ both effects are more or less constant, and the change in rate is thus almost negligible.

However, not all reactions of [Mn(R)(CO)$_5$] with phosphines give the expected products; thus mention should be made of three somewhat unusual reactions. (a) Rosen et al. (133) reacted [Mn(CH$_3$)(CO)$_5$] with [Cp-Fe(CO)$_2$(PPh$_2$)] and isolated a heterobimetallic acetyl-bridged complex as the product [Eq. (33)]. (b) Vaughan et al. (134) reacted [Mn(CH$_3$)(CO)$_5$]

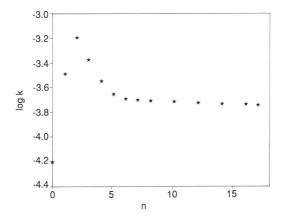

FIG. 3. Plot of n in $[Mn\{(CH_2)_nCH_3\}(CO)_5]$ versus log k for the reaction of $[Mn\{(CH_2)_nCH_3\}(CO)_5]$ with PPh_3 (from ref. 18).

with $PPh_2(SiMe_3)$ to form a novel ylide complex [Eq. (34)]. (c) Ressner *et al.* (*135*) reacted $[Mn(CH_3)(CO)_5]$ with $PPh_2CH_2Si(CH_3)_2CH_2PPh_2$ to give $[Mn(CO)_3\{CH_2Si(CH_3)_2(CH_2PPh_2)_2\}]$, formed by intramolecular activation of a C–H bond in a methylsilyl group by manganese.

$$[Mn(CH_3)(CO)_5] + [CpFe(CO)_2(PPh_2)] \xrightarrow[22°C]{THF} Cp(CO)Fe \overset{\overset{Ph_2}{P}}{\underset{C=O}{\diagup\diagdown}} Mn(CO)_4 \quad (33)$$

$$[Mn(CH_3)(CO)_5] + PPh_2(SiMe_3) \rightarrow (CO)_4Mn \overset{OSiMe_3}{\underset{Ph_2}{\diagdown C \diagup CH_3}} \quad (34)$$

3. Reaction of [Mn(R)(CO)₅] with Carbon Monoxide

Reactions of $[Mn(R)(CO)_5]$ compounds with carbon monoxide have been studied extensively. In this case only one product, namely, the acyl species, $[Mn(COR)(CO)_5]$, is formed [Eq. (35)].

$$[Mn(R)(CO)_5] + CO \rightarrow [Mn(COR)(CO)_5] \quad (35)$$

Early studies (*97,98*) concentrated on the reaction of $[Mn(CH_3)(CO)_5]$ with labeled CO (^{13}CO and ^{14}CO) in order to probe the mechanism and stereochemistry of the alkyl migration reaction. It was those two studies which were instrumental in proving conclusively that the carbonylation reaction takes place via alkyl migration rather than by direct CO substitution. Subsequent studies involved the measurement of carbonylation rates in order to determine thermodynamic parameters (e.g., activation energies) for the alkyl migration process (95,136). The variation in rate with change (both steric and electronic) in R for the reaction represented by Eq. (35) was investigated (11,98,112). The following order was found with n-propyl being the fastest: R = n-Pr > Et > $CH_2C_6H_{11}$ > C_6H_5 > CH_3 > CH_2OCH_3 > $CH_2C_6H_5$ > CH_2CO_2H, and R = CH_3 > CH_2F > CF_3.

The carbonylation of [MN(CH_3)(CO)$_5$] at high pressure (320 atm CO, 67°C in tetradecane) was reported (137); the only product isolated was [Mn(COCH$_3$)(CO)$_5$]. In another study (136), [Mn(CH_3)(CO)$_5$] was reacted with CO at 70°C to give [Mn(COCH$_3$)(CO)$_5$] with a rate constant k of 1.9×10^{-5} second^{-1}. This was compared to the rates for the reaction of cis-[Mn(CH_3)(CO)$_4$(PR$_3$)] with CO to give cis-[Mn(COCH$_3$)(CO)$_4$(PR$_3$)] [PR$_3$ = P(OPh)$_3$, k = 3.6×10^{-5} second^{-1}; PR$_3$ = P(OMe)$_3$, k = 2.3×10^{-5}-second^{-1}]. The similarity in the three rate constants implies a very small ligand effect, which was taken to indicate a transition state with very little unsaturation. More recent studies (21,23) involved the carbonylation of [CO$_5$Mn(CH$_2$)$_n$Mn(CO)$_5$] (under 1 atm CO at 25°C) to give [(CO)$_5$MnCO(CH$_2$)$_n$COMn(CO)$_5$] (n = 4–6).

4. Reaction of [Mn(R)(CO)$_5$] with Isocyanides and Amines

Another fairly common and well-documented reaction of [Mn(R)(CO)$_5$] compounds is that with alkyl isocyanides, RNC (which are isoelectronic with CO), or alkyl amines, RNH$_2$. The usual product is the isocyanide- or amine-substituted acyl species, [Mn(COR)(CO)$_4$(L)] (formed by alkyl migration). The first study on this type of reaction was that of Keblys and Filbey (138) who reacted [Mn(R)(CO)$_5$] (R = CH$_3$, C$_6$H$_5$) with NH$_3$ and aliphatic and aromatic amines [(C$_6$H$_{11}$)NH$_2$, Me$_2$NH, (C$_6$H$_5$)NH$_2$] to give cis-[Mn(COR)(CO)$_4$(amine)]. Subsequent studies by Mawby et al. (122), using [Mn(CH$_3$)(CO)$_5$] gave the same results except for the reaction with N-methylcyclohexylamine, which did not go to completion. They also made the observation that the reaction was solvent dependent. Kraihanzel and Maples (126) also repeated these results and made the additional observation that the amine-substituted acyl products, [Mn(COR)-(CO)$_4$(amine)], were difficult to decarbonylate. Kuty and Alexander (139) reported the reactions of [Mn(R)(CO)$_5$] (R = CH$_3$, CH$_2$C$_6$H$_5$ and p-O$_2$-NC$_6$H$_4$CH$_2$) with alkylated isocyanides, R'NC (R' = CH$_3$, t-C$_4$H$_9$, and C$_6$H$_{11}$) to give the acyl isocyanide adduct [Mn(COR)(CO)$_4$CNR')] (mixture of cis and trans), except for [Mn(p-O$_2$NC$_6$H$_4$CH$_2$)(CO)$_5$], which did not undergo any reaction.

A more recent study by Motz *et al.* (*140*) involved the interesting reaction of [Mn(p-XC$_6$H$_4$CH$_2$)(CO)$_5$] (X = Cl, OCH$_3$) with p-tolylisocyanide (p-CH$_3$C$_6$H$_4$CH$_2$NC) in the presence of PdO to give the double isocyanide inserted products [Mn{C(=CH(C$_6$H$_4$-p-Cl))N(p-tolyl)(NH-p-tolyl)}(CO)$_4$] for X = Cl (which is the product of intramolecular attack by an imino nitrogen atom on a coordinated p-tolylisocyanide) and [Mn{C(=N-p-tolyl)C(=N-p-tolyl)CH$_2$C$_6$H$_4$-p-X}(CO)$_4$] for X = Cl and OCH$_3$.

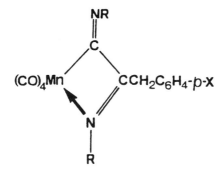

R = p-tolyl; X = Cl, OMe

5. Reaction of [Mn(R)(CO)$_5$] with Other Neutral Nucleophiles

Besides the numerous studies already mentioned on the reactions of [Mn(R)(CO)$_5$] with phosphines, amines, isocyanides, and carbon monoxide, there have been extensive investigations into the reactions of [Mn(R)(CO)$_5$] compounds with other neutral donor ligands. A brief description of these now follows.

a. *Reaction of [Mn(R)(CO)$_5$] with Alkenes and Alkynes.*   Reaction of [Mn(R)(CO)$_5$] with alkenes and alkynes usually results in migration of R and subsequent insertion of the alkene or alkyne via addition of the acyl species across the olefinic (or alkynyl) linkage. An extensive study of this general reaction was performed by DeShong *et al.* (*141*) who reacted [Mn(R)(CO)$_5$] (R = CH$_3$, CH$_2$Ph, CH=CHCH$_3$) with a variety of substituted alkenes and alkynes to give substituted manganacycles [Eqs. (36) and (37)] (the structures of the products were confirmed by X-ray crystallography). The unsymmetrical alkenes and alkynes exhibited a high degree of regioselectivity. The reaction of $[Mn(R)(CO)_5]$ ($R = CH_3$, CH_2Ph) with norbornylene occurs in a similar manner, as did the reaction with cyclopentene (*142*) [Eqs. (38) and (39)].

$$[Mn(R)(CO)_5] + R_3\text{—}=\text{—}R_2 \rightarrow \qquad (36)$$

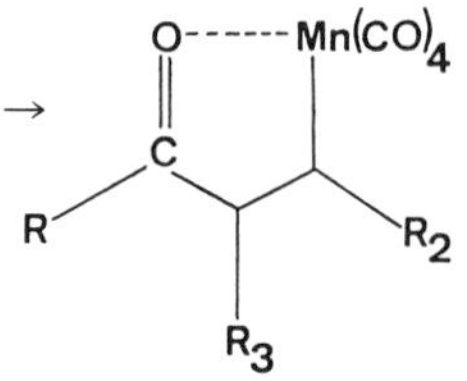

$$[Mn(R)(CO)_5] + R_5\text{—}\equiv\text{—}R_4 \rightarrow \qquad (37)$$

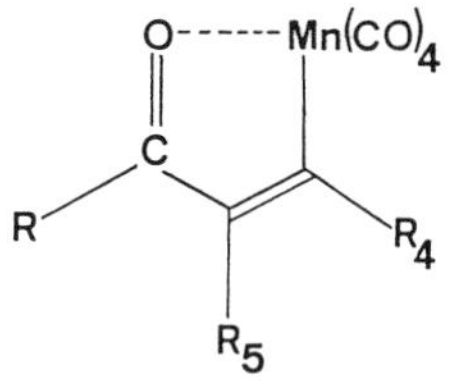

$$[Mn(R)(CO)_5] + \qquad \xrightarrow{\text{3850 atm}} \qquad (38)$$

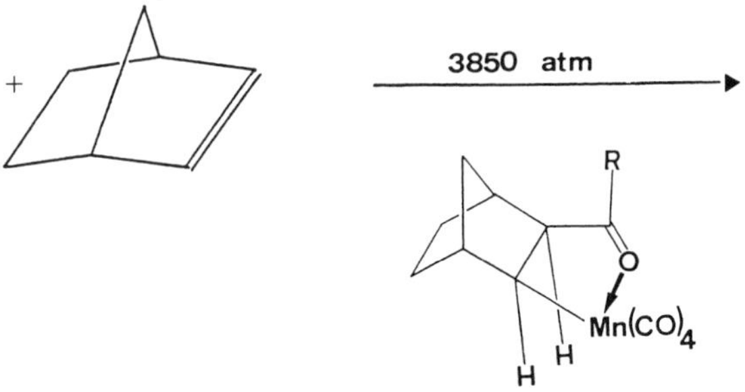

$$[Mn(CH_3)(CO)_5] + \qquad \xrightarrow{\text{3850 atm}} \qquad (39)$$

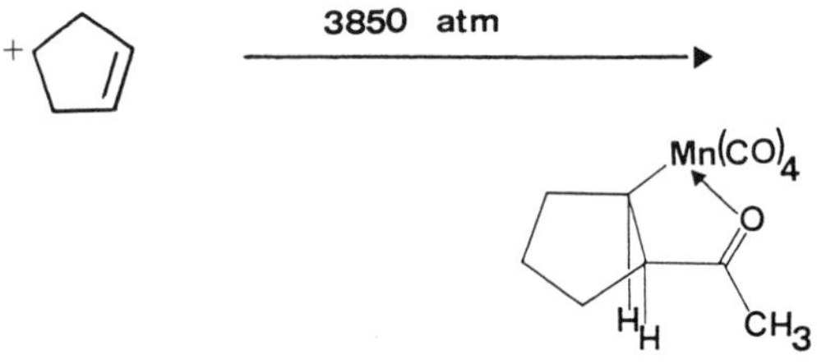

Booth *et al.* (*143*) carried out the reaction of $[Mn(CH_3)(CO)_5]$ with cyclopentadiene and found the reaction to be so slow that dimerization of cyclopentadiene occurred before reaction with $[Mn(CH_3)(CO)_5]$ [Eq. (40)]. The reaction of $[Mn(R)(CO)_5]$ ($R = CH_3$, Ph) with dicyclopentadiene directly, however, also gave a 1:2 adduct [Eq. (41)]. The reaction of $[Mn(CH_3)(CO)_5]$ with norbornadiene, which gave a mixture of the 1:1 and 1:2 adducts, was also reported in this study. Other studies with different alkenes and alkynes gave the same type of products (*144,145*).

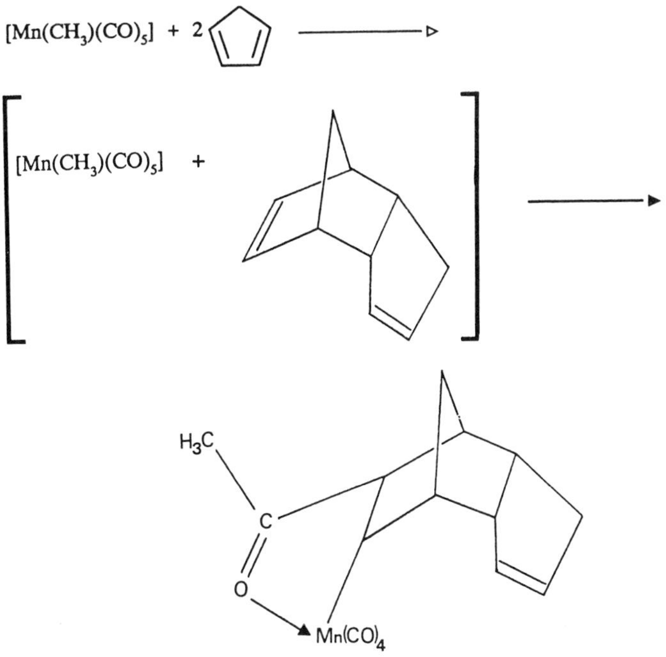

$$[Mn(CH_3)(CO)_5] + 2 \;\; \text{(cyclopentadiene)} \quad \longrightarrow$$

$$\left[[Mn(CH_3)(CO)_5] \; + \quad \right] \quad \longrightarrow$$

$$\overset{H_3C}{\underset{O}{\overset{|}{C}}} \qquad Mn(CO)_4 \qquad (40)$$

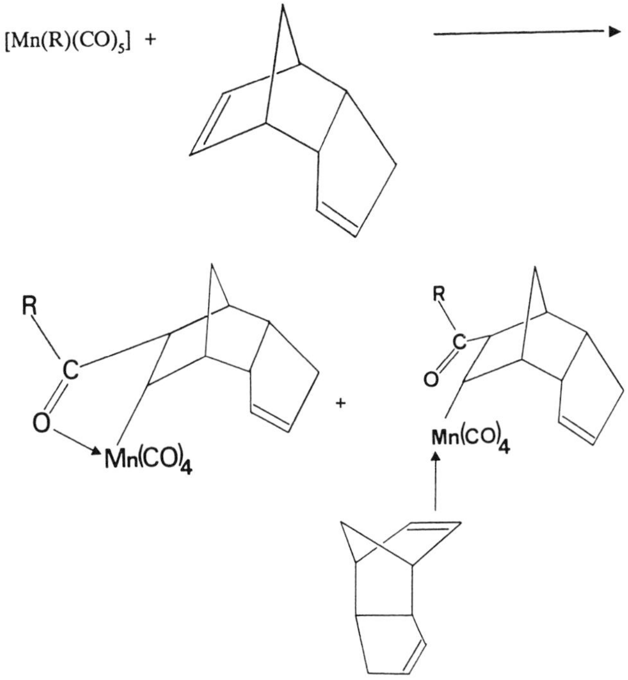

$$[Mn(R)(CO)_5] \; + \qquad \longrightarrow$$

$$\overset{R}{\underset{O}{\overset{|}{C}}} \; Mn(CO)_4 \qquad + \qquad \overset{R}{\underset{O}{\overset{|}{C}}} \; Mn(CO)_4 \qquad (41)$$

b. *Reaction of [Mn(R)(CO)₅] with Synthesis Gas.* The first report on the reaction of synthesis gas (CO/H_2) with alkyl manganesepentacarbonyl species can be considered to be that of King *et al.* in 1978 (*137*). Although they did not use synthesis gas as such (an equimolar mixture of CO and H_2), they reacted [Mn(CH₃)(CO)₅] with CO (320 atm, 67°C), resulting in the formation of the acyl compound, [Mn(COCH₃)(CO)₅]. They then reacted this with H_2 (313 atm, 95°C) (after releasing the CO pressure) to yield [Mn₂(CO)₁₀] and acetaldehyde (a very small amount of formaldehyde was also detected).

In 1979, Dombek (*94*) investigated the reactions of [Mn(R)(CO)₅] [R = H, CH₃, CH₂OCH₂CH₃, CH₂OC(O)CMe₃] with synthesis gas (6.8 atm H_2, 2.0 atm CO) in sulfolane at 75°C. Acetaldehyde was the only organic product from the reaction of [Mn(CH₃)(CO)₅] with synthesis gas; however, when [Mn(CH₃)(CO)₅] was reacted with H_2 only (6.8 atm), ethanol was the only organic product. A high yield of alcohol was also obtained from the reaction of [Mn(CH₂OCH₂CH₃)(CO)₅] with H_2, indicating that the hydroxymethylation of the alkyl group with H_2 under pressure is a general process and that the α-oxygen atom does not have a great influence on the reaction. When this reaction (i.e., with H_2 only) was repeated with the addition of CH₃CHO, the aldehyde was catalytically hydrogenated to CH₃CH₂OH by [Mn(CH₃)(CO)₅]. Dombek proposed that the reaction proceeded via a complexed aldehyde species, and the presence of CO inhibited the hydrogenation of this intermediate.

$$[Mn(COR)(CO)_5] \xrightarrow[\text{163 atm}]{H_2/CO} [Mn\{OC(O)CH_2R\}(CO)_5] \qquad (42)$$
$$R = CH_2Ph, CH_2CH_2Ph$$

Freudenberger and Orchin, in 1982 (*48*), found that syngas can be incorporated into [Mn(COR)(CO)₅] compounds in hexane to give alkoxycarbonyl species [Eq. (42)]. The reaction terminated at this point; the alkoxycarbonyl was inert toward further reaction with CO and H_2. In contrast to the reactions in hexane, the same reactions in sulfolane gave only the aldehydes, RCHO (*48*). The mechanism shown in Scheme 9 for the formation of the alkoxycarbonyl species was proposed. We believe, however, that the final step shown in the above scheme, namely,

$$RCH_2OMn(CO)_5 \xrightarrow{CO} RCH_2O\overset{\overset{\displaystyle O}{\|}}{-}C-Mn(CO)_5$$

is extremely unlikely. It is known that (a) some transition metals are oxophobic in nature, and thus tend not to bind to oxo ligands, and (b) a large, negative ρ^* is a fairly common feature of carbonyl insertion reactions (*112*), that is, the reaction is very susceptible to polar substituent

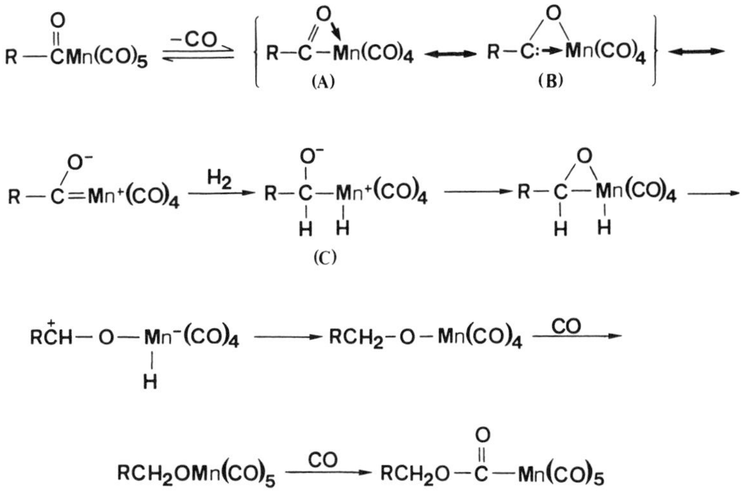

$$R-\overset{\overset{O}{\|}}{C}Mn(CO)_5 \;\underset{}{\overset{-CO}{\rightleftharpoons}}\; \left[R-\overset{\overset{O}{\|}}{C}-Mn(CO)_4 \;\longleftrightarrow\; R-C\overset{O}{\underset{}{\longrightarrow}}Mn(CO)_4 \right] \;\longleftrightarrow$$
$$\hspace{8cm}(A)\hspace{3cm}(B)$$

$$R-\overset{\overset{O^-}{|}}{C}=Mn^+(CO)_4 \;\overset{H_2}{\longrightarrow}\; R-\underset{\underset{H}{|}}{\overset{\overset{O^-}{|}}{C}}-Mn^+(CO)_4 \;\longrightarrow\; R-\underset{\underset{H}{|}}{\overset{\overset{O}{\diagup}}{C}}-Mn(CO)_4 \;\longrightarrow$$
$$\hspace{9cm}(C)$$

$$R\overset{+}{C}H-O-\underset{\underset{H}{|}}{Mn^-}(CO)_4 \;\longrightarrow\; RCH_2-O-Mn(CO)_4 \;\overset{CO}{\longrightarrow}$$

$$RCH_2OMn(CO)_5 \;\overset{CO}{\longrightarrow}\; RCH_2O-\overset{\overset{O}{\|}}{C}-Mn(CO)_5$$

SCHEME 9.

effects. Any oxo ligands, by virtue of the large electronegativity of the oxygen atom (higher than chlorine), will have a large, positive Taft σ^* value, making migration of this ligand a highly unfavorable process.

In sulfolane, intermediate **C** (Scheme 9) may be prevented from forming a manganese–oxygen bond by interaction with a molecule of solvent (structure **D**), collapsing instead to give the aldehyde. Reaction of the related alkyl species, $[Mn(R)(CO)_5]$ (R = CH_2Ph, CH_2CH_2Ph), gave similar results. For example, reaction of $[Mn(CH_2Ph)(CO)_5]$ with CO/H_2 in sulfolane gave $PhCH_2CHO$, and reaction of $[Mn(CH_2CH_2Ph)(CO)_5]$ with CO/H_2 in hexane gave $[Mn\{OC(O)(CH_2)_3Ph\}(CO)_5]$.

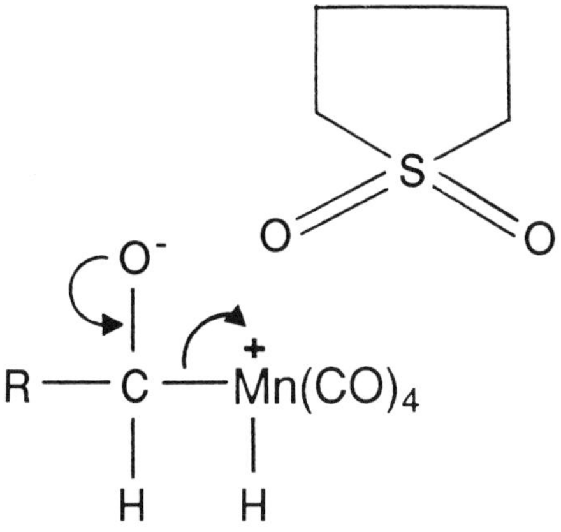

(D)

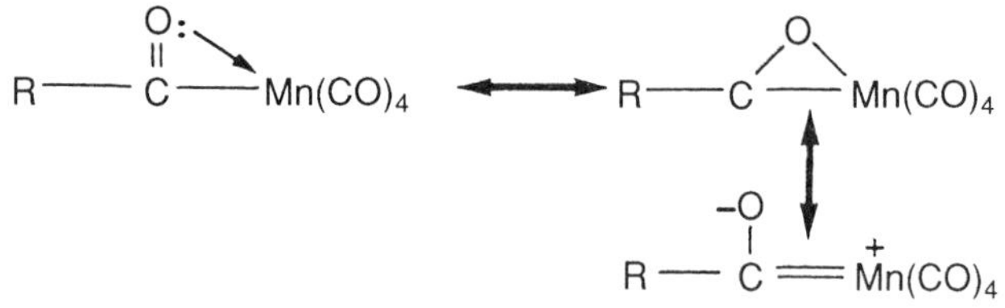

$$R \overset{O:}{\underset{}{\underset{}{C}}} Mn(CO)_4 \longleftrightarrow R \overset{O}{\underset{}{C}} Mn(CO)_4$$

$$R \overset{-O}{\underset{}{C}} \overset{}{=} \overset{+}{Mn}(CO)_4$$

SCHEME 10.

A later study of Orchin and co-workers in 1986 (22) extended this reaction to a wider range of $[Mn(COR)(CO)_5]$ compounds $[R = CH_3,$ $(CH_2)_4CH_3$, p-$CH_3C_6H_4]$, including the binuclear compounds $[(CO)_5Mn$-$(CH_2)_nMn(CO)_5]$ ($n = 4, 8$), and obtained similar results, that is, formation of an alkoxycarbonyl species in hydrocarbon solvents. They also carried out a labeling study with $[Mn(p$-$CH_3C_6H_4{}^{13}CO)(CO)_5]$ and found that the original acyl group in the reactant was reduced exclusively to the methylene group of the product. The formation of an acyl manganese-tetracarbonyl species was proposed as an intermediate, which could adopt a carbenoid structure as shown in Scheme 10.

In a study reported by Mapolie and Moss (23), $[(CO)_5Mn(CH_2)_4$-$Mn(CH)_5]$ was reacted with CO/H_2 (40 atm at 70°C in THF) to give the diol $HOCH_2(CH_2)_4CH_2OH$ as the only isolable organic product. Similar results were obtained from the reaction of synthesis gas with $[(CO)_5$-$Mn(CH_2)_5Mn(CO)_5]$ and $[(CO)_5Mn(CH_2)_6Mn(CO)_5]$, namely diol formation (24). A study by Ahmed et al. involved the reaction of $[Mn(CH_2$-$Ph)(CO)_5]$ with $PhC{\equiv}CPh$ in the presence of CO and H_2 to give an aldehyde and $[Mn_2(CO)_{10}]$ via an unstable acyl intermediate (146).

The most recent study is that of Andersen and Moss (42) which involved the reaction of synthesis gas with (a) $[Mn\{(CH_2)_8CH_3\}(CO)_5]$ in tetrahydrofuran (THF) (40 atm CO/H_2, 55°C) to give 1-decanol, CH_3-$(CH_2)_8CH_2OH$, and (b) $[Mn\{(CH_2)_{12}CH_3\}(CO)_5]$ in hexane (94 atm CO/H_2, 55°C) to give the alkoxycarbonyl compound $[Mn\{OC(O)CH_2(CH_2)_{12}CH_3\}$-$(CO)_5]$. The difference in products was accounted for by the difference in solvents, which could direct the reaction along alternative pathways.

c. SO_2 Insertion. Unlike CO insertion which occurs via alkyl migration, SO_2 insertion is believed to occur by the initial electrophilic attack of SO_2 at the α-carbon atom of the alkyl group. The $[Mn(CO)_5]$ moiety is displaced as a cationic species onto an oxygen atom of the sulfinato group. However, these oxygen-bonded sulfinate complexes easily isomerize to the more thermodynamically stable S-sulfinato complexes (147) [Eq. (43)]. SO_2 cannot be eliminated from these complexes (33), although

[Mn(SO$_2$R)(CO)$_5$] compounds can in turn react with a variety of ligands, resulting in carbonyl substitution (*148*) [Eq. (44)].

$$[Mn(R)(CO)_5] + SO_2 \rightarrow [Mn(SO_2R)(CO)_5] \qquad (43)$$
$$R = CH_3, C_2H_5, CH_2Ph, Ph$$

$$[Mn(SO_2R)(CO)_5] + PPh_3 \rightarrow [Mn(SO_2R)(CO)_4(PPh_3)] + CO \qquad (44)$$
$$R = CH_3, C_6H_5, CH_2C_6H_5$$

d. *Reaction of [Mn(R)(CO)$_5$] with Other Organometallic Species.* Reaction of [Mn(CH$_3$)(CO)$_5$] with η^5-(cyclopentadienyl)carbonyltriphenylphosphine rhodium and -iridium results in carbon–phosphorus bond cleavage in the triphenylphosphine complexes and isolation of bridged diphenylphosphidoheterodinuclear complexes (*149*) [Eq. (45)]. The reaction of [Mn(CH$_3$)(CO)$_5$] with acetylferrocene gave tetracarbonyl-2-(acetyl)ferrocenyl manganese (*150*).

$$[Mn(CH_3)(CO)_5] + [CpM(CO)(PPh_3)] \rightarrow [CpM\text{-}\mu\text{-}\{C(C_6H_5)O\}\text{-}\mu\text{-}\{CMeO\}\text{-}\mu\text{-}$$
$$\{PPh_2\}Mn(CO)_5] + [Cp\{M\text{-}\mu\text{-}(PPh_2)Mn(CO)_4\}] \qquad (45)$$
$$M = Rh, Ir$$

A study by Hart *et al.* (*151*) reported the reaction of [Mn(CH$_3$)(CO)$_5$] with [CpW{≡C(p-MeC$_6$H$_4$)}(CO)$_2$] to give a heterodinuclear complex [Eq. (46)]. The combination of the acyl and alkylidyne moieties took place through carbon–carbon bond coupling to give the bridging C(p-MeC$_6$H$_4$)C(O)Me ligand, which formally donates six electrons.

$$[Mn(CH_3)(CO)_5] + [CpW\{\equiv C(p\text{-MeC}_6H_4)\}(CO)_2] \rightarrow$$

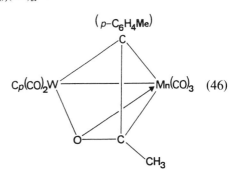

(46)

e. *Miscellaneous Reactions of [Mn(R)(CO)$_5$] with Other Nucleophiles.* There are numerous other reactions of [Mn(R)(CO)$_5$] compounds with other neutral nucleophiles not mentioned so far which have been reported. However, these reactions do not strictly fall within the scope of this article, and hence only brief mention will be made of them. They include (i) the reaction of $[Mn(R)(CO)_5]$ (R = Me, Ph, p-Tol) with CS_2 to give dithiocarboxylate complexes (152,153); (ii) the reaction of $[Mn(CH_3)(CO)_5]$ with boranes to produce hydrocarbons (154); (iii) the reaction of $[Mn(CH_3)(CO)_5]$ with tetrafluoroethylene to give the fluorocarbon-inserted product (155); (iv) the reaction of $[Mn(CH_3)(CO)_5]$ with benzyl methylketone to give $[Mn(CO)_3\{\eta^5\text{-}CHC_6H_5CMeCH(CH_2C_6H_5)O\}]$ (156); and (v) the reaction of $[Mn(CH_3)(CO)_5]$ with N-sulfinylsulfonamides and sulfur bis(sulfonyl)imides to give the inserted chelate products (157).

C. *Reactions of Rheniumpentacarbonyl Alkyls*

Compared to manganese, little work has been carried out on analogous alkyl complexes of rhenium, although that which has may be divided into the same categories, namely, (1) reaction with transition metal hydrides, for example $[CpW(H)(CO)_3]$; (2) reaction with Lewis acids, for example, $AlCl_2Et$; (3) orthometallation reactions; (4) reaction with cationic species, for example, H^+; (5) reaction with anionic nucleophiles, for example, BF_4^-; (6) reaction with neutral nucleophiles, for example, PPh_3.

1. *Reaction of $[Re(R)(CO)_5]$ with Transition Metal Hydrides*

Ethyl rheniumpentacarbonyl has been reacted with various metal hydrides in acetonitrile (158). The observed products were heterobimetallic compounds, although a solvated rheniumtetracarbonyl acyl complex was detected [Eq. (47)]. If the metal hydride is in excess, the rate-determining step is formation of the propionyl complex. The reaction was subsequently found to be first order in both the propionyl complex and the metal hydride. The second-order rate constants were measured and were found to be the reverse of the order of the acidities of the transition metal hydrides, which implies that the hydrides react as nucleophiles with the propionyl complex. In a separate experiment, $[Re(COEt)(CO)_5]$ was found to react with $[Re(H)(CO)_5]$ only after carbonyl dissociation, implying that the metal and not the acyl carbonyl is the site of nucleophilic attack by transition metal hydrides on acyl complexes.

$$[Re(Et)(CO)_5] \xrightarrow{\ CH_3CN\ } [Re(COEt)(CO)_4(CH_3CN)] \xrightarrow{\ M-H\ }$$

$$[Re(M)(CO)_4(CH_3CN)] + EtCHO \xrightarrow{\ CO\ } [Re(M)(CO)_5] + CH_3CN \qquad (47)$$

$$M-H = [Re(H)(CO)_5], [Mn(H)(CO)_5], [CpW(H)(CO)_3], [Os(H)_2(CO)_4]$$

Another study by the same group reported the above reaction in benzene instead of acetonitrile (50). In this case, the solvated species was not observed, and the reaction proceeded directly to the binuclear compounds.

2. Reaction of [Re(R)(CO)₅] with Lewis Acids

McKinney and Stone have isolated a rhenium hydroxycarbene species from the reaction of $[Re(R)(CO)_5]$ with a Lewis acid (*159*) [Eq. (48)]. The hydroxycarbene compound was found to function as a catalyst for olefin metathesis.

$$[Re(R)(CO)_5] + 2AlCl_2Et \rightarrow cis\text{-}[Re(Cl)\{C(R)(OH)\}(CO)_4] + \text{other products} \quad (48)$$
$$R = CH_3, C_6H_5$$

3. Orthometallation Reactions

Methyl rheniumpentacarbonyl reacts with anthraquinone to give an orthometallated product (*160*) [Eq. (49)]. Reaction of methyl rheniumpen-

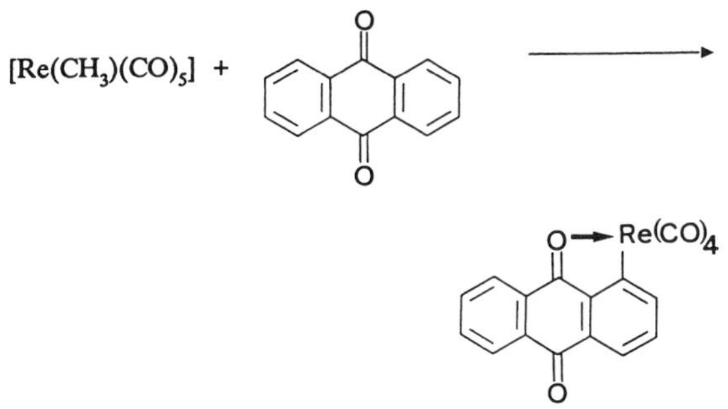

$$(49)$$

tacarbonyl with substituted aceto- and benzophenones also gave ortho-metallated products (*160*), namely,

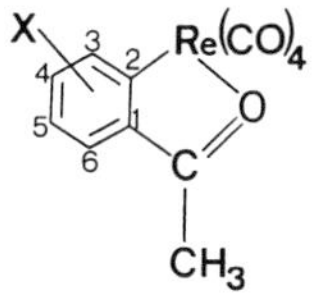

X = 3-, 4-, or 5-CH₃,
3-, 4-, or 5-OCH₃, 4-Cl

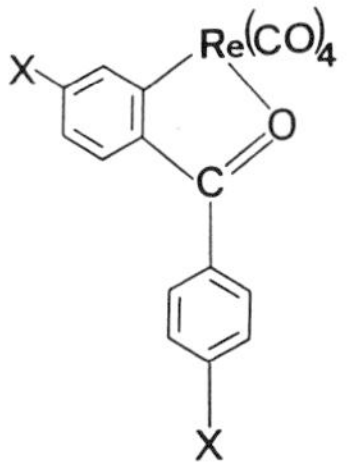

X = H, CH₃,
OCH₃, F, Cl

McKinney and Kaesz reacted $[Re(CH_3)(CO)_5]$ with orthometallation products of alkyl manganesepentacarbonyls to yield bis(tetracarbonyl-metal) secondary metallation products (*161*), as shown in Scheme 11.

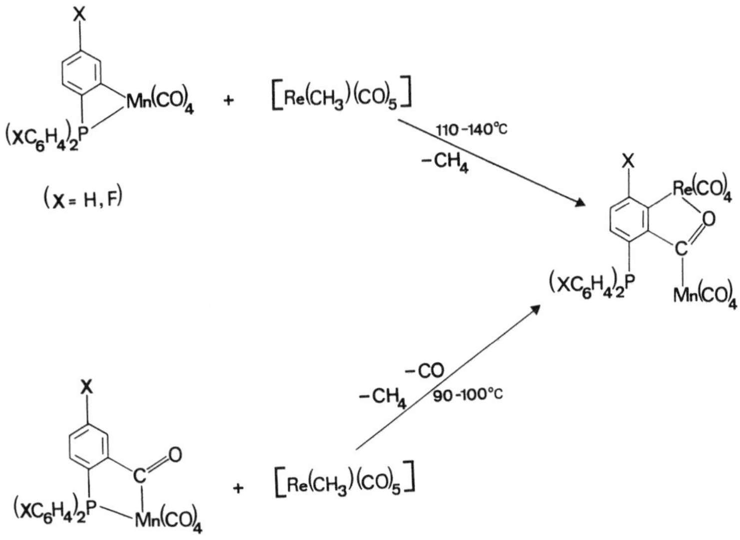

SCHEME 11.

4. Reaction of [Re(R)(CO)$_5$] with Cationic Species

The gase-phase reaction of [Re(CH$_3$)(CO)$_5$] with various proton donors has been reported (85) [Eqs. (50a) and (50b)]. The products of the reaction depend on the nature of B.

$$[Re(CO)_5]^+ + CH_4 + B \qquad (50a)$$

$$[Re(CH_3)(CO)_5] + BH^+$$

$$[Re(H)(CH_3)(CO)_5]^+ + B \qquad (50b)$$

B = NH$_3$, 2-MeTHF, Et$_2$O, THF, Me$_2$CO, Me$_2$CCH$_2$, Me$_2$O, MeCHO, MeOH, MeCHCH$_2$, F$_2$CHCH$_2$OH, HCN, H$_2$S, H$_2$O)

5. Reaction of [Re(R)(CO)$_5$] with Anionic Nucleophiles

Methyl rheniumpentacarbonyl has been subjected to three tetrafluoroboranation studies, all by Beck and co-workers (20,162,163). The first study (162) involved treating the tetrafluoroborate derivative (formed by reaction of [Re(CH$_3$)(CO)$_5$] with triphenylcarbenium tetrafluoroborate) with a variety of σ and π donors, L, forming the salts [Re(L)(CO)$_5$]$^+$BF$_4^-$ [Eq. (51)]. The second study involved the formation of binuclear rhenium compounds (20) [Eq. (52)]. The third study used the tetrafluoroboranation reactions as a route to a polynuclear rhenium compound (163) [Eq. (53)].

$$[Re(CH_3)(CO)_5] + Ph_3CBF_4 \rightarrow [Re(FBF_3)(CO)_5] + H_3CCPh_3 \xrightarrow{L}$$
$$[Re(L)(CO)_5]^+BF_4^-$$ (51)
$$L = Me_2CO, CH_3CN, CO, \text{ethylene, propylene, pentene}$$

$$[Re(CH_3)(CO)_5] + Ph_3CBF_4 \rightarrow [Re(FBF_3)(CO)_5] + H_3CCPh_3 \xrightarrow{CH_2CHR}$$
$$[Re\{CH_2CH(R)\}(CO)_5]^+BF_4^- \xrightarrow{[Re(CO)_5]^-} [(CO)_5Re\{CH_2CH(R)\}Re(CO)_5]$$ (52)
$$R = H, CH_3$$

$$[Re(CH_3)(CO)_5] + Ph_3CBF_4 \rightarrow [Re(FBF_3)(CO)_5] + H_3CCPh_3 \xrightarrow[H_2O]{OH^-}$$
$$[Re(CO)_4(COOH)]_n \xrightarrow[-H_2O]{acetone}$$

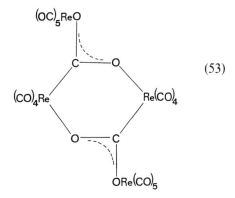

(53)

Reaction with germyllithium reagents can yield CO inserted products (*92*) [Eq. (54)]. Thus methyl rheniumpentacarbonyl reacts with germyl-lithium compounds to give anionic acyl complexes which can be isolated as tetraethylammonium salts or alkylated to give carbene complexes.

$$[Re(CH_3)(CO)_5] + R_3GeLi \longrightarrow Li[Re(COCH_3)(R_3Ge)(CO)_4]$$

Et$_4$NCl   Et$_3$CBF$_4$

$$Et_4N[Re(COCH_3)(R_3Ge)(CO)_4] \qquad [Re\{C(OEt)CH_3\}(R_3Ge)(CO)_4]$$ (54)
$$R_3 = Ph_3 \text{ or MePh-1-naphthyl}$$

6. Reaction of [Re(R)(CO)$_5$] with Neutral Nucleophiles

Alkyl rheniumpentacarbonyl species are known to be reluctant to undergo alkyl migration reactions (*43*); third row *d*-block transition metals (i.e., *5d*) are usually less inclined to undergo alkyl migration than the *3d*

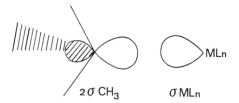

FIG. 4.   Bonding orbitals in [M(CH₃)(CO)₅].

and $4d$ metals of the same subgroup *(164)*. This has been attributed to the higher strength of the metal–carbon bond for $5d$ metals *(49)*.

The increase in bond strength going from $3d$ to $4d$ to $5d$ correlates with an increase in the bonding overlaps between the 2σ CH_3 orbital and the (p, d) hybrid σ MLn orbital on the metal center (in [M(CH₃)(CO)₅]), as shown in Fig. 4. The M–C(methyl) bond dissociation energies in [M(CH₃)(CO)₅] (M = Mn, Re) have been measured and were found to be 117 kJ/mol for M = Mn and 223 kJ/mol for M = Re, indicating a much stronger Re–C(alkyl) bond *(49)*. More recent calculations have been reported by McQuillan *et al.* *(65)*; these results were discussed in Section III.

a. *Reaction of [Re(R)(CO)₅] with Carbon Monoxide.* The reluctance of [Re(CH₃)(CO)₅] to undergo alkyl migration was demonstrated by King *et al.*, who reacted it with CO (320 atm at 200°C) but isolated only [Re₂(CO)₁₀] and unreacted [Re(CH₃)(CO)₅] from the reaction mixture *(137)*. Ethyl rheniumpentacarbonyl is also very resistant to CO-induced alkyl migration; the reaction of [Re(C₂H₅)(CO)₅] with CO was not complete after 1 hour at 100°C and 100 atm of CO *(14)*. However, the reaction of the binuclear species [(CO)₅Re(CH₂)₄Re(CO)₅] was partially successful and gave some acyl product; after reaction with CO (40 atm, 70°C) over 24 hours, 30% conversion to the acyl species, [CO₅ReCO(CH₂)₄CORe(CO)₅], was observed *(23)*.

b. *Reaction of [Re(R)(CO)₅] with SO₂.* SO₂ is known to insert into rhenium–alkyl bonds *(105,165)*; the kinetic products of SO₂ insertion are the O-bonded sulfinate derivatives, [Re{OS(O)R}(CO)₅] *(105)*. The mechanism of this reaction is thought to be the same as for the analogous [Mn(R)(CO)₅] compounds *(43)*. For example, Hartman and Wojcicki *(165)* have isolated the species [Re{S(O)₂R}(CO)₅] [Eq. (55)].

$$[Re(R)(CO)_5] + SO_2 \rightarrow [Re\{S(O)_2R\}(CO)_5] \qquad (55)$$
$$R = CH_3, CH_2C_6H_5$$

c. *Reaction of [Re(R)(CO)₅] with Tertiary Phosphines.* In contrast to the reactions of $[Mn(R)(CO)_5]$ compounds with tertiary phosphines, which yield acyl products, reactions of the analogous $[Re(R)(CO)_5]$ compounds with phosphines give only the phosphine-substituted alkyl species; in other words, the rhenium compounds undergo CO substitution as opposed to alkyl migration. For example, the reaction of the butanediyl compound $[(CO)_5Re(CH_2)_4Re(CO)_5]$ with tertiary phosphines gives the binuclear phosphine-substituted alkyl compounds (*23*) [Eq. (56)].

$$[(CO)_5Re(CH_2)_4Re(CO)_5] + 2PR_3 \xrightarrow[66°C]{THF} cis,cis\text{-}[(PR_3)(CO)_4Re(CH_2)_4Re(CO)_4(PR_3)] \quad (56)$$

$$PR_3 = PPh_3, PPh_2Me, PPhMe_2$$

Anisimov *et al.* (*166*) reported the reaction of $[Re(p\text{-}ClC_6H_4)(CO)_5]$ with PPh₃ to give the aryl product [Eq. (57)]. Wang and Angelici (*167*) have reported the reaction of $[Re(CH_3)(CO)_5]$ with PPh₃ using $[Pt(PPh_3)_4]$ as a catalyst [Eq. (58)]. The phosphine-substituted product was obtained in 86% yield with almost quantitative recovery of the catalyst.

$$[Re(p\text{-}ClC_6H_4)(CO)_5] + PPh_3 \xrightarrow{50–60°C} cis\text{-}[Re(p\text{-}ClC_6H_4)(CO)_4(PPh_3)] \quad (57)$$

$$[Re(CH_3)(CO)_5] + PPh_3 \xrightarrow[benzene]{[Pt(PPh_3)_4]} cis\text{-}[Re(CH_3)(CO)_4(PPh_3)] \quad (58)$$

The most recent study on the reactions of $[Re(R)(CO)_5]$ with a tertiary phosphine is that of Andersen and Moss (*16,42*). This involved reaction of the homologous series of compounds $[Re(R)(CO)_5]$ (R = CH₃ to $n\text{-}C_{18}H_{37}$) with triphenylphosphine [Eq. (59)]. Kinetic studies on the reaction were attempted and showed, on a qualitative basis, that (i) only a slight dependence of reaction rate on alkyl chain length was observed; (ii) this dependence followed the same general pattern as that of the $[Mn(R)(CO)_5]$ system (*42*); (iii) estimates of the rate constants k indicated that they fell in the range of 0.15×10^{-4} to 0.3×10^{-4}/second, spanning a much narrower range than those measured for the $[Mn(R)(CO)_5]$ system (*16–18,42*); and (iv) no evidence was obtained for acyl species being formed in the course of the reactions. This, in addition to fact (i) above, implies that the reactions of $[Re(R)(CO)_5]$ compounds with PPh₃ may therefore be simple substitution reactions in which a terminal carbonyl group is replaced by PPh₃, which most probably proceeds via initial loss of a terminal carbonyl group. An alternative explanation is that acyl species are first formed but then are decarbonylated under the conditions of the experiment, although no evidence could be found for acyl intermediates. Thus, different products were obtained from the reactions of (a) manganese and (b) rhenium long-chain alkyl complexes. This may be largely a result of the much stronger rhenium–carbon (alkyl) bond compared to the manganese–carbon (alkyl) bond (49).

$$[\text{Re(R)(CO)}_5] + \text{PPh}_3 \xrightarrow[\text{toluene}]{94°\text{C}} \textit{cis}\text{-}[\text{Re(R)(CO)}_4(\text{PPh}_3)] \qquad (59)$$

$$\text{R} = \text{CH}_3 \text{ to } \textit{n}\text{-C}_{18}\text{H}_{37}$$

In the above studies (16,42), no evidence for $[\text{Re(C}_2\text{H}_5)(\text{CO})_5]$ undergoing any carbonylation reaction was obtained, which may explain why ethyl rheniumpentacarbonyl is considerably more stable than ethyl manganesepentacarbonyl.

d. *Miscellaneous Reactions.* The reaction of methyl rheniumpentacarbonyl with tetrafluoroethylene has been reported (168) [Eq. (60)]. The fluorocarbon-inserted product is formed.

$$[\text{Re(CH}_3)(\text{CO})_5] + \text{CF}_2\text{CF}_2 \rightarrow [\text{Re(CF}_2\text{CF}_2\text{CH}_3)(\text{CO})_5] \qquad (60)$$

The reaction of methyl rheniumpentacarbonyl with other organometallic compounds is known. For example, reaction with η^5-(cyclopentadienyl)carbonyltriphenylphosphine rhodium resulted in the formation of a bridged diphenylphosphido heterodinuclear complex (149) [Eq. (61)]. The reaction of $[\text{Re(CH}_3)(\text{CO})_5]$ with amino- and phenylazocarboranes has been reported (169) [Eq. (62)]. The reaction of the binuclear complex $[(\text{CO})_5\text{Re(CH}_2)_4\text{Re(CO)}_5]$ with synthesis gas (40 atm, 70°C) was carried out but gave a mixture of unreacted starting material, with minor amounts of $[\text{CO}_5\text{ReCO(CH}_2)_4\text{CORe(CO)}_5]$, 1,6-hexanediol, and the corresponding aldehyde, $\text{OHC(CH}_2)_4\text{CHO}$ (24).

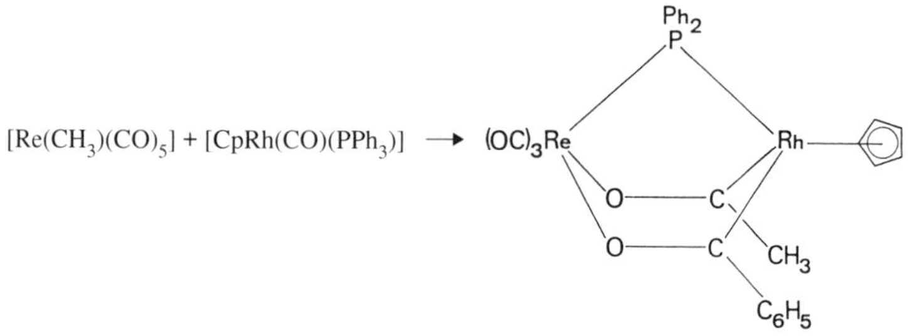

$$[\text{Re(CH}_3)(\text{CO})_5] + [\text{CpRh(CO)(PPh}_3)] \longrightarrow (\text{OC})_3\text{Re} \overset{\overset{\displaystyle \text{Ph}_2}{\underset{\displaystyle}{\text{P}}}}{\diagup} \diagdown \text{Rh} \qquad (61)$$

$$[\text{Re(CH}_3)(\text{CO})_5] + \underset{\text{B}_{10}\text{H}_{10}}{\diagup\diagdown} \overset{\text{H} \quad \text{CH}_2\text{NEt}_2}{\diagdown\diagup} \xrightarrow{100°\text{C}} \underset{\text{B}_{10}\text{H}_9 \longrightarrow \text{Re(CO)}_4}{\diagup\diagdown} \overset{\text{H} \quad \text{CH}_2\text{NEt}_2}{\diagdown\diagup} + \text{CH}_4 + \text{CO} \qquad (62)$$

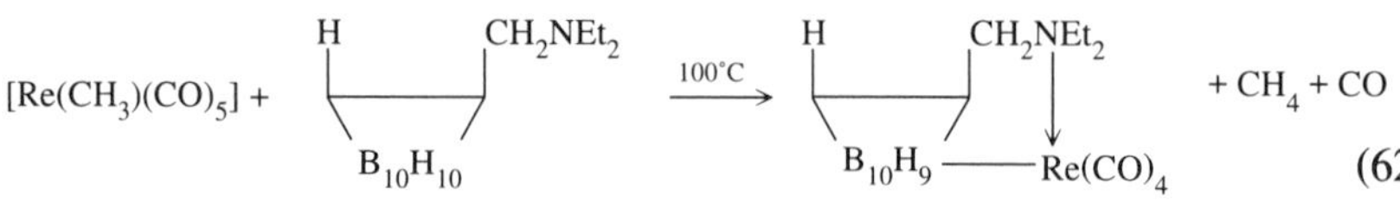

V

CONCLUDING REMARKS

A wide variety of alkyl derivatives of manganese- and rheniumpenta-carbonyl have now been prepared, many of them as recently as the early 1990s. This can partly be attributed to the instability of $[Mn(C_2H_5)(CO)_5]$, which may have deterred researchers from investigating this field any further. The synthetic routes to these alkyl derivatives vary, but the most widely used pathways involve treating $Na[M(CO)_5]$ (M = Mn, Re) with the appropriate alkyl halide, or, in cases where the metalpentacarbonyl anion will not react with the alkyl halide in question, thermal decarbonylation of the acyl derivative.

Studies have reported the synthesis of two extensive homologous series of n-alkyl derivatives $[M(R)(CO)_5]$ (M = Mn, Re; R = CH_3 to n-$C_{18}H_{37}$), and the higher n-alkyl derivatives were found to be relatively stable, both thermally and to air. This would seem to refute the assumptions that the shorter chain manganesepentacarbonyl alkyl compounds were unstable owing to β-hydride elimination. It has been suggested that the reason for the instability of the short-chain alkyl species is in fact an extremely facile carbonylation process (16,18). The analogous rhenium short-chain alkyl compounds are more stable than the manganese analogs, and this may be due to a rapid carbonylation process not being a viable mode of decomposition. This is probably partly due to the rhenium–carbon bond being considerably stronger than the manganese–carbon bond.

Reactions of these alkyl derivatives are now extensive, the most common being reaction with a neutral nucleophile in order to induce alkyl migration. Rates have been measured for the reaction of the series $[M(R)(CO)_5]$ (M = Mn, Re; R = CH_3 to n-$C_{18}H_{37}$) with PPh_3. The results obtained show a dependence of the rates of the reactions on the nature (both steric and electronic) of R, the alkyl group. The effect is much greater for M = Mn than for M = Re. These reactions may have relevance to catalytic reactions such as the Fischer–Tropsch synthesis and hydroformylation (e.g., in a mixture of α-olefins, those of a particular chain length could perhaps be selectively hydroformylated). Thus, reactions of $[M(R)(CO)_5]$ compounds with synthesis gas may have important applications. Partly because of the weaker Mn–C(alkyl) bond compared to the Re–C(alkyl) bond, manganesepentacarbonyl alkyl derivatives have been more widely studied in terms of reactivity. Thus, several new n-alkyl derivatives of manganesepentacarbonyl and rheniumpentacarbonyl have been prepared, the majority of which are stable and exhibit interesting reactivity patterns.

This revisit to the alkyl compounds of the manganese group shows that considerable progress, most of it in the 1990s, has been made since the pioneering paper of Closson *et al.* in 1957. For roughly 30 years, only three *n*-alkyl complexes of manganesepentacarbonyl (and two of rheniumpentacarbonyl) were known. Now, however, the situation has changed, with eighteen *n*-alkyl complexes of each of the two metal carbonyl systems being known. With this in mind, there is considerable scope for the further development of this area of organometallic chemistry. Looking to the future, one can certainly anticipate a variety of exciting and innovative studies on this topic.

REFERENCES

(1) Collman, J. P.; Hegedus, L. S.; Norton, J. R.; Finke, R. G. "Principles and Applications of Organotransition Metal Chemistry"; 2nd ed.; 1987; p.94, University Science Books, Mill Valley, California.
(2) Parshall, G. W.; Ittel, S. D. "Homogeneous Catalysis"; 2nd ed.; Wiley(Interscience): New York, 1992.
(3) Henrici-Olivé, G.; Olivé, S. "Catalyzed Hydrogenation of Carbon Monoxide"; Springer-Verlag; Berlin, 1984.
(4) Roth, J. F.; Craddock, J. H.; Hershman, A.; Paulik, F. E. *Chem. Technol.* **1971**, *1*, 600; Grove, H. D. *Hydrocarbon Process.* **1972**, 76.
(5) Henrici-Olivé, G.; Olivé, S. *Transition Met. Chem. (London)* **1976**, *1*, 77.
(6) Closson, R. D.; Kozikowski, J.; Coffield, T. H. *J. Org. Chem.* **1957**, *22*, 598.
(7) Pope, W. J.; Peachey, S. J. *Proc. Chem. Soc.* **1907**, *23*, 86.
(8) Mendeleyev, D. I. quoted in Hein, F. *Chem. Ber.* **1921**, *54*, 1905; Zeltner, J. *J. Prakt. Chem. Ber.* **1908**, *77*, Chapter 2, 394.
(9) Mulliken, R. S. *J. Phys. Chem.* **1952**, *56*, 295.
(10) Jaffé, H. H.; Doak, G. O. *J. Chem. Phys.* **1953**, *21*, 196.
(11) Calderazzo, F.; Cotton, F. A. *Abstr. Int Conf. Coord. Chem., Stockholm* **1962**, paper 6H7.
(12) Green, M. L. H.; Nagy, P. L. I. *J. Organomet. Chem.* **1963**, *1*, 58.
(13) Hieber, W.; Braun, G. *Z. Naturforsch.* **1959**, *14B*, 132.
(14) Davison, A.; McCleverty, J. A.; Wilkinson, G. *J. Chem. Soc.* **1963**, 1133.
(15) Ziegler, T.; Tschinke, V.; Becke, A. *J. Am. Chem. Soc.* **1987**, *109*, 1351.
(16) Andersen, J. M. Ph.D. Thesis, University of Cape Town, South Africa, 1993.
(17) Andersen, J. M.; Moss, J. R. *J. Organomet. Chem.* **1992**, *439*, C25.
(18) Andersen, J. M.; Moss, J. R. *Organometallics* **1994**, in press.
(19) Beck, W.; Olgemöller, B. *J. Organomet. Chem.* **1977**, *127*, C45.
(20) Raab, K.; Nagel, U.; Beck, W. *Z. Naturforsch.* **1983**, *38B*, 1466.
(21) Mapolie, S. F.; Moss, J. R.; Scott, L. G. *J. Organomet. Chem.* **1985**, *297*, C1.
(22) Sheeran, D. J.; Arenivar, J. D.; Orchin, M. D. *J. Organomet. Chem.* **1986**, *316*, 139.
(23) Mapolie, S. F.; Moss, J. R. *J. Chem. Soc. Dalton Trans.* **1990**, 299.
(24) Mapolie, S. F. Ph.D. Thesis, University of Cape Town, South Africa, 1988.
(25) Lindner, E.; Pabel, M. *J. Organomet. Chem.* **1991**, *414*, C19.
(26) Lindner, E.; Pabel, M.; Fawzi, R.; Mayer, H. A.; Wurst, K. *J. Organomet. Chem.* **1992**, *435*, 109.
(27) Hieber, W.; Braun, G.; Beck, W. *Chem. Ber.* **1960**, *93*, 901.
(28) Noack, K.; Schaerer, U.; Calderazzo, F. *J. Organomet. Chem.* **1967**, *8*, 517.

(29) Nesmeyanov, A. N.; Anisimov, K. N.; Kolobova, N. E.; Ioganson, A. A. *Dokl. Akad. Nauk SSSR* **1967,** *175,* 1293.

(30) Nesmeyanov, A. N.; Anisimov, K. N.; Kolobova, N. E.; Ioganson, A. A. *Dokl. Akad. Nauk SSSR* **1967,** *175,* 358.

(31) Nesmeyanov, A. N.; Anisimov, K. N.; Kolobova, N. E.; Ioganson, A. A. *Dokl. Akad Nauk SSSR, Ser. Khim.* **1968,** 395.

(31) Nesmeyanov, A. N.; Anisimov, K. N.; Kolobova, N. E.; Ioganson, A. A. *Dokl. Akad, Nauk SSSR, Ser. Khim.* **1968,** 892.

(33) Stewart, R. P.; Treichel, P. M. *J. Am. Chem. Soc.* **1970,** *92,* 2710.

(34) Hieber, W.; Wagner, G. *Ann. Chem.* **1958,** *618,* 24.

(35) Hieber, W.; Lindner, E. *Chem. Ber.* **1962,** *95,* 273.

(36) Calderazzo, F., 1990, personal communication.

(37) King, R. B. *Adv. Organomet. Chem.* **1964,** *2,* 157.

(38) Dessy, R. E.; Pohl, R. L.; King, R. B. *J. Am. Chem. Soc.* **1966,** *88,* 5121.

(39) Coffield, T. H.; Kozikowski, J.; Closson, R. D. *J. Org. Chem.* **1957,** *22,* 598.

(40) Motz, P. L.; Sheeran, D. J.; Orchin, M. D. *J. Organomet. Chem.* **1990,** *383,* 201.

(41) Blumer, D. J.; Barnett, K. W.; Brown, T. L. *J. Organomet. Chem.* **1979,** *173,* 71.

(42) Andersen, J. M.; Moss, J. R. submitted for publication.

(43) Treichel, P. M.; Boag, N. M.; Kaesz, H. D. In "Comprehensive Organometallic Chemistry"; Wilkinson, G.; Stone, F. G. A.; Abel, E. W., Eds.; Pergamon: Oxford, 1982; Vol. 4.

(44) Cotton, F. A.; Wilkinson, G. "Advanced Inorganic Chemistry"; 5th ed.; Wiley (Interscience), New York, 1988; p.1120.

(45) Wozniak, P.; Ruddick, J. D.; Wilkinson, G. *J. Chem. Soc. A* **1971,** 3116.

(46) Gismondi, T. E.; Rausch, M. D. *J. Organomet. Chem.* **1985,** *284,* 59.

(47) Bullock, R. M.; Rappoli, B. J. *J. Am. Chem. Soc.* **1991,** *113,* 1659.

(48) Freudenberger, J. H.; Orchin, M. *Organometallics* **1982,** *1,* 1408.

(49) Brown, D. L. S.; Connor, J. A.; Skinner, H. A. *J. Organomet. Chem.* **1974,** *81,* 403.

(50) Warner, K. E.; Norton, J. R. *Organometallics* **1985,** *4,* 2150.

(51) Seip, H. M.; Seip, R. *Acta Chem. Scand.* **1970,** *24,* 3431.

(52) Rankin, D. W. H.; Robertson, A. *J. Organomet. Chem.* **1976,** *105,* 331.

(53) Andrews, M.; Eckert, J.; Goldstone, J. A.; Passel, L.; Swanson, B. *J. Am. Chem. Soc.* **1983,** *105,* 2262.

(54) Evans, S.; Green, J. C.; Green, M. L. H.; Orchard, A. F.; Turner, D. W. *Discuss. Faraday Soc.* **1969,** *54,* 112.

(55) Lichtenberger, D. L.; Sarapur, A. C.; Fenske, R. F. *Inorg. Chem.* **1973,** *12,* 702.

(56) Hall, M. B. *J. Am. Chem. Soc.* **1975,** *97,* 2057.

(57) Guest, M. F.; Higginson, B. R.; Lloyd, D. R.; Hillier, I. H. *J. Chem. Soc. Faraday Trans. 2* **1975,** 902.

(58) Avanzino, S. C.; Jolly, W. L. *J. Am. Chem. Soc.* **1976,** *19,* 6505.

(59) Avanzino, S. C.; Chen, H.-W.; Donahue, C. J.; Jolly, W. L. *Inorg. Chem.* **1980,** *19,* 2201.

(60) Ricco, A. J.; Bakke, A. A.; Jolly, W. L. *Organometallics* **1982,** *1,* 94.

(61) Huang, Y.; Butler, I. S.; Bilson, D. F. R.; LaFleur, D. *Inorg. Chem.* **1991,** *30,* 117.

(62) Orgel, L. E. *Inorg. Chem.* **1962,** *1,* 25.

(63) Beck, W.; Hieber, W.; Tengler, H. *Chem. Ber.* **1961,** *94,* 862.

(64) Long, C.; Morrisson, A. R.; McKean, D. C.; McQuillan, G. P. *J. Am. Chem. Soc.* **1984,** *106,* 7418.

(65) McQuillan, G. P.; McKean, D. C.; Long, C.; Morrisson, A. R.; Torto, I. R. *J. Am. Chem. Soc.* **1986,** *108,* 863.

(66) Gang, J.; Pennington, M.; Russel, D. K.; Davies, P. B.; Hansford, G. M.; Martin, N. A. *J. Chem. Phys.* **1992**, *97*, 3885.
(67) Drew, D.; Darensbourg, M. Y.; Darensbourg, D. J. *J. Organomet. Chem.* **1975**, *85*, 73.
(68) Calderazzo, F.; Lucken, E. A. C.; Williams, D. F. *J. Chem. Soc. A* **1967**, 154.
(69) Webb, M. J.; Graham, W. A. G. *J. Organomet. Chem.* **1975**, *93*, 119.
(70) Mann, B. E.; Taylor, B. F. "¹³C NMR Data for Organometallic Compounds"; Academic Press: London, 1981.
(71) Mays, M. J.; Simpson, R. N. F. *J. Chem. Soc. A* **1967**, 1936.
(72) Schwartz, H. *Acc., Chem. Res.* **1989**, *22*, 282.
(73) Halpern, J. *Acc. Chem. Res.* **1982**, *15*, 322.
(74) Nappa, M. J.; Santi, R.; Halpern, J. *Organometallics* **1985**, *4*, 34.
(75) Butts, S. B.; Strauss, S. H.; Holt, E. M.; Stimson, R. E.; Alcock, N. W.; Shriver, D. F. *J. Am. Chem. Soc.* **1980**, *102*, 5093.
(76) Richmond, T. G.; Basolo, F.; Shriver, D. F. *Inorg. Chem.* **1982**, *21*, 1272.
(77) Nolan, S. P.; de la Vega, R. L.; Hoff, C. D. *J. Am. Chem. Soc.* **1986**, *108*, 7852.
(78) Correa, F.; Nakamura, R.; Stimson, R. E.; Burwell, R. L.; Shriver, D. F. *J. Am. Chem. Soc.* **1980**, *102*, 5112.
(79) Shusterman, A. J.; Tamir, I.; Pross, A. *J. Organomet. Chem.* **1988**, *340*, 203.
(80) Cameron, A.; Smith, V. H.; Baird, M. C. *Organometallics* **1983**, *2*, 465.
(81) Bennett, R. L.; Bruce, M. I.; Matsuda, A. *Aust. J. Chem.* **1975**, *28*, 1265.
(82) McKinney, R. J.; Firestein, G.; Kaesz, H. D. *Inorg. Synth.* **1989**, *26*, 155.
(83) Robinson, N. P.; Main, L.; Nicholson, K. *J. Organomet. Chem.* **1992**, *430*, 79.
(84) Johnson, R. W.; Pearson, R. G. *Inorg. Chem.* **1971**, *10*, 2091.
(85) Stevens, A. E.; Beauchamp, J. L. *J. Am. Chem. Soc.* **1979**, *101*, 245.
(86) Butts, S. B.; Richmond, J. G.; Shriver, D. F. *Inorg. Chem.* **1981**, *20*, 278.
(87) Calderazzo, F.; Noack, K. *J. Organomet. Chem.* **1965**, *4*, 250.
(88) Fischer, E. O.; Aumann, R. *J. Organomet. Chem.* **1967**, *8*, 1.
(89) Moss, J. R.; Green, M.; Stone, F. G. A. *J. Chem. Soc., Dalton Trans.* **1973**, 975.
(90) Casey, C. P.; Cyr, C. R. *Organomet. Chem.* **1973**, *57*, C69.
(91) Gibson, D. H.; Ahmed, F. U.; Hsu, W. L. *J. Organomet. Chem.* **1981**, *215*, 379.
(92) Carré, F.; Cerveau, G.; Colomer, E.; Corriu, R. J. P. *J. Organomet. Chem.* **1982**, *229*, 257.
(93) Wang, P.; Atwood, J. D. *J. Am. Chem. Soc.* **1992**, *114*, 6424.
(94) Dombek, B. D. *J. Am. Chem. Soc.* **1979**, *101*, 6466.
(95) Calderazzo, F.; Cotton, F. A. *Inorg. Chem.* **1962**, *1*, 30.
(96) Bent, T. L.; Cotton, J. D. *Organometallics* **1991**, *10*, 3156.
(97) Bird, C. W. *Chem. Rev.* **1962**, *62*, 283, quoting unpublished results obtained by T. H. Coffield and co-workers.
(98) Noack, K.; Calderazzo, F. *J. Organomet. Chem.* **1967**, *10*, 101.
(99) Flood, T. C.; Jensen, J. E.; Statler, J. A. *J. Am. Chem. Soc.* **1981**, *103*, 4410.
(100) Horton-Mastin, A.; Poliakoff, M.; Turner, J. D. *Organometallics* **1986**, *5*, 405.
(101) McHugh, T. M.; Rest, A. J. *J. Chem. Soc., Dalton Trans.* **1980**, 2323.
(102) Axe, F. U.; Marynick, D. S. *J. Am. Chem. Soc.* **1987**, *110*, 572.
(103) Alonso, F. J. G.; Liamazares, A.; Riera, V.; Vivanco, M.; Diaz, M. R.; Granda, S. G. *J. Chem. Soc., Chem. Commun.* **1991**, 1058.
(104) Noack, K.; Ruch, M.; Calderazzo, F. *Inorg. Chem.* **1968**, *7*, 345.
(105) Wojcicki, A. *Adv. Organomet. Chem.* **1973**, *11*, 87.
(106) Wax, M. J.; Bergman, R. G. *J. Am. Chem. Soc.* **1981**, *103*, 7028.
(107) Cotton, J. D.; Kroes, M. M.; Markwell, R. D.; Miles, E. A. *J. Organomet. Chem.* **1990**, *388*, 133.

(108) Cotton, J. D.; Markwell, R. D. *Organometallics* **1985**, *4*, 937.
(109) Cotton, J. D.; Dunstan, P. R. *Inorg. Chim. Acta* **1984**, *88*, 223.
(110) Tolman, C. A. *Chem. Rev.* **1977**, *77*, 313.
(111) Calderazzo, F.; Noack, K. *Coord. Chem. Rev.* **1966**, *1*, 118.
(112) Cawse, J. N.; Fiato, R. A.; Pruett, R. L. *J. Organomet. Chem.* **1979**, *172*, 405.
(113) Andersen, J. M.; Moss, J. R., 1994. Unpublished results.
(114) Cotton, J. D.; Markwell, R. D. *J. Organomet. Chem.* **1990**, *388*, 123.
(115) Cotton, J. D.; Kimlin, H. A.; Markwell, R. D. *J. Organomet. Chem.* **1982**, *232*, C75.
(116) Cotton, J. D.; Crisp, G. T.; Daly, V. A. *Inorg. Chim. Acta* **1981**, *47*, 165.
(117) Brinkman, K. Ph.D. Thesis, University of Utah, Salt Lake City, 1984.
(118) Dobson, G. R.; Ross, E. P. *Inorg. Chim. Acta* **1971**, *5*, 199.
(119) Clark, H. C.; Hauw, T. C. *J. Organomet. Chem.* **1972**, *42*, 429.
(120) Connor, J. A.; Zafarani-Moattar, M. T.; Bickerton, J.; Saied, N. I.; Suradi, S.; Carson, R.; Taknin, G.; Skinner, H. A. *Organometallics* **1982**, *1*, 1166.
(121) Saddei, D.; Freund, H. F.; Hohlneicher, G. *J. Organomet. Chem.* **1980**, *186*, 63.
(122) Mawby, R. J.; Basolo, F.; Pearson, R. G. *J. Am. Chem. Soc.* **1964**, *86*, 3994.
(123) Bannister, W. D.; Booth, B. L.; Green, M.; Haszeldine, R. N. *J. Chem. Soc. A* **1969**, 698.
(124) Calderazzo, F.; Cotton, F. A. *Chim. Ind. (Milan)* **1964**, *46*, 1165.
(125) Kraihanzel, C. S.; Maples, P. K. *J. Am. Chem. Soc.* **1965**, *87*, 5267.
(126) Kraihanzel, C. S.; Maples, P. K. *Inorg. Chem.* **1968**, *7*, 1806.
(127) Kraihanzel, C. S.; Maples, P. K. *J. Organomet. Chem.* **1969**, *20*, 269.
(128) Mawby, R. J.; Morris, D.; Thorsteinson, E. M.; Basolo, F. *Inorg. Chem.* **1966**, *5*, 27.
(129) Kraihanzel, C. S.; Maples, P. K. *J. Organomet. Chem.* **1976**, *117*, 159.
(130) Motz, P. L.; Ho, D. M.; Orchin, M. *J. Organomet. Chem.* **1991**, *407*, 259.
(131) King, R. B.; Zinich, J. A.; Cloyd, J. C. *Inorg. Chem.* **1975**, *14*, 1554.
(132) King, R. B.; Kapoor, P. N.; Kapoor, R. N. *Inorg. Chem.* **1971**, *10*, 1841.
(133) Rosen, R. P.; Hoke, J. B.; Whittle, R. R.; Geoffroy, G. L.; Hutchinson, J. B.; Zubieta, J. B. *Organometallics* **1984**, *3*, 846.
(134) Vaughan, G. D.; Krein, K. A.; J. A. Gladysz, *Angew. Chem., Int. Ed. Engl.* **1984**, *23*, 245.
(135) Ressner, J. M.; Werneth, P. C.; Kraihanzel, C. S.; Rheingold, A. L. *Organometallics* **1988**, *7*, 1661.
(136) Ruszcyk, R. J.; Huang, B. L.; Atwood, J. D. *J. Organomet. Chem.* **1986**, *299*, 205.
(137) King, R. B.; King, A. D.; Iqbal, M. Z.; Frazier, C. C. *J. Am. Chem. Soc.* **1978**, *100*, 1687.
(138) Keblys, K. A.; Filbey, A. H. *J. Am. Chem. Soc.* **1960**, *82*, 4204.
(139) Kuty, D. W.; Alexander, J. J. *Inorg. Chem.* **1978**, *17*, 1489.
(140) Motz, P. L.; Alexander, J. J.; Ho, D. M. *Organometallics* **1989**, *8*, 2589.
(141) DeShong, P.; Sidler, D. R.; Rybczynski, P. J.; Slough, G. A.; Rheingold, A. R. *J. Am. Chem. Soc.* **1988**, *110*, 2575.
(142) DeShong, P.; Slough, G. A. *Organometallics* **1984**, *3*, 636.
(143) Booth, P. L.; Gardner, M.; Haszeldine, R. N. *J. Chem. Soc., Dalton Trans.* **1975**, 1856.
(144) Booth, B. L.; Hargreaves, H. G. *J. Chem. Soc. A* **1969**, 308.
(145) Booth, B. L.; Hargreaves, H. G. *J. Chem. Soc. A* **1969**, 2766.
(146) Ahmed, I.; Alam, F. R.; Azam, K. A.; Kabir, S. E.; Ullah, S. S. *J. Bangladesh Acad. Sci.* **1992**, *16*, 31.
(147) Hartman, F. A.; Wojcicki, A. *Inorg. Chem.* **1968**, *7*, 1504.
(148) Hartman, F. A.; Wojcicki, A. *Inorg. Chim. Acta.* **1968**, *2*, 351.
(149) Blickensderfer, J. R.; Kaesz, H. D. *J. Am. Chem. Soc.* **1975**, *97*, 2681.

(150) Crawford, S. S.; Kaesz, H. D. *Inorg. Chem.* **1977,** *16,* 3193.
(151) Hart, I. J.; Jeffrey, J. C.; Lowry, R. M.; Stone, F. G. A. *Angew. Chem., Int. Ed. Engl.* **1988,** *27,* 1703.
(152) Lindner, E.; Grimmer, R. *J. Organomet. Chem.* **1970,** *25,* 493.
(153) Lindner, E.; Grimmer, R.; Weber, H. *J. Organomet. Chem.* **1970,** *23,* 209.
(154) Stimson, R. E.; Shriver, D. F. *Organometallics* **1982,** *1,* 787.
(155) Wilford, J. B.; Stone, F. G. A. *Inorg. Chem.* **1965,** *4,* 93.
(156) Bennett, R. L.; Bruce, M. I. *Aust. J. Chem.* **1974,** *28,* 1141.
(157) Leung, T. W.; Christoph, G. G.; Gallucci, J.; Wojcicki, A. *Organometallics* **1986,** *5,* 366.
(158) Martin, B. D.; Warner, K. E.; Norton, J. R. *J. Am. Chem. Soc.* **1986,** *108,* 33.
(159) McKinney, R. J.; Stone, F. G. A. *Inorg. Chim. Acta* **1980,** *44,* 227.
(160) McKinney, R. J.; Firestein, G.; Kaesz, H. D. *Inorg. Chem.* **1975,** *14,* 2057.
(161) McKinney, R. J.; Kaesz, H. D. *J. Am. Chem. Soc.* **1975,** *97,* 3066.
(162) Raab, K.; Olgemöller, B.; Schloter, K.; Beck, W. *J. Organomet. Chem.* **1981,** *214,* 81.
(163) Beck, W.; Raab, K. *Inorg. Synth.* **1990,** *28,* 15.
(164) Calderazzo, F. *Angew. Chem., Int. Ed. Engl.* **1977,** *16,* 299.
(165) Hartman, F. A.; Wojcicki, A. *Inorg. Chem.* **1968,** *7,* 1504.
(166) Anisimov, K. N.; Kolobova, N. E.; Ioganson, A. A. *Izv. Akad. Nauk SSSR, Ser. Khim.* **1969,** 749.
(167) Wang, S.; Angelici, R. J. *Inorg. Chem.* **1988,** *27,* 3233.
(168) Wilford, J. B.; Stone, F. G. A. *Inorg. Chem.* **1965,** *4,* 93.
(169) Kalinin, V. N.; Usatov, A. V.; Zakharkin, L. I. *Obshch. Khim.* **1981,** *51,* 2151.

ADVANCES IN ORGANOMETALLIC CHEMISTRY, VOL. 37

σ,π-*Bridging Ligands in Bimetallic and Trimetallic Complexes*

SIMON LOTZ, PETRUS H. VAN ROOYEN, and RITA MEYER

Department of Chemistry
University of Pretoria
Pretoria 0001, South Africa

I

INTRODUCTION AND SCOPE OF REVIEW

The activation of simple organic molecules by more than one transition metal constitutes an area of research which has grown in interest since the mid 1970s. Transition metal clusters, homonuclear and later heteronuclear compounds, have been widely studied because of the potential for application in catalysis. The chemistry of such complexes has been the topic of many review articles, some of which appeared in this series (*1–5*). The concept of utilizing binuclear complexes with metal–metal bonds for ligand activation has been recognized (*6–8*), and, as stated by Casey and Audett, "there is some hope that a smooth transition from the chemistry of mononuclear compounds to dinuclear compounds to metal clusters to metal surfaces may be found" (*9*).

Parallel to the development of the chemistry of dinuclear transition metal complexes with metal–metal bonds, studies related to bimetallic complexes, hetero- and homonuclear, without metal–metal bonds are appearing in the literature at ever increasing frequency (*10–13*). In these complexes, two or more metals are separated by a common ligand which acts as a bridge between them. Two metal centers acting in a joint fashion could enhance the activation of an organic substrate considerably and in such a manner that their unique effects lead to a transformation that is unachievable by either metal functioning alone. Furthermore, the hypothesis that two metals kept in close proximity by a bridging ligand could react cooperatively with substrate molecules has initiated a very broad interest in ligand systems able to lock two metal centers in such a position (14–16).

Transition metal complexes with bridging ligands and without metal–metal bonds can be conveniently divided into three classes based on the mode of coordination of the ligand to metal. The bridging ligand may be attached to the metal centers by (i) σ,σ, (ii) σ,π, or (iii) π,π bonds. Synthetic challenges associated with the preparation of heterobimetallic complexes are presently being addressed and overcome, adding to novelty and momentum in this area of chemistry. New and unique reaction pathways can be expected when disparate transition metal atoms are bridged by organic substrates (17). Beck and co-workers (18), who are very active in this area, have given a comprehensive account of the status of heterobimetallic complexes and the synthetic methods available for the preparation of compounds incorporating hydrocarbons as bridges between transition metal atoms.

As new synthetic methods of preparing heterobimetallic complexes emerge, selective and regiospecific activation of organic substrates by two or more metals becomes possible, creating opportunities for the synthesis of targeted compounds with useful properties. The importance and exploitation of σ,π coordination of organic moieties have been recognized in many intermediates during conversions or in catalytic processes. Whereas the fluxional behavior of ligands at metal centers has long been investigated (19–21), novel migrations relating to $\sigma \rightarrow \pi$ exchange processes or metal coordination site preference for bimetallic complexes with σ,π-bridging acetylene (22) and σ,π-thiophene ligands have also been observed (23). Another area of significant importance relates to structural features of the bridged organic substrate resulting from steric and electronic properties inflicted by the metal fragments on the bridging ligand (24,25). In many instances the bridging ligand has a conjugated π system, thereby enabling the σ-bonded metal to communicate with the π-bonded metal through π-resonance effects. Heterobimetallic compounds with unsaturated bridging ligands, "electronic bridges" (26,27), have great potential in application to the development of novel organometallic materials which exhibit conducting and optical properties (28,29).

The scope of this article is confined to the synthesis and properties of bimetallic and trimetallic complexes linked by bridging organic substrates via σ- and π-bonding modes. Thus, ligand activation is achieved both in the σ frame and π frame of the molecule. A section of this article dealing with hydrocarbon substrates acting as σ,π-bridging ligands overlaps with the review of Beck and co-workers (*18*). Unlike Beck, who chose to organize material according to reaction type, we have arranged complexes according to the nature of compounds represented.

Bimetallic and trimetallic compounds with σ,π-bridging ligands encompass a general theme in chemistry, which is far too wide to be justifiably dealt with in a single article. The range of complexes, which qualifies for inclusions, is very diverse and often unrelated. For this reason we have decided to limit artificially the scope of the article by excluding certain areas which have been reviewed extensively. Therefore, the following selection criteria were enforced when material was collected.

First, only transition metal complexes are considered, thereby excluding bimetallic and trimetallic compounds with σ- or π-bonded main group metals, for example, $(\eta^5\text{-}C_5H_4ER_3)M(CO)_nX$ (M = Fe, Mo; E = Ge, Sn, Pb; X = Me) (*30*), as well as lanthanides and actinides (*31*).

Second, complexes with metal–metal bonds are excluded from this review article. The presence or absence of a metal–metal bond is not always obvious, and its inclusion or omission is often decided by comparing bond distances and/or counting electrons. However, complexes with metal–metal bonds which relate closely to material used in the text are referred to but not discussed, such as complexes with $\mu_2\text{-}\sigma,\pi$ vinyl (*6*) or $\mu_2\text{-}\sigma,\pi$ acetylide ligands (*32*). Examples where either the σ-bound or π-bound transition metal fragment contains a fragment with a metal–metal bonded species, for instance, $(\eta^5\text{-}C_5H_5)Fe\{\eta^5\text{-}C_5H_4\text{—}CCo_3(CO)_9\}$ (*33*) or $W\{\mu\text{-}\eta^1{:}\eta^2{:}\eta^2\text{-}C(N(CH_2CH{=}CH_2)_2)C{\equiv}CR\{Co_2(CO)_6\}\}(CO)_5$ (*34*), are omitted.

Third, heteroatoms may form part of the π bond, but both the coordinating functionalities must involve a carbon atom. Bimetallic and trimetallic complexes with η^1- and η^2-bridging ligands without carbon atoms, such as X=X dichalcogene ligands (X = S, Se, Te) (*35,36*), diphosphenes, and higher congeners (X = PR, AsR; R = alkyl or aryl) (*37,38*); X≡X (X = N, P, As, Sb, etc.) (*39,40*); pentaphosphaferrocenes such as $(\eta^5{:}\eta^1{:}\eta^1\text{-}cyclo\text{-}P_5\{Cr(CO)_5\}_2)Fe(Cp^*)$ (*41*); metal-containing heterocycles as bridging ligands such as $Cp(CO)_2M\{P(Ph)_2E(R)\}Cr(CO)_5$ (M = Cr, Mo, W; E = N, P, As, Sb; R = alkyl and aryl substituents) (*42*); $Cp^*Rh(\mu,\eta^2{:}\eta^2\text{-}PP_2P)RhCp(CO)$ (*43*), etc., are not considered. [Cp, cyclopentadienyl; Cp′, methylcyclopentadienyl; Cp*, pentamethylcyclopentadienyl.]

Fourth, metallocycles where the metal fragment forms part of a pseudoaromatic ring which is π-bonded to a second transition metal fragment are omitted on the grounds of a metal–metal bond. This group includes metallometallocenes, metalloarenes, ferroles, etc. (*44,45*).

Although selective reference is made to older work, we have concentrated on the literature after 1980 (*45*). The review focuses on various aspects which were brought about by the distinct and selective influences of transition metals on bridging ligands. We hope to demonstrate the richness and potential of the chemistry of bimetallic and trimetallic complexes with σ,π-bridged ligands.

Complexes are arranged according to the σ bond of the bridging ligand. Three classes are recognized for σ-bonded carbon ligands: (i) metal–carbon single bonds (alkyl ligands) (**A**), (ii) metal–carbon double bonds (carbene ligands) (**B**), and (iii) metal–carbon triple bonds (carbyne ligands) (**C**). These classes are subdivided according to the bonding charac-

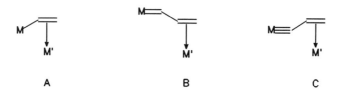

A   B   C

teristics of the π ligand, and complexes are listed according to (a) increasing bond order, (b) relative positions of the σ bond with respect to the π bond, (c) dimerization of metal fragments with σ,π functionalities, and (d) number of π attachments to one metal fragment. Alkene ligands (η^{2n}-coordination mode, $n = 1, 2, \ldots$) precede allylic ligands (η^{2n+1}-coordination mode, $n = 1, 2, \ldots$), which in turn are discussed before alkyne ligands. Rings without a conjugated π system ("open") are separated from those with conjugated π bonds and discussed with the linear counterparts. According to nomenclature used by Davies *et al.* (*46*), rings with conjugated π clouds are termed "closed" and arranged according to increasing ring size.

Metal–carbon π bonds

(a) Bond order

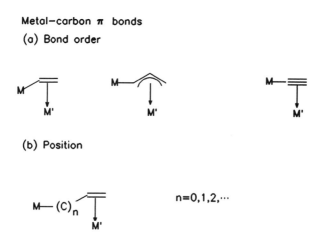

(b) Position

M— (C)$_n$ $n = 0, 1, 2, \cdots$

(c) Dimers

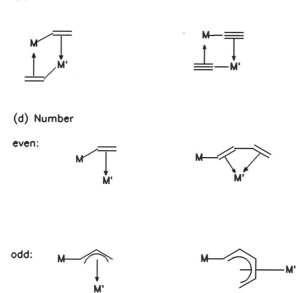

(d) Number

even:

odd:

Bridging ligands with heteroatoms (X) are grouped according to (a) heteroatom attachments as an additional σ-donor ligand (**A**), (b) heteroatoms as part of the π-bonded ligand (**B** and **C**), and (c) heteroatoms

Heteroatom(X) donor ligands

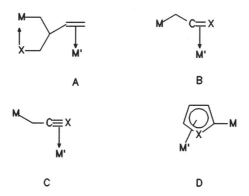

A   B

C   D

which form part of a conjugated π ring system (**D**). Within these guidelines, bimetallic complexes are listed before trimetallic compounds. Numbers given to compounds in the text may be representative of a specific compound or a class of analogous compounds.

II

BRIDGING LIGANDS WITHOUT HETEROATOMS

A. σ-Alkyl/π-Alkene Complexes

Vinyl ligands that bridge two metal centers of bimetallic and trimetallic complexes derived from cluster precursors almost invariably retain metal–metal bonding (**A**) (*6–8,47*). Other arrangements of bimetallic complexes with bridging ethylene ligands, which are also excluded from this review, are vinylidene complexes (**B**) (*48*) and 1,2-η^1:η^1-alkene bridged compounds (**C**) (*49*). In contrast, only a few simple bimetallic complexes without metal–metal bonds and with μ-σ,π-bonded vinyl ligands of types **D** and **E** have been recorded.

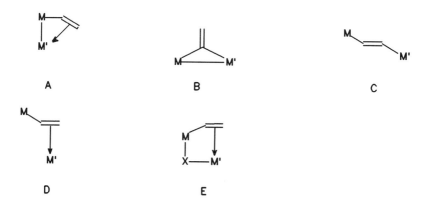

The complex $[Re(CO)_5(\mu$-η^1:η^2-CH=CH$_2)Re(CO)_5]^+BF_4^-$ (**1**) was formed by reacting $Re(CO)_5FBF_3$ with $Me_3SiCH=CH_2$ and probably proceeds through the formation of a σ-vinyl complex $Re(CO)_5\{CH=CH_2\}$, which reacts with a $[Re(CO)_5]^+$ species in a subsequent reaction (*50*). Treatment of $Fe(\eta^5$-C$_5$H$_5)(CO)_2(CH=CH_2)$ with the Lewis acid precursor $[Fe(\eta^5$-C$_5$H$_5)(CO)_2(THF)]^+BF_4^-$ affords a stable bimetallic complex with a bridging σ,π-vinyl ligand, $[Fe_2(\eta^5$-C$_5$H$_5)_2(CO)_4(\mu$-η^1:η^2-CH=CH$_2)]^+BF_4^-$ (**2**) (*51*). Spectral data indicate that the positive charge is delocalized over both iron centers and that **2** is fluxional at room temperature, with the μ-vinyl ligand oscillating between the two metals. Heterobimetallic complexes **3a** and **3b** with a bridging σ,π-ethenyl ligand were synthesized by reacting $W(\eta^5$-C$_5$H$_5)(CO)_3\{CH=CHC(O)Ph\}$ and $Re(CO)_5\{CH=CHC-(O)Me\}$, respectively, with $Fe_2(CO)_9$ [Eq. (1)] (*7,52*). Table I lists selected bimetallic complexes with σ,π-bridged vinyl ligands and Table II lists spectral data of some of these compounds.

TABLE I

SELECTION OF BIMETALLIC COMPLEXES[a] WITH σ,π-BRIDGED VINYL LIGANDS,
$L_nM^1\{\mu\text{-}\eta^1:\eta^2\text{-}(C(R^1)\text{=}CHR^2)\}M^2L_n$

	Complex					
No.	M^1	M^2	R^1	R^2	Method of characterization	Ref.
1	Re	Re	H	H	IR; ^{1}H NMR	(50)
2	Fe	Fe	H	H	IR; ^{1}H, ^{13}C NMR	(51)
3	W	Fe	H	$R^bC(O)$	IR; ^{1}H NMR	(52)
4a	Zr	Zr	H	Ph	IR; ^{1}H, ^{13}C NMR; X-ray	(53)
4c	Zr	Zr	Me	Me	^{1}H NMR; X-ray	(54)
5a,c	Rh	Rh	$C_6H_4R^b$	$C_6H_4R^b$	^{1}H, ^{31}P NMR; X-ray	(56)
6a	Rh	Rh	H	H	^{1}H, ^{13}C, ^{31}P NMR; X-ray	(57)
7	Rh	Rh	H(Et)	Et(H)	^{1}H NMR	(58)
8	Re	Re	H	R^b	MS; IR; ^{1}H NMR	(60)

[a] M^2 represents π-bonded metal.
[b] R represents hydrogen, alkyl, or phenyl substituent.

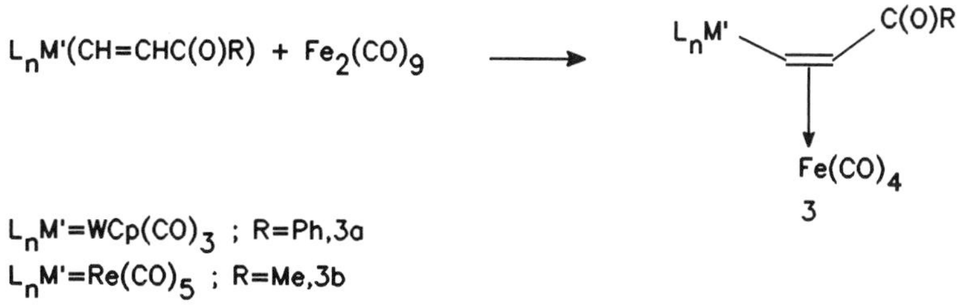

$$L_nM'(CH\text{=}CHC(O)R) + Fe_2(CO)_9 \longrightarrow$$

$$(1)$$

$$L_nM'=WCp(CO)_3 ; R=Ph, 3a$$
$$L_nM'=Re(CO)_5 ; R=Me, 3b$$

The bridging μ-alkenyl linkage is often aided by a second supporting bridging ligand such as H, Cl, and vinylidene ligands. Erker and co-workers (53) studied the reactions of vinylzirconocene complexes with zirconium hydrides. During a slow reaction of (β-styryl)zirconocene chloride with the oligomeric hydridozirconocene chloride, a binuclear complex with bridging σ,π-alkenyl and chloro ligands, $\{ZrCp_2\}_2(\mu\text{-}Cl)(\mu\text{-}\eta^1:\eta^2\text{-}CH\text{=}CHPh)$ (4a), was formed (53). The reactive Zr–C bond of 4a inserts carbon monoxide with the preservation of a binuclear compound. Based on spectroscopic evidence the product that formed on irradiation of $Cp_2(Ph)Zr\{CH\text{=}CHPh\}$ in benzene was formulated as $Cp_2(Ph)Zr\{\mu\text{-}\eta^1:\eta^2\text{-}CH\text{=}CHPh\}ZrCp_2$ (4b). Important features of the molecular structure of 4a are an almost planar bridging olefinic ligand, a nonbonding Zr–Zr distance of 3.498(1) Å, and a very short [2.196(7) Å] metal–carbon σ-bond length. The insertion of $^tBuC\text{≡}CH$ into a Zr–Me bond of Cp_2ZrMe_2 afforded a bimetallic complex (4c) with a $\mu\text{-}\eta^1\text{-}C\text{≡}C^tBu$ and a structurally

TABLE II

SPECTROSCOPIC DATA[a] OF σ,π-BRIDGED VINYL MOIETIES IN SELECTED BIMETALLIC
COMPLEXES, $L_nM^1\{\eta^1:\eta^2-C_1H_1C_2(H_a)H_b\}M^2L_n$

Complex			IR (cm^{-1})		^{1}H/^{13}C NMR (δ, ppm)						
No.	M^1	M^2	ν(C=C)	Medium	H$_1$	C$_1$	H$_a$	C$_2$	H$_b$	Solvent	Ref.
1	Re	Re	1535	KBr	6.81		5.15		5.63	CD$_3$NO$_2$	(50)
2	Fe	Fe	—	—	7.53		3.37		4.67	CD$_3$NO$_2$	(51)
						133.2		63.3		(CD$_3$)$_2$CO	(51)
3b	Re	Fe	1410	KBr	8.85		4.69		4.69	CDCl$_3$	(52)
4a	Zr	Zr	—	—	7.70	170.3	4.56	84.9	4.56	C$_6$D$_6$	(53)
6a	Rh	Rh	—	—	9.48	195.6	4.36	79.7	6.37	C$_6$D$_6$	(57)
8a	Re	Re	—	—	7.18		4.46		6.26	CDCl$_3$	(60)

[a] Assignment according to authors; M^2 represents π-bonded metal.

distorted σ,π-bridging alkene ligand. The structure of **4c** indicates a planar 1,2-dimetalloalkene ligand, in which one Zr center interacts with the methyl substituent and is σ-bonded to the alkene and the other is π-bonded to the carbons of the double bond (54).

Mononuclear metal complexes that catalyze alkyne hydrogenation yield cis-alkenes, and the formation of trans-alkenes is ascribed to a subsequent isomerization step. In contrast, Muetterties and co-workers (55) showed that a catalytic alkyne hydrogenation reaction effected at two or more metal centers has the potential for trans-alkene formation. Diarylalkynes and dialkylalkynes react with $(\mu$-H)(Rh{P(O^{i}Pr)$_3$}$_2$)$_2$ in the absence of hydrogen to form Rh$_2$\{P(O^{i}Pr)$_3$\}$_4$(μ-H)$_2$(μ-η^2-RC$_2$R) and \{Rh$_2$\{P(O^{i}Pr)$_3$\}$_4$(μ-H)$_3$\}(H) (56). In a subsequent reaction of the tetrahydrido complex and the appropriate alkyne, μ-σ,π-vinyl complexes with supporting hydrogen bridging ligands, Rh$_2$\{P(O^{i}Pr)$_3$\}$_4$(μ-H)\{μ-η^1:η^2-C(R)=C(H)R\} (**5**) [R = Me (**5a**), Ph (**5b**), MeC$_6$H$_4$ (**5c**)], were isolated. Similarly, the reaction of the electron-rich binuclear complex \{Rh-(dippe)\}$_2$(μ-H)$_2$ (dippe is 1,2-bis(diisopropylphosphino)ethane] with ethylene afforded a bimetallic rhodium complex with bridging hydride and σ,π-vinyl ligands (**6a**) (57). Reactions of the binuclear rhodium hydride with propene, 1-hexene, cis-2-pentene, and trans-butene generate the analogous σ,π-alkenyl hydrides \{(dippe)Rh\}$_2$(μ-H)\{μ-η^1:η^2-CR^1=CHR^2) (**6**) [R^1 = H, R^2= H (**6a**), Me (**6b**), C$_4$H$_9$ (**6c**), C$_3$H$_7$ (**6d**); R^1, R^2 = Me, Et (**6e**), H, Et (**6f**), Me, Me (**6g**)] in at least two structurally isomeric forms.

The crystal structure determination of Rh$_2$(dippe)$_2$(μ-H)(μ-η^1:η^2-CH=CH$_2$) (**6a**) reveals a disordered vinyl group and an asymmetrically bridging hydride ligand [Rh(1)–H distance 1.84(4) Å and Rh(2)–H dis-

TABLE III

SELECTED X-RAY DATA[a] FOR σ,π-BRIDGED VINYL MOIETIES IN BIMETALLIC COMPLEXES

$$L_nM^1 \underset{M^2L_n}{\overset{C_1=C_2}{\diagup}}$$

Complex	Bond distances (Å)						Bond angles (°)		
No. M^1 M^2	M^1–M^2	M^1–C_x	M^2–C_1	M^2–C_2	M^2C_2–M^2C_1	C_1–C_2	M^1–C_1–M^2	α^d	Ref.
4a Zr Zr	3.498	2.196	2.462	2.440	−0.022	1.42	97.2	143.5	(53)
4c Zr Zr	3.485	2.365	2.486	2.168	−0.318	1.336	91.8	—	(54)
5a Rh Rh	2.889	2.049	2.217	2.294	0.077	1.40	85.2	—	(56)
5c Rh Rh	2.936	2.076	2.226	2.292	0.066	1.40	86.0	—	(56)
6a[b] Rh Rh	2.866	2.091	2.210	2.295	0.085	1.362	83.5	—	(57)
		2.004	2.216	2.314	0.098	1.33			
9c[c] Ru Zr	—	2.095	2.185	2.564	0.359	1.304	—	144	(64)

[a] Standard deviations omitted.
[b] Disordered vinyl group; two positions.
[c] Ligand σ,σ-bridging ethylene, but with agostic Zr–H interaction.
[d] $\alpha = M^1$–C_1–C_2

tance 1.66(4) Å]. Even though complexes **5a**, **5c**, and **6a** are formally without metal–metal bonds, crystal structure data indicate short Rh–Rh distances of 2.889(2), 2.936(1), and 2.866(1) Å, respectively (Table III). The reaction of excess buta-1,3-diene with the coordinated unsaturated binuclear rhodium hydride complex $Rh_2\{P(O^iPr)_2CH_2CH_2P(O^iPr)_2\}_2(\mu\text{-}H)_2$ gave a mixture of the but-1-enyl hydrides $Rh_2\{P(O^iPr)_2CH_2CH_2P\text{-}(O^iPr)_2\}_2(\mu\text{-}H)(\mu\text{-}R^1CH{=}CHR^2)\}$ (**7**) [$R^1 = H$, $R^2 = Et$ (**7a**); $R^1 = Et$, $R^2 = H$ (**7b**)], which were characterized spectroscopically (58). The dimers, in the presence of excess butadiene, subsequently decomposed to a mixture of syn and anti isomers of the mononuclear rhodium allyl complexes.

It has been postulated that the formation of a $\mu\text{-}\eta^1{:}\eta^2$-vinyl ligand in binuclear intermediates involves the activation of a coordinated olefin by one metal center followed by an intramolecular oxidative addition of a

$$L_nM\text{---}M'L_n \xrightarrow{\text{oxidative addition}} L_nM\diagdown_H^{M'L_n} \qquad (2)$$

vinylic C–H bond to an adjacent, coordinatively unsaturated site [Eq. (2)] (59,60). Photolysis of $Re_2(CO)_{10}$ in the presence of substituted alkenes afforded the bimetallic complexes $Re_2(CO)_8(\mu-\eta^1:\eta^2-CH=CHR)(\mu-H)$ (8) [R = H (8a), Me (8b), Et (8c), n-C_4H_9 (8d)], which were formulated with Re–Re bonds (60). Metal–metal bond fissions by fluorinated alkenes have also been recorded, and the insertion of tetrafluoroethylene into an iron–iron bond was studied by Ibers and co-workers (61). They suggested that the thermal rearrangement of $\mu-\eta^1-CF_2=CF_2$ to $\mu-\eta^1-CFCF_3$ proceeds via a $\mu-\eta^1:\eta^2-CF=CF_2$ intermediate, and a related μ-vinyl bridged cation (9a) was obtained after fluoride abstraction with boron trifluoride etherate [Eq. (3)].

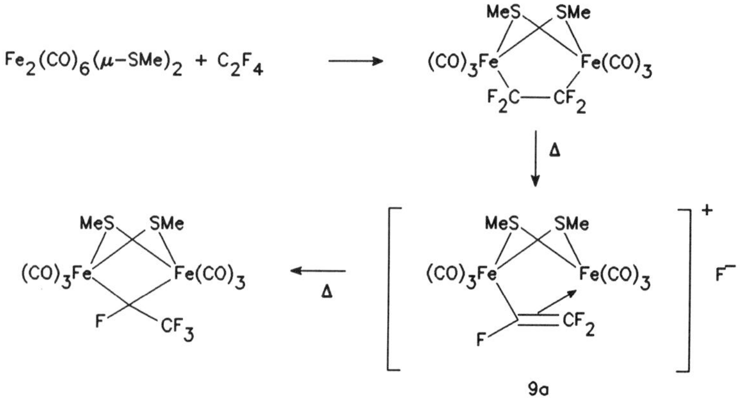

$$Fe_2(CO)_6(\mu-SMe)_2 + C_2F_4 \longrightarrow$$

(CO)$_3$Fe⟨MeS–SMe⟩Fe(CO)$_3$, F$_2$C—CF$_2$

Δ

(3)

$\left[\text{(CO)}_3\text{Fe}\langle\text{MeS–SMe}\rangle\text{Fe(CO)}_3, \text{F}–=\text{CF}_2 \right]^+$ F$^-$

9a

→ Δ → (CO)$_3$Fe⟨MeS–SMe⟩Fe(CO)$_3$, F–CF$_3$

In many instances, based on spectroscopic evidence, bridging $\eta^1:\eta^2$-olefinic intermediates or products have been postulated. The ^1H and ^{13}C NMR data reveal two nonequivalent Fp groups [Fp is CpFe(CO)$_2$] and a bridging ketene moiety for the 1,4-dimetalla-2-butanones, $FpCH_2C(O)ML_n$ [ML_n = CpFe(CO)$_2$, CpMo(CO)$_2$L, Mn(CO)$_5$, CpNi(CO), Co(CO)$_3$L] (62). The infrared data are consistent with a significant contribution of the π complex formulated as A in describing the structure of the heterobimetallic $\mu-\eta^1$-(C,C)-ketene complexes. Complexes containing a C_2 bridge between an electron-rich metal and an electron-deficient metal often exhibit unusual structural and reactivity features. The reaction of $Cp_2Zr(H)Cl$ polymer with $Cp(Me_3P)_2Ru\{CH=CH_2\}$ and $Cp(Me_3P)_2$-$Ru\{C\equiv CH\}$ gave the dimetalloalkane $Cp(Me_3P)_2Ru(\mu-CH_2CH_2)ZrCp_2Cl$ (9b) and the dimetalloalkene $Cp(Me_3P)_2Ru(\mu-CH=CH)ZrCp_2Cl$ (9c) respectively (63,64). Structural and spectroscopic data reveal that these complexes have a three-center, two-electron agostic interaction between the Zr and the proton of the carbon β to the zirconium.

Diallylplatinum reacts with dry hydrogen chloride to give polymeric

A

chloro(allyl)platinum, the crystal structure of which reveals a tetranuclear cyclic assembly of alternating nonplanar σ,π-bonded divinyl and dichloro bridging ligands (65). The reaction of $\{Pt(\eta^3\text{-}C_3H_5)Cl\}_4$ with thallium(I) acetylacetonate gives the dimer $Pt_2(acac)_2(\mu\text{-}\sigma,\pi\text{-}C_3H_5)_2$ (**10**), where each Pt atom is σ-bonded to one and π-bonded to the second vinyl ligand [acac, acetylacetonate.] The structure of **10** reveals a nonbonding Pt–Pt distance of 3.495(2) Å and a significant trans influence in the Pt–O bond lengths with respect to the σ- and π-bonded vinyl group. The Pt–O distances are 2.07(2) Å trans to σ-vinyl and 1.98(2) Å trans to π-vinyl ligand.

Hegedus and Tamura (66) postulated that reactions of acyliron, acylnickel, and cobalt carbonyl anions with η^3-allyl complexes of palladium proceed via unstable bimetallic intermediates with bridging σ,π-alkenyl ligands, $L_n(RC(O))M\{\eta^1\text{:}\eta^2\text{-}CH{=}CHR\}PdL_n$. Products isolated from the acylation of π-allyl ligands were α,β- and α,γ-unsaturated ketones.

The reaction of nucleophiles with cationic polyolefinic complexes has been widely investigated, and addition reactions occur according to the Davies–Mingos–Green rules (46). Beck and co-workers (18,67) have shown that the addition to π-coordinated unsaturated hydrocarbons in cationic complexes of carbonyl metallates, instead of common organic nucleophiles, affords a versatile entry to the direct synthesis of bimetallic and trimetallic complexes with σ,π-bridging ligands. Nucleophilic attack of anionic transition metal carbonyl complexes on cationic allyl complexes was employed to synthesize a large number of stable bimetallic and trimetallic complexes with bridging σ,π-propenyl ligands, $L_nM\{(\mu\text{-}\eta^1\text{:}\eta^2\text{-}C_3H_5)M'L_n\}_m$ (m = 1 or 2). Heterobimetallic complexes $(CO)_5M\{\mu\text{-}\eta^1\text{:}\eta^2\text{-}C_3H_5\}M'L_n$ (**11**) [M = Re, M'L$_n$ = W(CO)$_5$ (**11a**), Fe(CO)$_4$ (**11b**), Ru(CO)$_4$ (**11c**), Os(CO)$_4$ (**11d**), MnCp(CO)$_2$ (**11e**), ReCp(CO)$_2$ (**11f**), CoCp(CO) (**11g**), IrCp*(η^2-C$_2$H$_4$) (**11h**), MoCp(CO)(NO) (**11i**) MoCp'(CO)(NO) (**11j**); M = Mn, M'L$_n$ = ReCp(CO)$_2$ (**11k**); M = FeCp(CO)$_2$, M'L$_n$ = IrCp*(η^2-C$_2$H$_4$) (**11l**)] were directly synthesized from anionic metal substrates ML$_n$ and the cationic allyl precursors L_nM'(allyl)$^+$ [Eq. (4)] (67,68). In contrast to organic nucleophiles which attack the central carbon of the allyl ligand and yield metallocyclobutanes (46,69), organometallic nucleophiles attack the terminal carbon of the allyl ligand, affording bimetallic complexes with bridging σ,π-propenyl ligands (18). The cationic complex $[IrCp^*(\eta^2\text{-}C_2H_4)(\eta^3\text{-}C_3H_5)]^+$ has three different π-coordinated ligands. Whereas the Davies–Green–Mingos rules (46) predict that nucleophilic attack in this system should take place on the ethylene ligand, attack occurs only on the allyl ligand (68). The reaction of carbonylmetallates was shown to proceed stereospecifically, affording only the exo isomer (67,68). A ratio of 1:3.5 for the endo:exo isomers was recorded for the complexes $L_nM\{\mu\text{-}\eta^1{:}\eta^2\text{-}C_3H_5\}IrCp^*(\eta^2\text{-}C_2H_4)$ $[ML_n = Re(CO)_5, FeCp(CO)_2]$ (68).

The NMR data for **11** displayed no evidence of any fluxional behavior for the complexes in solution. Structural features associated with the μ-σ,π-propenyl ligand of **11h** and **11j** are consistent with a strong metal–carbon σ bond, a typical sp^3–sp^2 carbon–carbon single bond distance, and a significantly shorter olefinic carbon–carbon bond which is η^2-bonded to the second metal fragment (67,68).

$$L_nM \overbrace{}^{} R \quad\Bigg]^+ \; + \; M'(CO)_5 \Bigg]^- \longrightarrow \quad L_nM \overbrace{}^{R}\!\!\!\!\!\!\!\!\!\!\!\!\!\!\! M'(CO)_5 \qquad (4)$$

$$\textbf{11}$$

The complex where one fluoro substituent was displaced by a $Mn(CO)_5$ fragment, by the reaction of manganese pentacarbonyl anion and perfluoronorbornadiene, reacts with $Pt(PPh_3)_4$ to afford the Pt–Mn bimetallic complex $(CO)_4(PPh_3)Mn\{\mu\text{-}\eta^1{:}\eta^2\text{-}C_7F_7\}\{Pt(PPh_3)_2\}$ (**9b**) which has a σ,π-bonded bridging fluorinated norbornadiene ligand (70).

The complexes $Fp\{CH_2C(R){=}CH_2\}$ react rapidly with $HgCl_2$ in tetrahydrofuran (THF) to yield 1:1 adducts which convert to the respective PF_6^- salts, $[Fp\{\eta^2\text{-}CH_2{=\!\!=}C(R)CH_2HgCl\}]PF_6$ (**11**) (R = H (**11m**), Me (**11n**)] (71). The reaction of $Fp\{CH_2C{\equiv}CPh\}$ with $HgCl_2$ was slower and was investigated in greater detail. The cleavage of the iron–carbon σ bond was ascribed to an initial attack of the triple bond on the electrophilic $HgCl_2$, affording a σ,π-allene-bridged iron intermediate which rearranges to the final products.

King and Bisnette (72) abstracted a hydride with $Ph_3C^+PF_6^-$ from the β-carbon atom of the σ,σ-propadiyl bridge in the complex $FpCH_2CH_2CH_2Fp$ to give the cationic σ,π-bonded propenyl-bridged complex $[Fp_2\{\mu\text{-}\eta^1{:}\eta^2\text{-}CH_2CH{=}CH_2\}]PF_6$ (**12**). Whereas the spectroscopic data of complexes **11** did not display any fluxional behavior in solution, the corresponding cationic bis(iron) complex **12** belonged to the general class of dinuclear fluxional complexes. Infrared and variable-temperature ^1H and ^{13}C NMR studies of **12** and related complexes indicate asymmetrical hybrid structures (21,73), and an X-ray crystal structure determination reveals a nonclassically bonded rigid π-allylic bridging ligand in the solid state (74,75). The manganese precursor $\{Mn(CO)_5\}_2\{\mu\text{-}1,3\text{-}CH_2CH_2CH_2\}$, on treatment with triphenyl hexafluorophosphate, reacted differently and gave only oxidized products without bridging σ,π-propenyl ligands. Complex **12** unexpectedly formed in high yield from the reaction of 2 equiv of $Ph_3C^+PF_6^-$ with the triiron precursor $\{Cp(CO)_2FeCH_2\}_3CH$ (76). As very little triphenylmethane was detected, an alternative mechanism involving radical species was proposed for the formation of **12**. The stabilized cat-

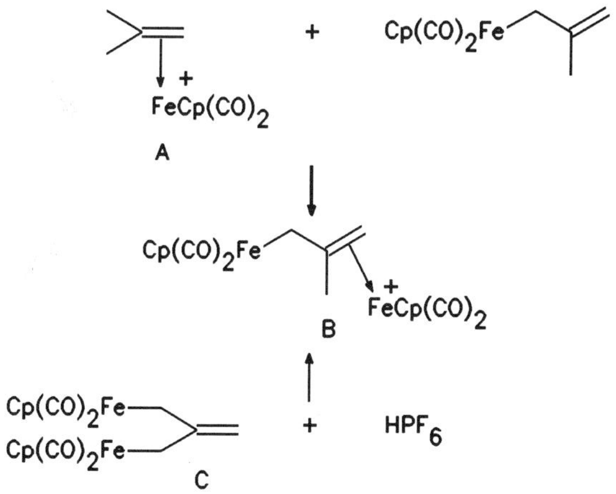

$$(5)$$

ionic complex **B** was isolated by utilizing the less stable precursor **A** as a metal fragment transferring agent [Eq. (5)] (77). Protonation of **C** gives **B**, and the reaction of **A** with **C** affords a cationic trimetallic complex with a σ,σ,π-bonded 2-methylene-1,3-propanediyl bridging ligand whereby the iron fragment of **A** was transferred to the double bond of **C** (76). Other examples of fluxional dinuclear cyclopentadienyliron complexes where **A** was used as a source of Fp^+ are those with σ,π-bound 3-cyclopentene, 3-cyclohexene, and 4-metoxypenta-1,3-diene ligands (77).

The iron methylidene complexes $Cp(CO)_2FeCH_2^+$ were used as alkylating agents toward uncoordinated double bonds in allylic ligands (78). In the reaction of $Cp(CO)_2Fe\{CH_2CH=CH_2\}$ the bimetallic complex $Fp_2(\mu\text{-}\eta^1:\eta^2\text{-}CH_2CH_2CH=CH_2)$ was obtained in low yield. Hydride abstraction from β-carbon atoms with triphenyl hexafluorophosphate of σ,σ-hydrocarbon-bridged bimetallic complexes, $\{FeCp(CO)_2\}_2\{\mu\text{-}(CH_2)_n\}$ ($n = 4–6$), transformed the σ,σ ligand into a σ,π bridge and gave the complexes $[\{FeCp(CO)_2\}_2\{\mu\text{-}\eta^1:\eta^2\text{-}(C_nH_{2n-1})\}]PF_6$ (79). An alternative route to the BF_4^- salt of the cation where n is 5 has been reported for the olefin-coupling reaction of $[FeCp(CO)_2\{CH_2=CH_2\}]^+$ with $FeCp(CO)_2\{CH_2CH=CH_2\}$ (80).

Olefin-coupling reactions of η^1-allyliron complexes with a variety of cationic iron–olefin complexes (ethylene, propene, styrene, etc.) were utilized to give cationic bimetallic complexes with σ,π-hydrocarbon bridges *(80,81)*. The condensation of simple [Fp(olefin)]$^+$ substrates with Fp(allyl) precursors was extended to the reaction with Fp(1,3-butadiene)$^+$. Initial attack at C-1 or C-4 leads to the formation of dinuclear complexes with σ-coordinated and π-coordinated Fp fragments, which by subsequent intramolecular condensation could give either cyclohexenyl or cyclopentenyl intermediates. Attack at C-2 yields a dinuclear complex incapable of further intramolecular reaction [Eqs. (6-8)].

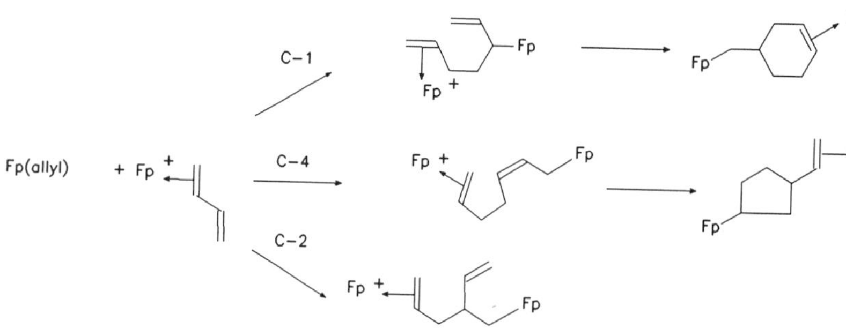

Beck and co-workers studied the reactions of dianionic osmiumtetracarbonyl, [Os(CO)$_4$]$^{2-}$, with 2 equiv of [ML$_n$(η^3-allyl)]$^+$, thereby extending the method of preparing binuclear complexes with alkyl–alkene bridges to trinuclear complexes with two σ,π-hydrocarbon bridges, {L$_n$M(η^1:η^2-CH$_2$CH=CH$_2$)}$_2$Os(CO)$_4$ (13) [ML$_n$ = MoCp(CO)(NO) (13a), Fe(CO)$_4$ (13b), Ru(CO)$_4$ (13c), MnCp(CO)$_2$ (13d), ReCp(CO)$_2$ (13e), IrCp*(η^2-C$_2$H$_4$) (13f)] *(18,68,82)*. The metal–propenyl ligands (L) in the trimetallic complexes 13 are in the cis positions of Os(CO)$_4$L$_2$ and are generally less stable than the bimetallic counterparts. Complex 13a is an unstable oil, and 13b and 13c are orange solids which decompose at 39 and 41°C, respectively. The proton resonances in the NMR spectra for the two σ,π-propenyl bridges gave two separate sets of signals.

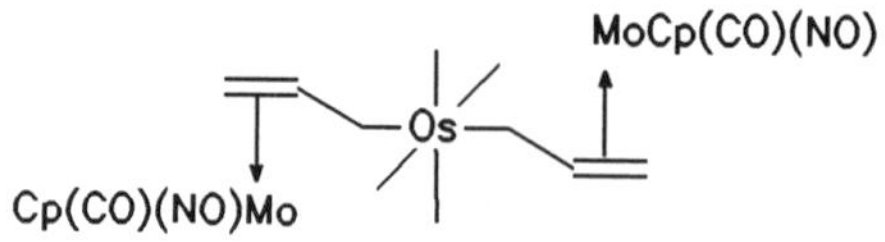

13a

A new mode of bridging for a buta-1,3-dienyl ligand was observed in Rh$_2$(dippe)$_2$(μ-H)(μ-η^1:η^4-C$_4$H$_5$) (14), the product from the reaction of buta-1,3-diene with $Rh_2(dippe)_2(\mu-H)_2$ (*58*). The structure of **14** was confirmed by an X-ray crystal structure determination and reveals a butadienyl unit which bridges the two rhodium centers via an η^4-cis attachment to one of the metals geometrically designated as square pyramidal with the μ-H in the apical position.

A spontaneous electron transfer reaction was observed by Lehmann and Kochi (*83*) on the addition of the cationic electrophilic complex $[Fe(\eta^5$-cyclohexadienyl)(CO)$_3]^+$ to $[Mo(\eta^5$-C$_5$H$_5$)(CO)$_3]^-$. Radical intermediates formed and combined to give the dimeric complexes $Fe_2\{\mu-\eta^4:\eta^4$-(C$_6$H$_7$—C$_6$H$_7$)\}(CO)$_6$ and $Mo_2Cp_2(CO)_6$, which were isolated in almost quantitative yields. In contrast, the open chain analogs $[Fe(\eta^5$-penta/hexadienyl)(CO)$_3]^+$ lead, under identical reaction conditions, to the nucleophilic adducts $Cp(CO)_3Mo(\mu-\eta^1:\eta^4$-CH$_2$-CH=CH—CH=CHR)Fe(CO)$_3$ (**15**) [R = H (**15a**), Me (**15b**)]. A bimetallic intermediate with a bridging $\eta^1:\eta^4$-cyclohexadienyl ligand could be characterized spectroscopically at low temperatures for the cyclohexadienyl precursor but on warming converted to the molybdenum and cyclohexadienyliron dimers.

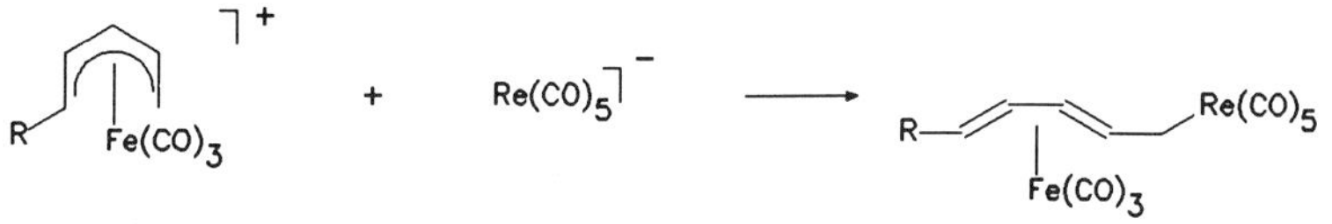

R=H a,Me b

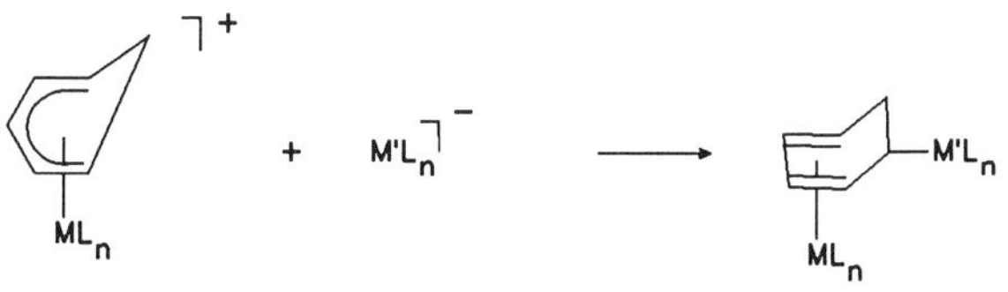

ML_n=Fe(CO)$_3$,M'L$_n$=Re(CO)$_5$ a,Mn(CO)$_5$ b,WCp(CO)$_3$ c
ML_n=Ru(CO)$_3$,M'L$_n$=Re(CO)$_5$ d,e,Mn(CO)$_5$ f,WCp(CO)$_3$ g
ML_n=CoCp,M'L$_n$=Ru(CO)$_5$ h

Beck and co-workers (*84,85*) utilized the anionic carbonylmetallates $[M(CO)_5]^-$ (M = Mn, Re) and $[WCp(CO)_3]^-$ in nucleophilic addition reactions with cationic hexadienyl, cyclohexadienyl, cycloheptadienyl, and cyclooctadienyl complexes of iron and ruthenium to give heterobimetallic complexes with $\eta^1:\eta^4$-hydrocarbon bridges. The reaction of $[Fe(\eta^5$-

$C_5H_6R)(CO)_3]^+$ with $[Re(CO)_5]^-$ afforded the open chain, η^4-butadiene bimetallic complexes $(CO)_5Re\{\mu$-η^1:η^4-CH_2CH=CH—CH=$CHR\}Fe$-$(CO)_3$ (16) [R = H (16a), Me (16b)] [Eq. (9)]. The corresponding cyclo-hexadienyl precursor $[M(\eta^5$-$C_6H_7)(CO)_3]^+$ with $M'L_n^-$ gave the analogous cyclic η^4-hexadiene complexes $L_nM'\{\mu$-η^1:η^4-$C_6H_7\}M(CO)_3$ (17) [M = Fe, $M'L_n$ = $Re(CO)_5$ (17a), $Mn(CO)_5$ (17b), $WCp(CO)_3$ (17c); M = Ru, $M'L_n$ = $Re(CO)_5$ (17d,e), $Mn(CO)_5$ (17f), $WCp(CO)_3$ (17g)] [Eq. (10)]. Interestingly, the σ-bonded rhenium fragment of the bridging cyclohex-adienyl ligand of 17d migrates to a sp^2 carbon to give the thermodynami-cally more stable, more highly substituted diene 17e (85). The complex $(CO)_5Re\{\mu$-η^1:2-5-η^4-$C_6H_7\}CoCp$ (17h) was synthesized by the same method using $CpCo(\eta^5$-cyclohexadienyl$)^+$ (86).

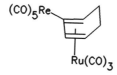

17e

Whereas the bimetallic complex $(CO)_5Re\{\mu$-η^1:2-5-η^4-cyclohepta-diene$\}Fe(CO)_3$ was the major product from the reaction of the cyclohepta-dienyl precursor of iron and pentacarbonylrhenate, only redox products, namely, $Mn_2(CO)_{10}$ and $(\mu$-η^4:η^4-C_7H_8—$C_7H_8)\{Fe(CO)_3\}_2$, were obtained from the similar reaction with $[Mn(CO)_5]^-$ (85). The cationic cyclohepta-dienyl compounds $[Fe(\eta^5$-$C_7H_9)(\eta^4$-diene)(CO)]$^+$ and $[CoCp(\eta^5$-$C_7H_9)]^+$ react with $[Re(CO)_5]^-$ to afford the corresponding η^1:2-5-η^4-cyclohepta-diene-bridged bimetallic complexes $(2$-5-η^4-$C_7H_9\{Re(CO)_5\})ML_n$ (17) [ML_n = Fe(diene)(CO), diene being 2,3-dimethylbutadiene (17i), 1,3-cyclohexadiene (17j), ML_n = CoCp (17k)] (68). Again, the same proce-dure with $[Mn(CO)_5]^-$ yielded only η^4:η^4-coordinated diiron and dicobalt compounds with C–C coupled rings and $Mn_2(CO)_{10}$ (86). The reaction of cationic 1-3-η^3:5-η^2-cyclooctadienyl complexes $[M(\eta^2$:η^3-$C_8H_{11})(CO)_3]^+$ [M = Fe (a), Ru, (b)] with $[Re(CO)_5]^-$ afforded bimetallic complexes 18 with the 3-η^1:1,5-η^4-cyclooctadienyl bridging ligand [Eq. (11)]. The crys-tal structure of 18b indicates a highly distorted boat conformation for the bridging COD ligand (COD is cyclooctadiene). The same regioselectivity whereby attack occurs on the terminal carbon of the allyl function was found for $[Fe\{2$-4-η^4:6-η^2-bicyclooctadienyl\}(CO)_3]^+$ (68,85). The π-bonded cationic bicyclo[5.1.0]octadienyl precursors reacted with $[Re(CO)_5]^-$ to give bicyclic products 18c–f the structure of which reveals that the cyclopropane ring stayed intact and is in the 2 position, which is adjacent to the σ-bonded Re fragment and one of the two π bonds of the $Fe(CO)_3$ unit.

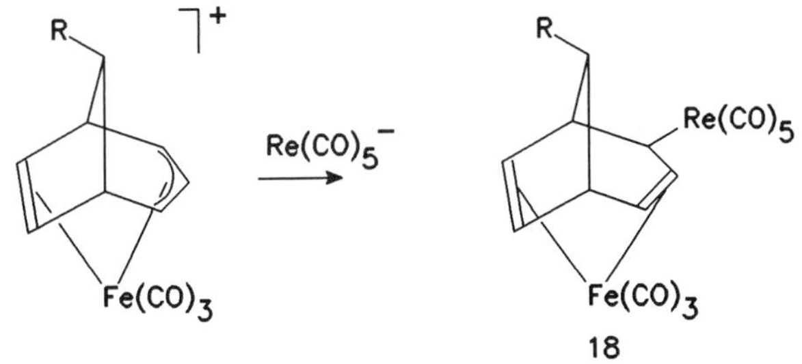

(11)

R=H c,MeC(O) d,EtC(O) e,PhC(O) f

Trimetallic complexes with two identical σ,π-bridging ring ligands having different ring sizes and numbers of π attachments have been studied. The complexes with two bridging cyclohexadienyl (**19a**), cycloheptadienyl (**19b**), and cycloheptatrienyl (**19c–f**) ligands were synthesized by the nucleophilic addition reactions of dianionic carbonylmetallates, $[M(CO)_4]^{2-}$ (M = Os, Ru), and two equivalents of the appropriate cationic metal allyl substrates (*82*).

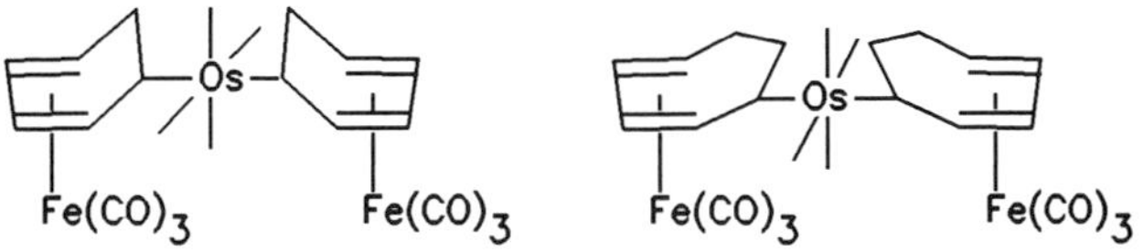

M=Os,M'=Cr c,Mo d
M=Ru,M'=Cr e,Mo f

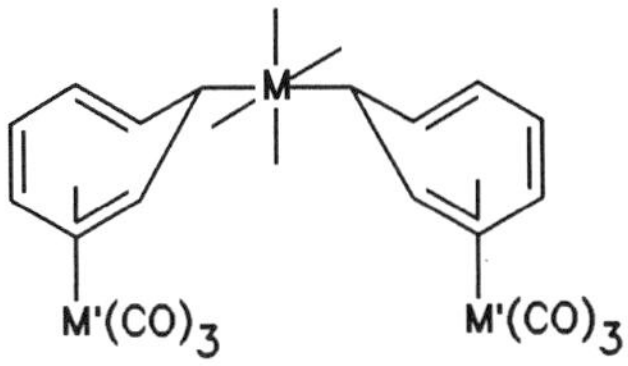

B. σ-Alkyl/π-Allyl Complexes

As far as we are aware, no examples of bimetallic complexes of the structural type **G** exist, where a C_3 allyl ligand acts as an $\eta^1:\eta^3$-C_3H_4 bridging ligand with two transition metals but without metal–metal bonds. Methods for the preparation of such complexes are not obvious and may pose a synthetic challenge. In contrast, many examples of μ-allyl (**B**), μ-

alkylidene (C) and μ-alkenyl (D) complexes with metal–metal bonds have been recorded (6). In many instances unambiguous classification of a bridging ligand such as B, being represented by only one of the structural representations C–E, is very difficult. Nevertheless, such classifications may be useful in describing different aspects of the structures, spectroscopic data, and reactivity patterns of these complexes. Interestingly, one of the structural forms E is formulated without a metal–metal bond and represents an η^4-carbene–alkene ligand (87).

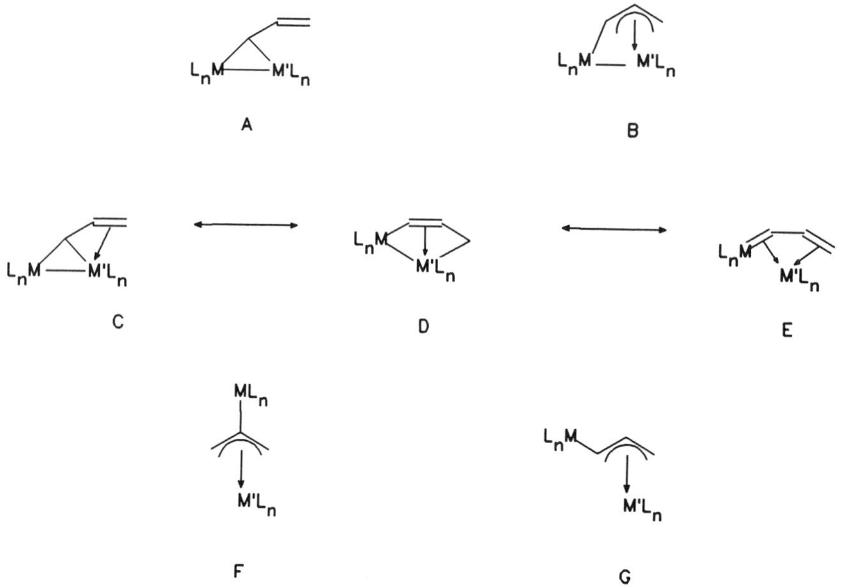

The conjugated alkylidene complexes $W(CO)_5\{=C(OR)CH=CHR'\}$ (R = Me, R' = H, Me, Ph) react with coordinatively unsaturated $W(CO)_5$ [generated from $(CO)_5WCPh_2$ (87) or $(CO)_5W(THF)$ (88)] to give the dinuclear complex $W_2(CO)_9\{\eta^1:\eta^3\text{-}C(OMe)CHCHR'\}$. Although it was not possible to detect or isolate the intermediate X, there is enough literature support for the formation of such a species. The synthesis of heterobimetallic μ-carbene complexes with metal–metal bonds through the addition of coordinatively unsaturated metal fragments to Fischer-type carbene complexes was introduced by Stone and co-workers (89,90) and has been extensively applied to the preparation of many novel cluster compounds. The disappearance of the characteristic carbene carbon signals in the ^{13}C NMR spectra of the tungsten–carbene precursors at around 305 ppm and the appearance of new signals at around 230 ppm are indicative of the formation of bridging alkylidene ligands (87). The $W_2Cp_2(CO)_6$ dimer reacts photochemically with 1,3-butadiene via CO elimination and a 1,4-H shift to give, in very low yield, a bimetallic complex with am η^1:η^3-CHCH=CHCH$_3$ bridging ligand (*91*). A single-crystal structure determination indicated a W–W distance of 3.11(1) Å, which corresponds to a W–W single bond.

X

Examples of bimetallic complexes with and without metal–metal bonds with bridging σ,π-2-allyl ligands (**F**) have been recorded (Tables IV and V) (*92,93*). The bimetallic complexes Cp(^{i}Pr$_3$P)Pd{μ-2-η^1:1-3-η^3-C$_3$H$_4$}Pd-(P^{i}Pr$_3$)X (**20**) [X = Cl (**20a**), Br (**20b**)] are formed through the insertion of allene into a Pd–Pd bond. The structure determination of **20b** reveals two Pd fragments bridged by a bent C$_3$H$_4$ unit and a Pd–Pd distance of 3.575(3) Å, which is significantly longer than the Pd–Pd distance of 2.61 Å found for the precursor. Powell and co-workers (*94*) obtained σ,π-2-allyl bridged Pt–Fe bimetallic complexes of the structure type **F** by adding Pt(PPh$_3$)$_2$(η^2-C$_2$H$_4$), with its labile ethylene ligand, to the allene complex FeCp(PRPh$_2$)(CO)(CH$_2$=C=CH$_2$) to give {η^3-2-C$_3$H$_5${FeCp(CO)(PR$_3$)}Pt(PPh$_3$)$_2$ (**20**) [PR$_3$ = PPh$_2$H (**20c**), PMe$_2$Ph (**20d**)]. The Pt–Fe distance of 3.794 Å in **20d** is clearly nonbonding, the plane of the allyl

TABLE IV

BIMETALLIC COMLEXES[a] WITH σ,π-BRIDGED ALLYL LIGANDS, L$_x$M^1{μ-(X)CR^{1}CR^{2}C(R^3)R^4}M^{2}L$_y$

			Complex					
No.	M^1	M^2	X	R^1	R^2	R^3	R^4	Method of characterization	Ref.
20b	Pd	Pd	H	H	M^{1}L$_y$	H	H	MS; ^{1}H NMR; X-ray	(*93*)
20d	Fe	Pt	H	H	M^{1}L$_y$	H	H	IR; ^{1}H, ^{13}C NMR; X-ray	(*94*)
22a,b	Mn	Mo	CH$_2$	H	H	H	H	IR; ^{1}H NMR	(*98*)
22c	Re	Mo	CH$_2$	H	H	H	Me	IR; ^{1}H NMR; X-ray	(*98*)
23	Re	Mo	H	H	CH$_2$M^{1}L$_y$	H	H	IR; ^{1}H, ^{13}C NMR	(*98*)
24	Pt	Pt	H	H	CH$_2$M^{1}L$_y$	H	H	^{1}H, ^{13}C, ^{31}P, ^{195}Pt NMR	(*101*)
25a	Pt	Pt	CPh	—	Ph	H	Ph	^{1}H NMR; X-ray	(*106*)
28	Fe	Fe	C(O)M^{2}L$_x$	Ph	M^{1}L$_y$	H	H	IR; ^{1}H , ^{13}C NMR; X-ray	(*111*)
29	Pd	Pd	C$_5$H$_8$M^{1}L$_x$	H	H	H	H	IR; ^{1}H NMR	(*112*)

[a] M^2 represents π-bonded metal.

TABLE V

SPECTROSCOPIC DATA[a] OF σ,π-BRIDGED ALLYL MOIETIES IN SELECTED BIMETALLIC COMPLEXES, $L_nM^1\{\mu\text{-}(X)C_1R^1C_2R^2C_3(R^3)R^4\}M^2L_n$

Complex			¹H/¹³C NMR[b] (δ, ppm)							
No.	M¹	M²	X / C₁	R¹	R² / C₂	R³	C₃	R⁴	Solvent	Ref.
20b	Pd	Pd	5.30	4.15	—	2.80		3.70		(93)
20d	Fe	Pt	4.50	3.71	—	2.96		3.61	CD₂Cl₂	(94)
			81.5		219		80.9		CD₂Cl₂	
22c	Re	Mo	2.83	1.92	1.64	1.92		2.83	CD₂Cl₂	(98)
			74.2		125.5		94.2		CD₂Cl₂	
24	Pt	Pt	3.46	2.49	2.46	2.49		3.46	CD₂Cl₂	(101)
			65.4		148		65.4		CD₂Cl₂	
28	Fe	Fe	—	—	—	2.00		2.75	CDCl₃	(111)
			140.2		73.2		54.7		CDCl₃	
29	Pd	Pd	—	4.0	5.43	3.0		4.0	CDCl₃	(112)

[a] Assignments according to authors; M² represents π-bonded metal.
[b] ¹³C NMR data are given under ¹H NMR data.

group is slanted with respect to the PtL_2 fragment, as is typical for Pt–allyl structures, and the iron is displaced out of the plane of the allyl ligand.

The dimetallocycle $Ru_2Cp_2(CO)(\mu\text{-}CO)\{\mu\text{-}C(O)C(Ph)=CPh_2\}$, which is readily obtained by the photochemical reaction of diphenylacetylene with the $Ru_2Cp_2(CO)_4$ dimer, has been reported to undergo quantitative diphenylacetylene exchange when heated with allene in toluene (95). Spectroscopic data suggest a bimetallic product, $Cp(CO)_2Ru\{\mu\text{-}2\text{-}\eta^1\text{:}\eta^3\text{-}CH_2C=CH_2\}RuCp(CO)$ (**21**), which contains an η^3-allyl complex of a $RuCp(CO)$ fragment, with the $RuCp(CO)_2$-fragment as a σ-bound 2-allyl substituent. Thus, in the process of coordinating to the two ruthenium metal atoms the metal–metal bonding is removed by the allene ligand. The reaction of **21** with HBF_4 yields a cationic dinuclear complex with a bridging methylvinyl ligand which, after subsequent treatment with $NaBH_4$, represents a convenient, high yield synthesis of a μ-dimethylcarbene complex, $\{RuCp(CO)\}_2(\mu\text{-}CO)(\mu\text{-}\eta^1\text{-}CMe_2)$.

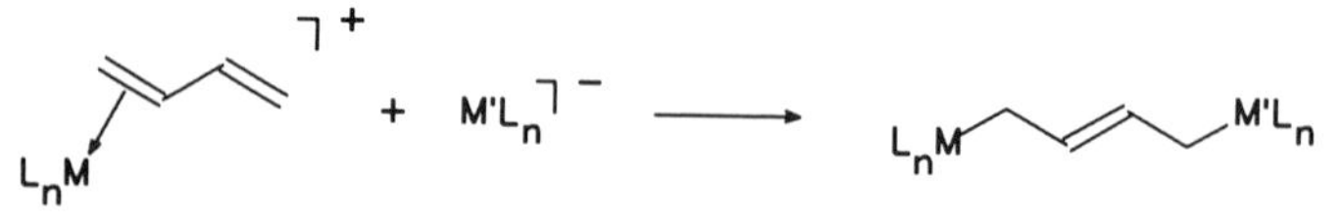

$$L_nM\text{-allyl}^+ + M'L_n^- \longrightarrow L_nM\text{-allyl-}M'L_n \quad (12)$$

A

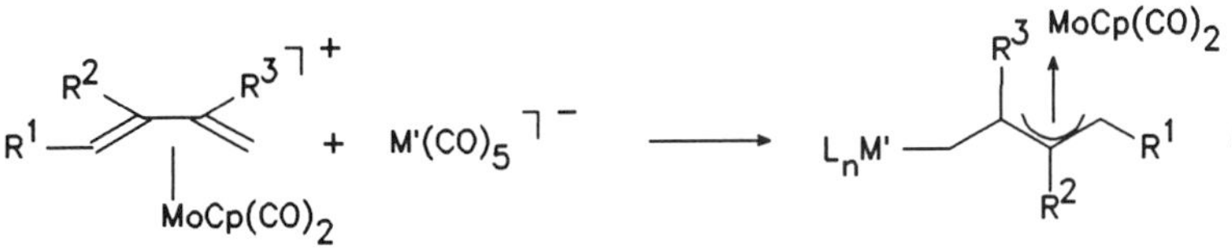

$$ \text{(13)} $$

22	M'	R^1	R^2	R^3
a	Mn	H	H	H
b	Re	H	H	H
c	Re	Me	H	H
d	Re	H	Me	H
e	Re	Me	Me	H

Where the addition of organometallic nucleophiles, $[M(CO)_5]^-$ (M = Mn, Re), to cationic η^2-coordinated 1,3-diene complexes afforded bimetallic complexes with σ,σ-bridging but-2-ene-1,4-diyl ligands (**A**) [Eq. (12)], η^4-coordinated 1,3-dienes yielded bridging $\eta^1{:}\eta^3$-butadienyl ligands (**B**) [Eq. (13)] (96–98). The heterobimetallic complexes $(CO)_5M\{\mu\text{-}\eta^1{:}\eta^3\text{-}CH_2CR^1CR^2CHR^3\}MoCp(CO)_2$ (**22**) [M = Mn (**22a**), Re (**22b**), $R^1 = R^2 = R^3 = H$; M = Re (**22c–e**), R^1, R^2, R^3 = H and/or Me] were synthesized from the reaction of $[Mo(Cp)(CO)_2(\text{diene})]^+$ and $[M(CO)_5]^-$. The crystal structure determination of **22c** (M = Re, $R^1 = R^2 = H$, $R^3 = Me$) indicates that the pentacarbonylrhenate(I) attacked a terminal methylene carbon atom of the diene, forcing it out of the plane of the allyl system and placing the metal fragment in an exo position in the solid state (98).

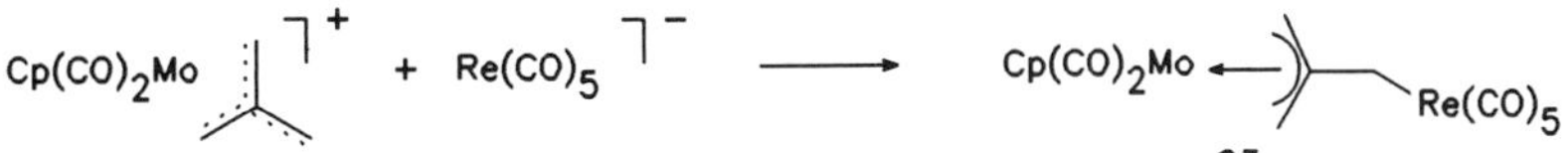

$$ \text{(14)} $$

Cyclopentadiene and 1,3-cyclohexadiene ligands react similarly with pentacarbonylrhenate(I) to give Mo–Re bimetallic complexes with cyclic ligands which are σ-bonded to the rhenium and π-bonded (allylic) to molybdenum (96,98,99). The σ,π-bridged C_4 iso-allyl analog **F**, $(CO)_5Re\{\mu\text{-}\eta^1{:}\eta^3\text{-}CH_2C(CH_2)CH_2\}MoCp(CO)_2$ (**23**), was synthesized by the addition of the organometallic nucleophile to the cationic precursor with a π-coordinated trimethylenemethane ligand [Eq. (14)]. The ^1H and ^{13}C NMR spectra indicate the formation of only one isomeric form for **23**. The mechanism of the reaction for the synthesis of $Fe(CO)_3\{\eta^4\text{-}C(CH_2)_3\}$ is considered to involve initial formation of a (2-chloromethylallyl)iron tricarbonyl chloride followed by the oxidative addition of the C–Cl bond to a coordinatively unsaturated $Fe(CO)_4$ species to give the σ,π-bonded intermediate $(CO)_4ClFe\{\eta^1:\eta^3\text{-}CH_2C(CH_2)CH_2\}Fe(CO)_3Cl$ (*100*). The complex **23** with a metal fragment as a σ-substituent in the 2 position of the allyl ligand has a greater stability in solution and in the solid state compared to the terminally substituted allyl complexes (**22**). Kemmit and Platt (*101*) studied reactions involving the reactive carbon–chloride bond in $Pt\{\eta^3\text{-}CH_2C(CH_2Cl)\!=\!CH_2\}(PPh_3)_2$. When treated with $Pt(C_2H_4)$-$(PPh_3)_2$ in the presence of KPF_6, the C–Cl bond was oxidatively cleaved, and a cationic dinuclear complex $[(Ph_3P)_2ClPt\{\eta^1:\eta^3\text{-}CH_2C(CH_2)\!=\!CH_2\}$-$Pt(PPh_3)_2]PF_6$ (**24**) formed.

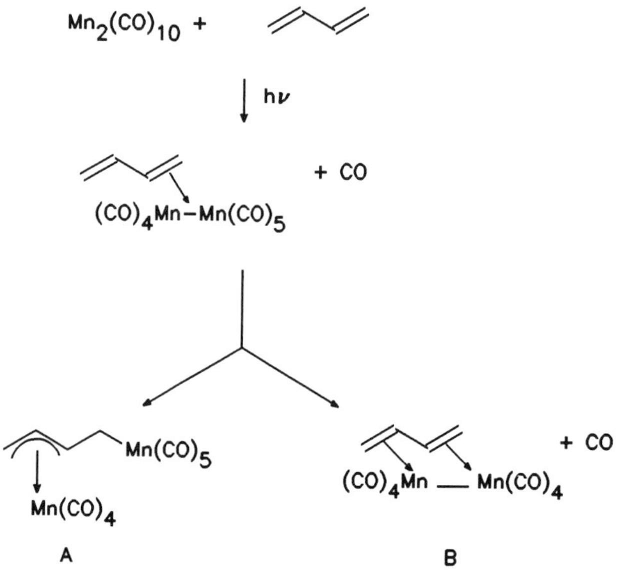

$$Mn_2(CO)_{10} \ +$$

$$\Big\downarrow \ h\nu$$

$$(CO)_4Mn\text{–}Mn(CO)_5 \qquad +\ CO$$

(15)

A   B

$$+\ CO$$

$$Mn(CO)_5$$
$$\Big\downarrow$$
$$Mn(CO)_4$$

$$(CO)_4Mn \ \text{—} \ Mn(CO)_4$$

The photochemical reaction of $Mn_2(CO)_{10}$ and butadiene gives as the major reaction products **A** and **B**, which could be separated by HPLC [Eq. (15)] (*102–104*). The splitting patterns and coupling constants of the six protons in the ^1H NMR spectrum of **A**, which reveal an η^3-enyl ligand with a methylene group in an *E* position, which together with the infrared and ^{13}C NMR data are consistent with the formulation given for **A**. Cleavage of the metal–metal bond by *trans*- or *cis*-1,3-pentadiene under photochemical reaction conditions gives two tetracarbonyl-η^3-enylmanganese complexes for each isomer (*105*). In both cases the major product, a mononuclear 2-4-η^3-pentadienyl complex, could be separated by HPLC from the 1-3-η-pentenyl derivatives. A dinuclear intermediate $(CO)_4Mn\{\mu\text{-}1\text{-}3\text{-}\eta^3\text{-}C_5H_8\}Mn(CO)_5$ with a σ,π-bridging 3-penten-2,5-diyl ligand was postulated, which, after β-hydrogen elimination, accounts for the formation of the pentadienyl complex and $HMn(CO)_5$.

TABLE VI

SELECTED X-RAY DATA[a] FOR σ,π-BRIDGED ALLYL MOIETIES IN
BIMETALLIC COMPLEXES

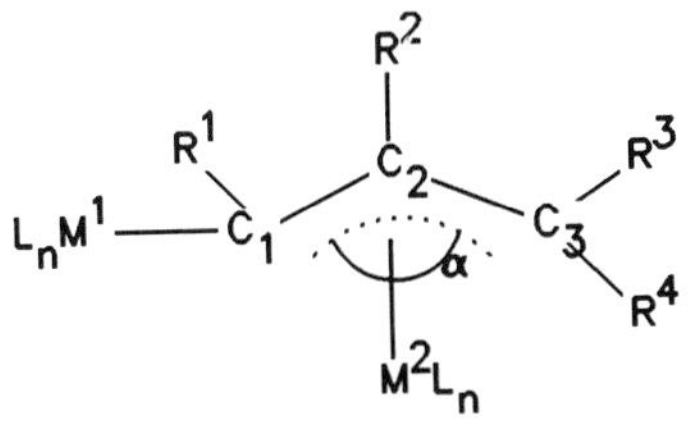

Complex		Bond distances (Å)								
No.	M^1	M^2	M^1–M^2	M^1–C_x	M^2–C_1	M^2–C_2	M^2–C_3	C_1–C_2	C_2–C_3	Bond angle α (°)	Ref.
20b	Pd	Pd	3.575	2.04	2.22	2.17	2.16	1.41	1.47	116.4	(93)
20d	Fe	Pt	3.794	1.967	2.151	2.272	2.165	1.421	1.464	—	(94)
22c	Re	Mo	—	2.268	2.378	2.210	2.433	1.430	1.488	126.4	(98)
25a	Pt	Pt	—	2.10	2.17	2.23	2.18	1.38	1.43	114	(106)
25b	Pt	Pt	—	2.09	2.18	2.23	2.18	1.41	1.47	113	(107)
28	Fe	Fe	—	—	2.110	2.194	2.133	1.434	1.417	111.5	(111)

[a] Standard deviations omitted; M^2 represents π-bonded metal.

The reaction of a cationic diplatinum μ-phenylethenylidene complex with an almost stoichiometric amount of diphenylacetylene and di-*p*-tolyl-acetylene under photochemical conditions gives the cationic diplatinum μ-η^1:η^3-butadienediyl complexes [(Et₃P)₂(PhC≡C)Pt{μ-η^1:η^3-C(Ph)= CC(R)CHR}Pt(PEt₃)₂]⁺ (25) [R = Ph (25a), *p*-tolyl (25b)] in very high yield (106,107). In this reaction the activated alkenylidene couples by C–C bond formation with the alkyne to give an expanded, unsaturated bridging butadienediyl ligand. The location of the two *p*-tolyl groups on adjacent carbon atoms of the butadienediyl skeleton is consistent with incorporation of the alkyne unit as an intact species (107). The crystal structure determination (Table VI) confirms a molecular formulation with the σ,π-bridging butadienediyl ligand between the two Pt moieties.

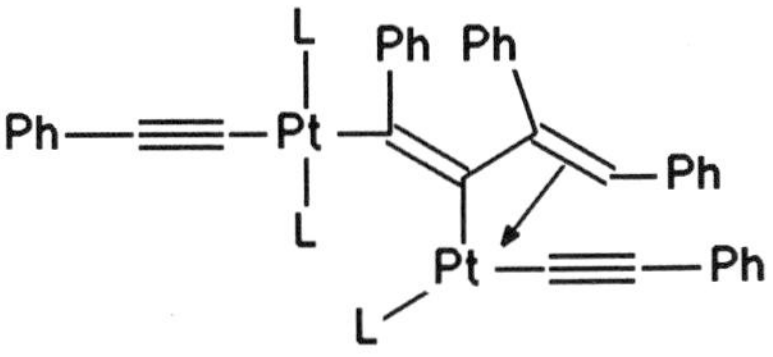

25a

The reaction of Ni(COD)(PMe$_3$)$_2$ with equimolar amounts of BrCH$_2$C$_6$H$_4$-o-Br affords $trans$-Ni(CH$_2$C$_6$H$_4$-o-Br)(PMe$_3$)$_2$Br, which oxidatively adds to a second equivalent of the Ni precursor to yield the binuclear $cis,trans$-Br(Me$_3$P)$_2$Ni{μ-η^1:η^1-CH$_2$-o-C$_6$H$_4$}Ni(PMe$_3$)$_2$ (108). This dinuclear Ni complex exists only in solution, and after removal of the solvent, resulting in the concomitant loss of a trimethylphosphine ligand, a dinuclear complex with a bridging ligand containing a σ-aryl and a π-pseudoallyl functionality (26) is formed. Complex 26 reacts with CO by insertion into the σ bond to give a σ,σ-bonded metallocyclic dinickel complex with an aroyl ligand. Whereas complex 26 reacts with dppm [dppm is bis(diphenylphosphino)methane] to displace a PMe$_3$ and a Br ligand to give a cationic dinuclear complex with both metal–ligand coordination modes intact, the less bulky dmpm [dmpm is bis(dimethylphosphino)methane] yields Br(dmpm)Ni{μ-η^1:η^1-CH$_2$-o-C$_6$H$_4$}Ni(dmpm)Br.

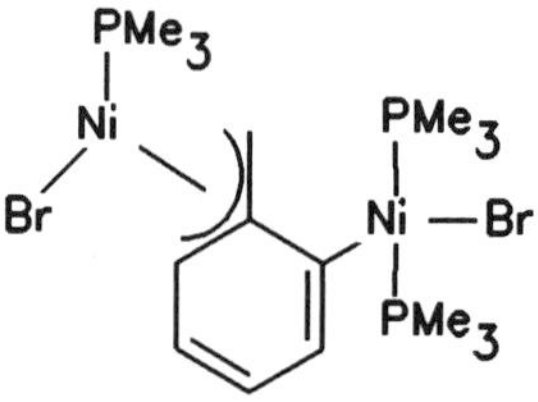

26

The tungsten and molybdenum propargyl complexes Cp(CO)$_3$M-{CH$_2$C≡CH} react with AlCl$_3$ to give products with β-chlorocarbonyl allyl ligands which were subsequently treated with [MCp(CO)$_2$L]$^-$. The reactions proceed via C–C bond formation to give dinuclear complexes of tungsten and molybdenum where the β-carbonylallyl ligand is σ-bonded to the WCp(CO)$_2$L fragment and η^3-bonded to the MCp(CO)$_2$ fragment (27) [M = W, L = CO (27a), P(OMe)$_3$ (27b); M = Mo, L = CO (27c)]. (109).

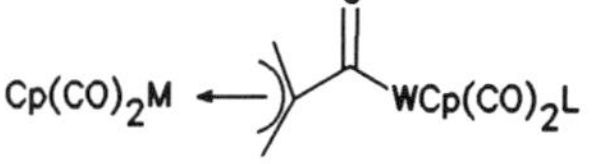

27 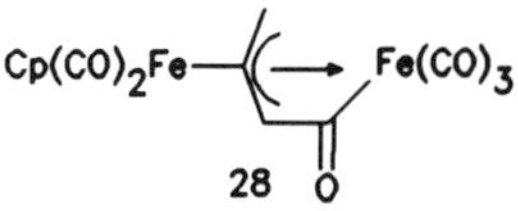   28

Dinuclear metal complexes with σ,π-bridging α-carbonylallylic ligands (28) have also been synthesized and form part of the rich chemistry of transition metal propargyl complexes (103). Propargyl ligands in the complexes L$_n$MCH$_2$C≡CR, are rapidly converted to allenyl ligands by electrophilic reagents (E$^+$) to give cationic complexes, [L$_n$MCH$_2$=C=

C(E)R]$^+$. This method was extended by Wojcicki and co-workers (*110,111*) to utilize coordinatively unsaturated metal fragments as electrophiles, whereby new synthetic methods for homo- and heterobinuclear and trinuclear metal–μ-allenyl complexes were developed. The difference between this approach and that of Stone and co-workers (*89,90*) is that the unsaturation is contained entirely within the ligand, which, as a result of this, often rearranges. Complex **28** was obtained in high yields from the reaction of the iron–propargyl precursor and diiron nonacarbonyl (*111*).

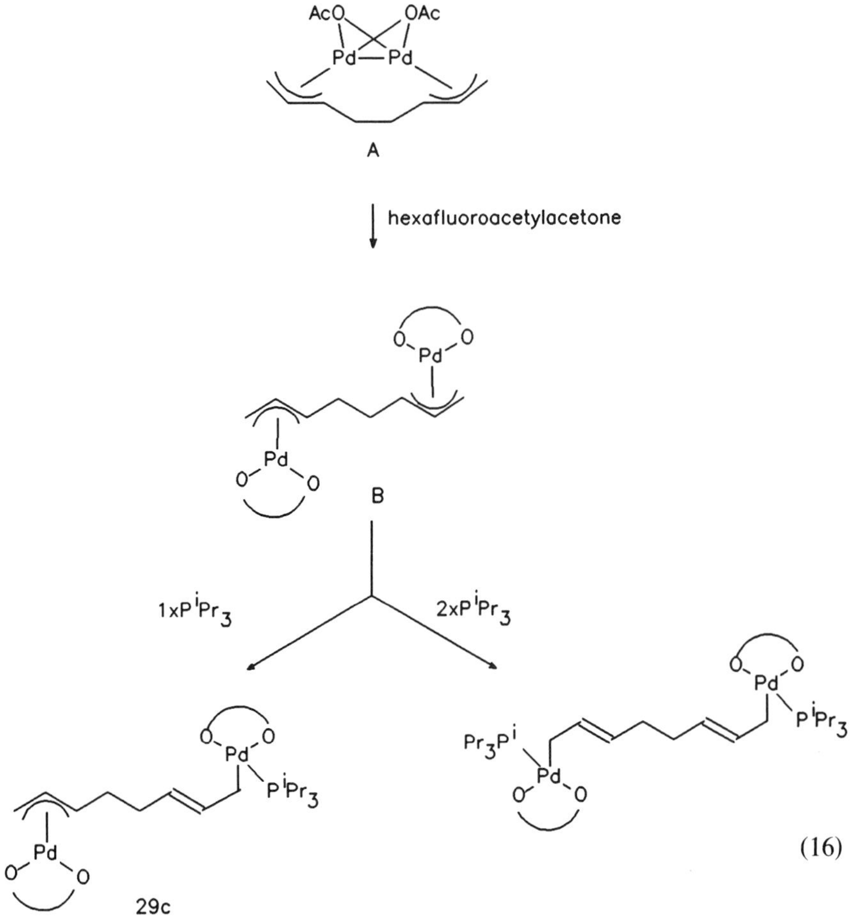

The telomerization reaction of acetic acid and 1,3-butadiene yielding acetoxyoctadienes with an (η^3-allyl)palladium acetate dimer was studied and a reaction mechanism proposed which was based on the key intermediate **A** (*112*). The reaction of bis(hexafluoroacetylacetonato)palladium in methanol with 1,3-butadiene or the reaction of **A** with hexafluoroacetone gives the μ-1-3-η:6-8-η-octadienyl-bridged diplatinum complex (**B**), as was confirmed by an X-ray structure determination. The reaction of **B** with 1 equiv of P^{i}Pr$_3$ leads to the addition of one phosphine ligand, and spectroscopic data indicate that one η^3-allyl ligand was converted to an η^1-allenyl ligand to give the η^1:η^3-octadienyl-bridged complex (**29c**) [Eq. (16)]. Two equivalents of triisopropylphosphine modifies both allyl ligands and affords a σ,σ-octadienyl-bridged dinuclear complex.

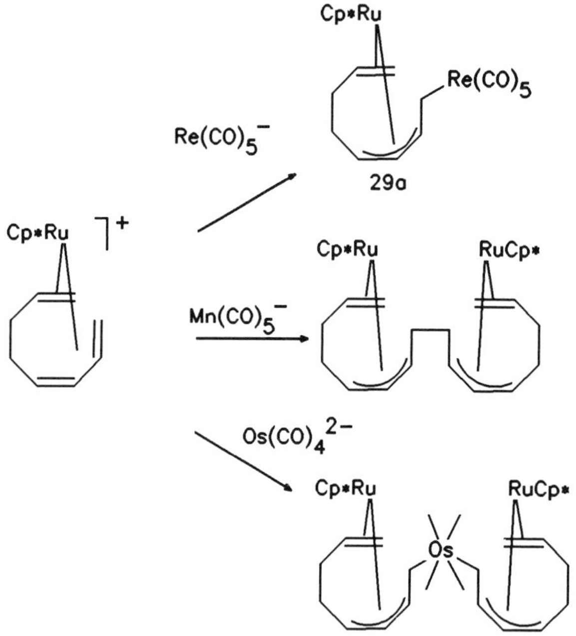

$$(17)$$

$$(18)$$

$$(19)$$

Other examples from the Beck laboratories where anionic metal substrates attack π-coordinated 1,3-dienes, giving bimetallic complexes with allyl ligands, were recorded for the reaction of [Cp*Ru{η^2:5-8-η^4-octatriene}]$^+$ with [Re(CO)$_5$]$^-$ (**a**) and [RuCp(CO)$_2$]$^-$ (**b**) to give L$_n$M{μ-η^1:2-4-η^3,7-η^2-C$_8$H$_{11}$}RuCp* (**29**) [Eq. (17)] (*86*). The corresponding reaction with [Mn(CO)$_5$]$^-$ again afforded only redox products with two Cp*Ru fragments bridged by a C–C coupled C$_{16}$ unit which is η^2:η^3-bonded to each Ru center [Eq. (18)]. The corresponding trimetallic complex was obtained from 2 equiv of triene precursor and [Os(CO)$_4$]$^{2-}$ [Eq. (19)] (*86*). Interestingly, for [Ru(η^6-C$_6$H$_6$){1-3-η^3:5-η^2-cyclooctadienyl}]$^+$ the attack of [Re(CO)$_5$]$^-$ occurs on the alkene and not on the allyl function of the C$_8$H$_{11}$ ring, to give the Ru–Re bimetallic complex with the σ-bonded Re(CO)$_5$ fragment in the 6 position of the ring (*68*).

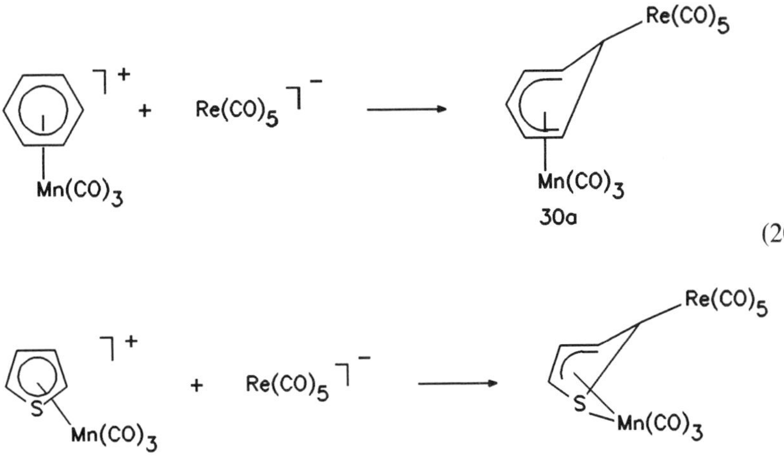

$$(20)$$

The addition of pentacarbonylrhenate to the π-coordinated benzene of the cationic complexes $[M(\eta^6\text{-}p\text{-}C_6H_4R_2)(CO)_3]^+$ gave σ,π-bridged fluxional cyclohexadienyl complexes $\eta^5\text{-}C_6H_4R_2\{Re(CO)_5\})M(CO)_3$ (**30a–g**) (M = Mn, Re; R = H, Me, Cl) [Eq. (20)] (*113,114*). Although only one signal was observed for the six hexadienyl protons in the ^1H NMR spectrum at room temperature, the anticipated four signals indicating the nonequivalence were observed in low-temperature studies. A crystal structure determination of **30a** (M = Mn, R = H) reveals that the plane described by the five carbon atoms attached to the $Mn(CO)_3$ fragment is distorted and a second plane containing the sixth carbon atom is displaced from the first by 34° (*106*). The rhenium fragment is in the exo position in the solid state. The $Re(CO)_5$ fragment is also found in the exo position in the complex $\eta^6\text{-}C_7H_7\{Re(CO)_5\})Mo(CO)_3$ obtained from the reaction of $[Mo(\eta^7\text{-}C_7H_7)(CO)_3]^+$ with $[Re(CO)_5]^-$ (*98*). Similarly, $[Re(CO)_5]^-$ reacts with the π-coordinated thiophene ligand of $(\eta^5\text{-thiophene})Mn(CO)_3^+$ by nucleophilic addition at a ring double bond to give the σ,π-thiophene-bridged complex $(2\text{-}4\text{-}\eta^3\text{-}S\text{-}SC_4H_3\{Re(CO)_5\}Mn(CO)_3$ (**30h**) (*114*).

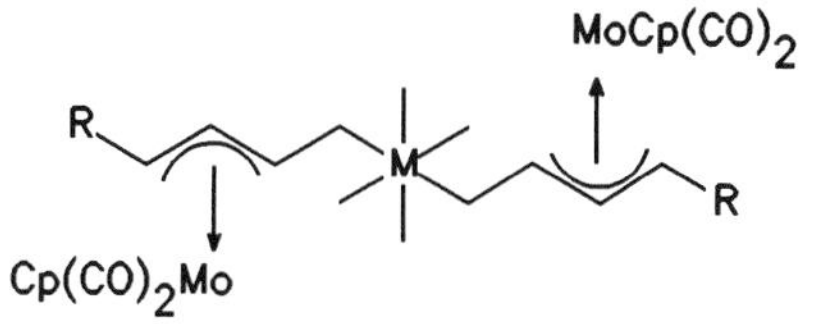

31

M=Os,R=H a,Me b

M=Ru,R=H c,Me d

The directed synthesis of trinuclear σ,σ,π-hydrocarbon-bridged complexes is achieved by adding 1 equiv of dianionic carbonylmetallates to 2 equiv of π-coordinated, unsaturated hydrocarbon ligands of cationic complexes. The addition of the organometallic nucleophiles $[M(CO)_4]^{2-}$ (M = Os, Ru) to 2 equiv of $[MoCp(CO)_2(butadiene)]^+$ leads to the formation of the heterotrimetallic $\eta^1:\eta^3$-hydrocarbon-bridged complexes (**31a–d**) (82). Trimetallic complexes with a central $Os(CO)_4$ fragment which is σ-bonded to two thienyl or two hexadienyl ligands in turn η^4-coordinated and η^5-coordinated to two $Mn(CO)_3$ fragments, $(CO)_4Os(\mu-\eta^1:\eta^3-S-SC_4H_3-\{Mn(CO)_3\})_2$ and $(CO)_4Os(\mu-\eta^1:\eta^5-C_6H_6\{Mn(CO)_3\})_2$, respectively, were prepared by the same method (82,115). Crystal structure determinations reveal the hydrocarbon bridges to be cis to one another, as is the case for the trimetallic complex $(CO)_4Os(\mu-\eta^1:\eta^6-C_7H_7\{Mo(CO)_3\})_2$.

C. σ-Alkyl/π-Alkyne Complexes

Many examples of dinuclear and trinuclear complexes with metal–metal bonds and acetylide ligands bridging the metal centers with σ,π, σ,σ,π, and σ,π,π bonds have been recorded, of which the bonding modes **A**, **B**, and **C** are representative (116–118). Multisite binding and activation of acetylides (the ethynyl anion $HC{\equiv}C^-$ is isoelectronic with CO) for synthetic exploitation in cluster and applied chemistry have been topics of current interest (32,119). Other arrangements of bimetallic complexes with bridging acetylene ligands, which are also not included in this review, are the $\mu-\eta^1$-ethylidyne complexes (**D**) (120) and μ-1,2-ethynediide complexes (**E**) (18,120,121). The formation of trimetallic complexes with bridged σ,σ,π-ethyne ligands was anticipated from the reactions of the dimetallated ethyne complex, $(CO)_5ReC{\equiv}CRe(CO)_5$, with $Pt(PPh_3)_2(\eta^2-C_2H_4)$, $Fe_2(CO)_9$, and $Co_2(CO)_8$, which have labile ligands (122). Although π-coordination was achieved, it was accompanied by metal–metal bond formation. The reaction of $(CO)_5ReC{\equiv}CRe(CO)_5$ and $[Cu(NCMe)_4]PF_6$ afforded $[\{Cu(NCMe)_2\}_2\{(CO)_5ReC{\equiv}CRe(CO)_5\}]^{2+}$, the structure of which was determined by X-ray diffraction (123). This complex, which contains a Cu–Cu bond, clearly shows the isolobal relationships between

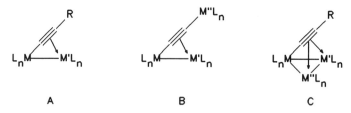

A   B   C

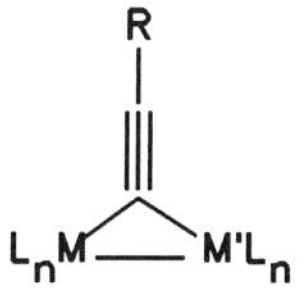

R
|
|||
L$_n$M———M'L$_n$ L$_n$M———≡———M'L$_n$

    D    E

L$_n$M———≡———R   L$_n$M———≡———R
        | | |
    M'L$_n$ X———M'L$_n$

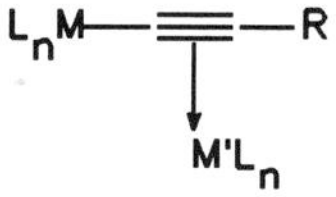

    F 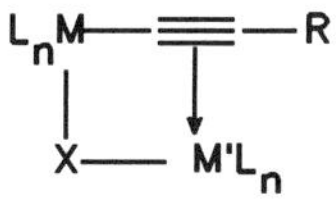   G

organic alkynes and $(CO)_5ReC{\equiv}CRe(CO)_5$. Complexes of classes **B** and **C** are, to our knowledge, unknown for compounds without metal–metal bonds. Thus, in this section we focus on bimetallic complexes of formulations **F** and **G**.

In the first type of nucleophilic reaction of the M—C≡CR group, interaction can occur via the π cloud of the acetylide with organometallic complexes which are coordinatively unsaturated. Spectroscopic data indicate that the addition of the Lewis acid $[Re(CO)_5]^+$ to $Me_3Sn\{C{\equiv}CMe\}$ affords the rhenium propynide complex $(CO)_5Re\{C{\equiv}CMe\}$ (*10,124*). In contrast, pentacarbonyl(tetrafluoroborato)rhenium reacts with ethynyltrimethylsilane presumably to give the corresponding pentacarbonylrhenium ethynide as a reactive intermediate. The latter, after subsequent reaction with a second equivalent of the rhenium precursor, gives the cationic σ,π-ethynide bridged complex $[(CO)_5Re\{\mu-\eta^1:\eta^2-C{\equiv}CH\}Re-(CO)_5]^+BF_4^-$ (**32**), which has been fully characterized (Table VII) (*124*). The π-coordination of the second Re fragment increases the acidity of the acetylide proton, which on deprotonation with sodium ethanolate gives the σ,σ-ethynediide-bridged dinuclear complex (**E**) [Eq. (21)]. Although the addition of HBF_4 to **E** regenerates **32**, the addition of other acids HX (X = Cl, OSO_2F) liberates acetylene and yields $Re(CO)_5X$. Two equivalents of triarylphosphinegold chlorides, $Au(PR_3)Cl$, react readily with acetylene in ethanolic sodium ethoxide to give the σ,σ-bonded ethynediyl complex $R_3PAu—C{\equiv}C—AuPR_3$ (*125*). There was no evidence for the formation of a dinuclear σ,π-bonded ethynyl intermediate in this case, nor during the reaction where the ethynide ligand was transferred to a Pt fragment in the reaction with PtL_2Cl_2.

Alternatively, the M—C≡CR complex can also react as a nucleophile with an electrophile at the β-carbon to give a cationic vinylidene complex

TABLE VII

Bimetallic Complexes[a] with σ,π-Bridged Acetylide Ligands,
$L_nM^1\{\mu\text{-}\eta^1\text{:}\eta^2\text{-}(C\equiv CR^1)\}_xM^2L_n$

	Complex					
No.	M^1	M^2	x	R^1	Method of characterization	Ref.
32	Re	Re	1	H	MS; IR; ^{1}H, ^{13}C NMR	(124)
33a	Re	Fe	1	Me	IR; ^{1}H NMR; X-ray	(127)
34	W	W	1	Ph	IR; ^{1}H NMR; X-ray	(130)
35	Fe	Fe, Mn	1	R^b	IR; ^{1}H, ^{13}C NMR; X-ray	(131,133)
36a	Ru	W	1	H	^{1}H NMR; X-ray	(134)
37–39	Mn, Ru	Cu, Ag	1, 2	^{t}Bu	IR; ^{1}H, ^{31}P NMR; X-ray	(135–138)
40	Zr	Zr	1	Me, Ph	IR; ^{1}H, ^{13}C NMR	(143)
41	Pt, Rh	Rh, Mo	1	Me	IR; ^{1}H, ^{31}P NMR; X-ray	(151,152)
42	Rh	Rh	1	R^b	IR; ^{1}H, ^{31}P NMR; X-ray	(153,154)
43	Pt	Ir	1	Ph	IR; ^{1}H, ^{31}P NMR	(155,156)
44a	Pt	Cu	2	Ph	^{31}P NMR; X-ray	(157)
46	Ti	Co, Ni	2	Ph	MS; IR; X-ray	(160,163)
46, 48	Ti	Fe, Ag	2	SiMe$_3$	MS; IR; X-ray	(161,162)
50	Hf	Ni, Co	2	SiMe$_3$	IR; ^{1}H, ^{13}C NMR; X-ray	(165,166)
51	Pt	Ti	2	^{t}Bu	IR; ^{1}H, ^{13}C, ^{19}F NMR; X-ray	(167)
56	Pt	Pt, Pd	2	R^b	IR; ^{1}H, ^{19}F NMR	(22,172)
57	Ir	Ir	2	SiMe$_3$	MS; ^{1}H NMR; X-ray	(174)
59	Zr	Zr	2	R^b	IR; ^{1}H, ^{13}C NMR	(143,177)
61	Ti	Ti	2	Ph	IR; ^{1}H NMR	(181)
62	Ti	Ti	2	SiMe$_3$	IR; ^{1}H, ^{13}C NMR; X-ray	(182)

[a] M^2 represents π-bonded metal.
[b] R represents alkyl or aryl groups.

(H) [Eq. (22)]. The protonation of σ-coordinated acetylides by strong protic acids results in the formation of cationic vinylidene complexes (126). Therefore, it was expected that the reaction of the Lewis acid [Re(CO)$_5$]$^+$ with metal acetylides could also lead to the formation of bimetallic vinylidene complexes (127). However, the reaction of [Re(CO)$_5$]$^+$ and FeCp(CO)$_2${C≡CR} [R = Ph (a), Me (b)] gives, during electrophilic substitution of the Fe fragment by the Re fragment, the σ,π-ethynide-bridged complexes [(CO)$_5$Re{(μ-η^1:η^2-C≡CR)FeCp(CO)$_2$}]$^+$BF$_4^-$ (33) [Eq. (23)]. The crystal structure of 33a (Fig. 1) reveals an asymmetrically π-coordinated Fe fragment, a nonlinear cis-distorted alkyne, and some evidence of electron delocalization to the σ-bonded Re atom when Re–C bond distances are compared in 33a [1.98(3) Å], Re-alkynyl [2.14(2) Å (121)], Re–carbene [2.09(2) Å (128)], and Re–vinylidene complexes [1.90(2) Å (129)].

$$L_nM \!-\!\!\!\equiv\!\!\!-H \; \rceil^+$$
$$\downarrow$$
$$M'L_n$$

F

32   (21)

NaOR $\downarrow$ $\uparrow$ HBF$_4$

$$L_nM \!-\!\!\!\equiv\!\!\!-M'L_n$$

E

HBF$_4$ $\downarrow$ $\uparrow$ Base

$$L_nM\!\!=\!\!=\!\!\smallsetminus_{M'L_n} \; \rceil^+$$ (22)
$$H$$

$$Cp(CO)_2Fe \!-\!\!\!\equiv\!\!\!-R \; + \; Re(CO)_5 \; \rceil^+ \longrightarrow \; (CO)_5Re \!-\!\!\!\equiv\!\!\!-R$$
$$\downarrow$$
$$Cp(CO)_2Fe^+BF_4^-$$ (23)

R=Ph a,Me b   **33**

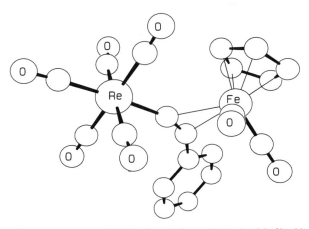

FIG. 1.   X-Ray structure of $[(CO)_5Re\{(\mu\text{-}\eta^1{:}\eta^2\text{-}C\!\equiv\!CR)FeCp(CO)_2\}]^+$ **33**) [Fritz *et al.* (*127*)].

The protonation of $Cp(CO)_3W\{\eta^2\text{-}C\equiv CPh\}$ by HBF_4 in dichlorome-thane gives a stable cationic binuclear complex of tungsten, $[Cp(CO)_3\text{-}W\{\mu\text{-}\eta^1:\eta^2\text{-}C\equiv CPh\}=WCp(CO)(\eta^2\text{-}CH\equiv CPh)]^+$ (**34**), with a σ,π-bridg-ing phenylacetylide ligand (*130*). A crystal structure determination of **34** reveals that the π-coordination does not affect the carbon triple bond distance significantly, which is the same as that for the uncoordinated acetylide, and suggests that the structure type with a π-bonded acetylide (**F**) dominates. Again the σ,π-bonded bridging ligand affords an asymmet-rical arrangement of atoms. However, the vinylcarbenium character of the σ-acetylide precursor was demonstrated by the formation of the phos-phonium cation $[Cp(CO)_3W\{C(PPh_3)=CHPh\}]BF_4$ on treatment with HBF_4/PPh_3.

Cationic diiron σ,π-acetylide complexes of the type $[Fp_2^{\#}\{\eta^1:\eta^2\text{-}(C\equiv CR)\}]^+$ (**35**) ($Fp_2^{\#}$ is Fp_2, $FpFp^*$, or Fp_2^*, with Fp being $(\eta^5\text{-}C_5H_5)Fe(CO)_2$ and Fp^* being $(\eta^5\text{-}C_5Me_5)Fe(CO)_2$; $Fp_2^{\#} = Fp_2^*$, R = H (**35a**) and R = Ph (**35b**)] were synthesized by the ligand exchange reaction of mononuclear acetylide complexes $Fp^{\#}(C\equiv CR)$ and the labile iron cations $[Fp^{\#}(THF)]BF_4$ (*131*). The fluxional behavior of the σ,π-acetylide diiron complexes was studied by means of variable-temperature NMR analyses (Table VIII). The fluxional behavior of $[Fp_2^{\#}(C\equiv CPh)]BF_4$ was ascribed to a flow of electron density from the π bond into the M–C σ bond, thereby creating a positive charge at the β position and generating a free cationic vinyl intermediate. Therefore, the motion of the bridging ligand with respect to the metal centers can be described by

$$\{(\eta^1(Fp):\eta^2(Fp')\} \rightleftharpoons [\{(\eta^1(Fp):\eta^1(Fp')\}]^{\#} \rightleftharpoons \{\eta^1(Fp'):\eta^2(Fp)\}.$$

On the other hand, the fluxional process for $[Fp_2^{\#}(C\equiv CH)]BF_4$ gave data consistent with a 1,2-proton shift where the H atom on the ethynyl ligand

TABLE VIII

SPECTROSCOPIC DATA[a] OF σ,π-BRIDGED ACETYLIDE MOIETIES IN SELECTED BIMETALLIC COMPLEXES, $L_nM^1\{\mu\text{-}\eta^1:\eta^2\text{-}(C_1\equiv C_2R^1)\}_xM^2L_n$

Complex			IR (cm^{-1})		$^1H/^{13}C^b$ NMR (δ, ppm)					
No.	M^1	M^2	$\nu(C\equiv C)$	Medium	x	R	C_1	C_2	Solvent	Ref.
32	Re	Re	1870	KBr	1	4.55	57.6	82.4	CD$_3$NO$_2$	(*124*)
35a	Fe	Fe	1809	KBr	1	3.76	124.2	72.2	CDCl$_3$	(*131*)
35c	Fe	Mn	1839	KBr	1	1.73	73.3	67.8	C$_6$D$_6$	(*131*)
47a	Ti	Ni	1857	CH$_2$Cl$_2$	2	7.3–7.7	187.2	130.1	CD$_3$Cl	(*162*)
50a	Hf	Ni	1880	Pentane	2	7.3–7.8	174.5	127.8	CD$_3$Cl	(*165*)

[a] M^2 represents π-bonded metal for $x = 1$.
[b] For ^{13}C NMR, discrimination between resonances was not always possible.

moves between the two ethynyl carbon atoms via a three-membered ring intermediate. Structural data reveal greater asymmetry in the geometry of the bridging acetylide ligand of **35b** compared to **35a**, which is ascribed to the steric bulk of the phenyl compared to the proton substituent and a small but significant contribution of the vinylidene structure, $L_nM=C=C(M'L_n)H$, for both complexes. Whereas deprotonation of the σ,π-acetylide complex $[Fp_2^*(C\equiv CH)]BF_4$ (compare **32**) with NaOMe in THF results in the formation of $Fp^*C\equiv CFp^*$, the analogous complex $[Fp_2(C\equiv CPh)]BF_4$ affords the metallocyclic adduct $Cp(CO)Fe\{C(Ph)=C(Fp)C(O)OMe\}$, suggesting initial attack from the nucleophile on a carbonyl ligand and insertion of the acetylide ligand into the newly formed Fe–acyl bond (*132*).

$$[Fe]\diagup\!\!\!=\!\!=[Mn] \qquad (24)$$

$$[Fe]\!-\!\!\equiv\!\!-[Mn]\;\rceil^{-} \quad\xleftarrow{\;H^+\;}$$

$$\xrightarrow{\;Me^+\;}$$

$$[Fe]\!-\!\!\equiv\!\!-Me \qquad (25)$$
$$\qquad\quad | $$
$$\qquad\quad [Mn]$$

$$\textbf{35c}$$

The electronic influences of the metal centers on the structure of a binuclear vinylidene $Cp'(CO)_2Mn=C=C(H)Fp^*$ or η^2-structure $Cp'(CO)_2Mn(\eta^1{:}\eta^2\text{-}C\equiv CR)Fp^*$ were discussed with respect to the iso-electronic dinuclear acetylide-bridged compounds $\{Cp^*M(CO)_2\}_2(\mu\text{-}C\equiv CR)$ (M = Fe^+ and/or Mn) (*133*). The deprotonation of $[Cp'-(CO)_2MnC\equiv CH]Fp^*$ with *n*-BuLi generates an anionic ethynediyl intermediate, $[Cp'(CO)_2MnC\equiv CFp^*]^-$, which can be protonated on the $C_\beta(Mn)$ position to give the bimetallic vinylidene complex $Cp'(CO)_2$-$Mn=C=CHFp^*$ or methylated on the $C_\alpha(Mn)$ position, owing to steric hindrance at the $C_\beta(Mn)$ position, to give the bimetallic complex with an η^2-coordinated Mn fragment, $Fp^*\{\mu\text{-}\eta^1{:}\eta^2\text{-}C\equiv CMe\}Mn(Cp')(CO)_2$ (**35c**) [Eq. (24), and (25)]. Protonation of $[Cp'(CO)_2MnC\equiv CH]Fp^*$, on the other hand, gives the cationic μ-vinylidene complex $Cp'(CO)_2Mn\{\mu\text{-}\eta^1{:}\eta^1\text{-}C=CH_2\}Fp^*$. The electrophilic addition (protonation) of neutral η^1-vinylidene complexes occurs readily at the β-carbon atom to give a cationic "carbyne" complex **X** (*133*). The transfer of electron density from $Fp^*(M'L_n)$ to $MnCp(CO)_2(ML_n)$ affords an $\eta^1{:}\eta^2$-ethenylidene intermediate, which after migration of Fp* along the π system ultimately results in the formation of the μ-vinylidene complex $[(\mu-\eta^1:\eta^1-C=CH_2-\{Fp^*\})MnCp'(CO)_2]BF_4$.

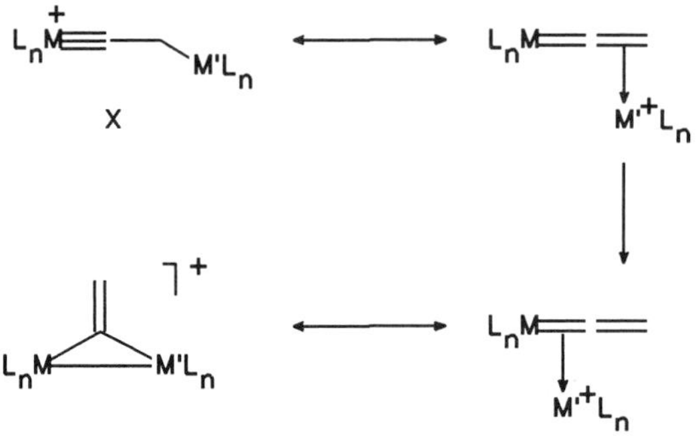

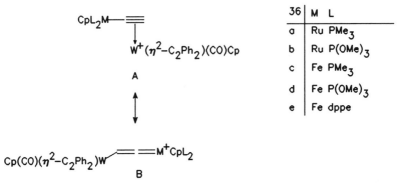

Frank and Selegue (134) found that the solid state structure of $(\mu-\eta^1:\eta^2-C(H)\equiv C\{RuCp(PMe_3)_2\})WCp(\eta^2-Ph_2C_2)(CO)$ (**36a**) closely resembles a cationic ruthenium complex bearing a tungsten-substituted vinylidene ligand (**B**) with virtually complete loss of the π-bonding interaction (**A**). Deprotonation of the μ-ethynyl complexes with phosphine and phosphite ligands with $NaN(SiMe_3)_2$ and $KOCMe_3$, respectively, yielded dinuclear complexes with bridging ethynediyl ligands.

CpL_2M—≡≡

$W^+(\eta^2-C_2Ph_2)(CO)Cp$

A

$Cp(CO)(\eta^2-C_2Ph_2)W$≡≡≡M^+CpL_2

B

36	M	L
a	Ru	PMe_3
b	Ru	$P(OMe)_3$
c	Fe	PMe_3
d	Fe	$P(OMe)_3$
e	Fe	dppe

The π-coordination of group 11 (IB) metals to σ-alkynyl complexes of manganese, $Mn(dppe)(CO)_3\{C\equiv C'Bu\}$ [dppe is bis(diphenylphosphino)ethane], in the presence of $P(C_6H_4-o-Me)_3$ in dichloromethane gave the cationic σ,π-acetylide complexes, $[(\eta^1:\eta^2-C\equiv C'Bu)\{Mn(dppe)-(CO)_3\}M\{P(C_6H_4-o-Me)_3\}]^+$ (**37**) [M = Cu (**37a**), M = Ag (**37b**), M = Au (**37c**)] (135,136). The neutral complexes $(\eta^1:\eta^2-C\equiv CR)\{Mn(dppe)(CO)_3\}-CuCl$ (**38**) [R = 'Bu (**38a**), Ph (**38b**), CH_2OMe (**38c**)] were isolated on reacting CuCl in the absence of the phosphine ligands (136,137). The

TABLE IX

SELECTED X-RAY DATA[a] FOR σ,π-BRIDGED ACETYLIDE MOIETIES IN
BIMETALLIC COMPLEXES

Complex			Bond distances (Å)					Bond angles (°)		
No.	M^1	M^2	M^1–M^2	M^1–C_x	M^2–C_1	M^2–C_2	C_1–C_2	α	β	Ref.
33a	Re	Fe	—	2.088	2.319	2.124	1.237	158.7	150.1	(127)
34	W	W	4.216	2.19	2.21	2.10	1.30	146	148	(130)
35a	Fe	Fe	4.038	1.912	2.422	2.114	1.226	162.0	152	(131)
35b	Fe	Fe	4.012	1.944	2.357	2.134	1.244	158.2	147.2	(131)
35c	Fe	Mn	3.919	1.955	2.281	2.144	1.223	156.9	155.1	(133)
36a	Ru	W	—	1.96	2.53	2.05	1.25	163	—	(134)
38b	Mn	Cu	3.443	2.023	2.036	2.030	1.226	171.1	165.6	(137)
39a	Mn, Mn	Cu	—	2.03	2.08	2.09	1.24	170	164	(136)
41a	Pt	Rh	3.066	1.95	2.22	2.46	1.21	177	166	(151)
41b	Pt	Rh	3.099	2.01	2.22	2.40	1.16	178	172	(151)
42c	Rh	Rh	3.155	2.043	2.209	2.616	1.182	168	172.9	(154)
44a	Pt	Cu, Cu	3.250	2.002	2.101	2.173	1.212	177.7	169.5	(157)

[a] Standard deviations omitted; M^2 represents π-bonded metal.

structure of **38b** was confirmed by an X-ray structure analysis (Table IX). Attempts to prepare silver and gold analogs of **38** failed, but the synthesis of $(\eta^1:\eta^2\text{-C}\equiv\text{CR})\{\text{Mn(dppe)(CO)}_3\}\text{AuC}_6\text{F}_5$ (**38**) [R = ^{t}Bu (**38d**), CH_2OMe (**38e**)] was successful (136). A crystal structure determination of $(\eta^1:\eta^2\text{-}$ C$\equiv$CPh)$\{$RuCp(PPh$_3$)$_2\}$CuCl (**38f**) was recorded and, in contrast to the structure of a related dimeric iron–copper complex, was monomeric (138). The trimetallic complexes $[\{$Mn(dppe)(CO)$_3$(μ-$\eta^1:\eta^2$-C$\equiv$ C^{t}Bu)$\}_2$M]X (**39**) [M = Cu, X = PF$_6$ (**39a**); M = Ag, X = BF$_4$ (**39b**); M = Au, X = PF$_6$ (**39c**)] can be obtained directly from the σ-alkynyl Mn(dppe)-(CO)$_3$(C$\equiv$C^{t}Bu) with CuCl, [Ag(NCMe)$_4$]BF$_4$, and Au(C$_4$H$_8$S)Cl combined with TlPF$_6$ (136). Reaction of CuCl with LiC$_5$Me$_5$ at low temperatures, which acts as a source of Cp*Cu$^+$, will add over the triple bond to give bimetallic and trimetallic complexes when treated with a metal–carbyne complex (139). The structure of **39a** reveals that the coordination around the copper is pseudotetrahedral, with four Cu–C bond distances which do not differ significantly (136). It is known that copper acetylides react with FpCl to give products in which the acetylide is π-bonded to the CuCl fragment and σ-bonded to the CpFe(CO)$_2$ fragment (*140*). A structural study has shown that these complexes are dimeric via copper-halogen bridges (*141*).

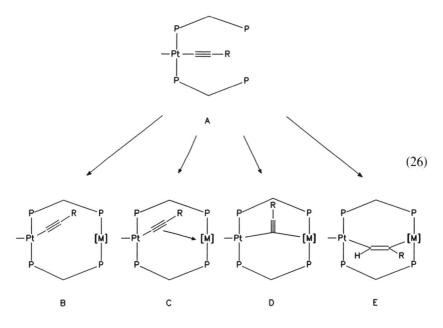

Utilization of a bidentate ligand to lock two metals in close proximity to one another has enabled the synthesis of heterobimetallic complexes with σ,π-bridging acetylide ligands (*142*). The complexes (R'Cp)$_2$Zr(C≡CR)Cl (R = Me, Ph, R' = H, Me) react with (R''Cp)$_2$Zr (R' = H, Me), which was generated *in situ*, to yield binuclear σ,π-alkynyl- and chloro-bridged complexes of the type (R'Cp)$_2$Zr(μ-Cl)(μ-C≡CR)Zr(R'Cp)$_2$ (**40**) [R' = R'' = H, R = Me (**40a**), Ph (**40b**); R' = R'' = Me, R = Me (**40c**), Ph (**40d**)] (*143*).

(26)

Shaw and co-workers *(144–150)* have extensively used bis(diphenyl-phosphino)methane as a bidentate metal fragment link to synthesize systematically a variety of complexes of the type $Pt(C\equiv CR)_2(\mu$-dppm)$_2ML_n$. Four classes of neutral and cationic bimetallic complexes without metal–metal bonds were found; those (i) without σ,π-bridging acetylide ligands (**B**) [M = Cd, Hg, Cu, Ag, Au *(144–147)*], (ii) with σ,π-bridging acetylides **C** (*vide infra*), (iii) with μ-η^1-bridging acetylide ligands (**D**) [W, Pt *(148,149)*] and (iv) with μ-1,2-σ,σ-alkene ligands (**E**) [Rh *(150)*] [Eq. (26)]. The reaction of the monodentate dppm precursors $Pt(dppm-P)_2(C\equiv CR)X$ (**A**) [R = Me, Ph, *p*-MeC$_6$H$_4$, CH$_2$CH$_2$Ph, C(Me)=CH$_2$] or $(RC\equiv C)(X)Pt(\mu$-dppm)$_2ML_n$ (**B**) (X = C$\equiv$CPh, R = Ph, ML$_n$ = HgCl$_2$, AgCl, CuI, Au; X = Cl, R = Me, Ph, *p*-tolyl, ML$_n$ = AgCl) with $Rh_2(CO)_4Cl_2$ afforded $[(RC\equiv C)Pt(\mu$-dppm)$_2(\mu$-σ,π-C$\equiv$CR)-Rh(CO)]Cl$ (**41**), which is readily converted to the corresponding PF$_6^-$ salt *(120,151)*. The bimetallic compounds with σ,π-bridging acetylene ligands are also directly accessible by treating $[Pt(\mu$-dppm-PP')$_2]Cl_2$ first with $Ag(O_2CMe)PhC\equiv CH$ and thereafter with $Rh_2(CO)_4Cl_2$. The neutral bimetallic complexes $(RC\equiv C)Pt(\mu$-dppm)$_2(\mu$-σ,π-C$\equiv$CR)RhCl$ (**C**) (R = *p*-tolyl, Ph) were prepared either by refluxing the μ-σ,π-bimetallic chlorosalts in toluene or by treating $Pt(C\equiv CR)_2(dppm-P)_2$ (**A**) with $Rh_2(cyclooctene)_4Cl_2$ *(151)*. Treatment of $[(CO)_5Mo(\mu$-dppm)$_2$-Rh(CO)$_2]PF_6$ with LiC$\equiv$CR in THF gave, after rapid evolution of CO, the Mo–Rh complexes $(CO)_3Mo(\mu$-dppm)$_2(\mu$-η^1:η^2-C$\equiv$CR)Rh(CO)$ (**41**) [R = Me (**41c**), Ph (**41d**), *p*-tolyl (**41e**)] *(152)*. The crystal structure of **41c** shows that the two metal centers are again fixed (3.1 Å apart) by the geometry of the frame of the bridging dppm ligands. Crystal structure determinations of $[XPt(\mu$-dppm)$_2(σ,π$-C$\equiv$CMe)Rh(CO)Cl]PF_6$ [X = Cl (**41a**), C$\equiv$CMe (**41b**)] were also recorded. In both structures **41a,b** the metal centers are linked by the two dppm bridges to give rise to an eight-membered Pt$_2$P$_4$C$_2$Rh ring in the boat conformation with non bonding Pt–Rh distances of around 3.1 Å, which are consistent with the intraligand P–P distances *(151)*.

The carboxylate complexes $[Rh_2(\mu$-O$_2$CR)(μ-dppm)$_2$(CO)$_2]ClO_4$ (R = CH$_3$, CF$_3$) react with terminal acetylenes, R'C$_2$H [R' = H, Ph, *t*Bu (R = CF$_3$ only)], to yield dirhodium complexes of the type $[Rh_2(\mu$-C$\equiv$CR)(μ-dppm)$_2$(CO)$_2]ClO_4$ (**42**) [R = H (**42a**), Ph (**42b**), *t*Bu (**42c**)] with σ,π-bridging acetylide complexes in which the acetylide ligand is fluxional, exhibiting an oscillatory type of motion *(153)*. The bridging acetylide ligands are accessible to nucleophilic attack by hydride to yield unstable μ-η^1-bridging vinylidene complexes. The geometry of the σ,π-bridging acetylide ligand of $Rh_2(\mu$-dppm)(μ-η^1:η^2-C$\equiv$C'Bu)(CO)$_2$ (**42c**) was shown to be similar to analogous bimetallic complexes which have been studied structurally (*154*). The complex has the typical A-frame geometry with the β-carbon atom further away from the Rh metal owing to the steric bulk of the ^{t}Bu substituent. Extension of these studies to the corresponding ethynide precursor $Pt(C{\equiv}CH)_2(dppm—P)_2$, which can be generated by treating $Pt(dppm—PP)Cl_2$ with $NaC{\equiv}CH$ and an excess of dppm in liquid ammonia, yielded the σ,π-bridging ethynide complexes $[(HC{\equiv}C)Pt(\mu\text{-}dppm)(\mu\text{-}\eta^1{:}\eta^2\text{-}C{\equiv}CH)Rh(CO)]^+$ (**43a**) and $(HC{\equiv}C)Pt(\mu\text{-}dppm)(\mu\text{-}\eta^1{:}\eta^2\text{-}C{\equiv}CH)Ir(CO)Cl$ (**43b**) after reaction with $RhCl_2(CO)_4$ and $Ir(PPh_3)_2(CO)Cl$, respectively (*155*).

The Pt-Ir complexes $(RC{\equiv}C)Pt(\mu\text{-}dppm)_2(\mu\text{-}\eta^1{:}\eta^2\text{-}C{\equiv}CR)IrCl$ (**43**) [R = Ph (**43c**), p-tolyl (**43d**)], $(RC{\equiv}C)Pt(\mu\text{-}dppm)_2(\eta^1{:}\eta^2\text{-}C{\equiv}CR)Ir(CO)Cl$ [R = Me (**43e**), Ph (**43f**), p-tolyl (**43g**)], $[(RC{\equiv}C)Pt(\mu\text{-}dppm)_2(\mu\text{-}\eta^1{:}\eta^2\text{-}C{\equiv}CR)Ir(CO)]X$ [R = Me, X = PF_6 (**43h**), BPh_4 (**43i**); R = p-tolyl, X = PF_6 (**43j**), BPh_4 (**43k**); R = Ph, X = BPh_4 (**43l**)], and $[(RC{\equiv}C)Pt(\mu\text{-}dppm)_2(\mu\text{-}\eta^1{:}\eta^2\text{-}C{\equiv}CR)(\mu\text{-}H)Ir(CO)H]PF_6$ [R = Me (**43m**), p-tolyl (**43n**)] were synthesized from $trans\text{-}Pt(C{\equiv}CR)_2(\mu\text{-}dppm\text{-}P)_2$ (**A**) and $Ir_2Cl_2(cyclooctene)_4$ or $Ir(PPh_3)_2(CO)Cl$ (*155,156*). Treatment of the bis(phenylacetylene) complex $Pt(dppy)_2(C{\equiv}CPh)_2$ (dppy is diphenylphosphinopyridine) with 2 equiv of $[M(MeCN)_4]^+$ in dichloromethane afforded the trimetallic complexes $[Pt(dppy)_2\{(\mu\text{-}\eta^1{:}\eta^2\text{-}C{\equiv}CPh)(Cu(MeCN)_2)\}_2]^{2+}$ (**44**) [M = Cu (**44a**), Ag (**44b**), Au (**44c**)] (*157*). Structural features of the Pt–Cu_2 complex indicate C–C (acetylene) distances not significantly different from those of uncoordinated acetylenes and Cu–C distances comparable to those of polymeric σ,π-bridging Cu acetylide complexes (*45*). Bruce and co-workers (*158*) reported the synthesis of silver–rhodium acetylide complexes, and a structure determination of $RhAg_2(C{\equiv}CC_6F_5)_5(PPh_3)_3$ (**45**) was recorded, showing each silver interacting with three of the acetylide ligands (*159*).

44

M=Cu a,Ag b

Organometallic complexes with two σ-acetylide ligands can act as a bidentate chelating ligand via η^2-coordination to a second metal fragment.

Typical building blocks of this kind are the "open clamshell" structure of titanocene diacetylides and the square planar structure of *cis*-platinum diacetylides. Lang and co-workers (*160–163*) have studied the synthesis, structural features, and reactivity of heterobimetallic complexes of titanium with two σ-bonded acetylide ligands, both of which are both π-bonded to a second, different metal center. The complex (η^5-C$_5$H$_4$Si-Me$_3$)$_2$Ti(C≡CR)$_2$ reacts with 1 equiv of FeCl$_2$ (*160*), 0.5 equiv of Co$_2$(CO)$_8$ (*161*), 1 equiv of CuCl (*160*), 1 equiv of NiCl or Ni(CO)$_4$ (*162*), 1 equiv of a silver salt AgX (*163*) and 1 equiv of Pt(η^2-C$_2$H$_4$)$_2$PR$_3'$ (*164*) to afford, in high yields, the bimetallic complexes (η^5-C$_5$H$_4$SiMe$_3$)$_2$Ti{η^1:η^2-C≡CR}$_2$ML$_n$ with ML$_n$ = FeCl$_2$ and R = SiMe$_3$ (**46a**), ML$_n$ = Co(CO) and R = Ph (**46b**), ML$_n$ = CuCl and R = SiMe$_3$ (**46c**), ML$_n$ = Ni(CO) and R = Ph (**47a**) or ML$_n$ = NiCl$_2$ and R = SiMe$_3$ (**47b**), ML$_n$ = AgX and R = SiMe$_3$ (**48**) [X = Cl (**48a**), CN (**48b**), SCN (**48c**), NO$_2$ (**48d**) and ClO$_4$ (**48e**)], and ML$_n$ = PtPR$_3'$ (**49**) [R' = P(cyclo-C$_6$H$_{11}$)$_3$ (**49a**), PMe$_2$Ph (**49b**), PMePh$_2$ (**49c**), PPh$_3$ (**49d**), P^{i}Pr$_2$Ph (**49e**)], respectively. The Ti–C(alkynyl) σ bonds of **47** are cleaved with HX or X$_2$ (X = Cl, Br) to yield titanocene dihalides. The carbonyl ligand of **47a** can be displaced by phosphine ligands to give (η^5-C$_5$H$_4$SiMe$_3$)$_2$Ti{η^1:η^2-C≡CPh}NiPR$_2$R' [R = Ph, R' = C≡CPh (**47c**); R = C≡CPh, R' = Ph (**47d**); R = R' = OMe (**47e**)].

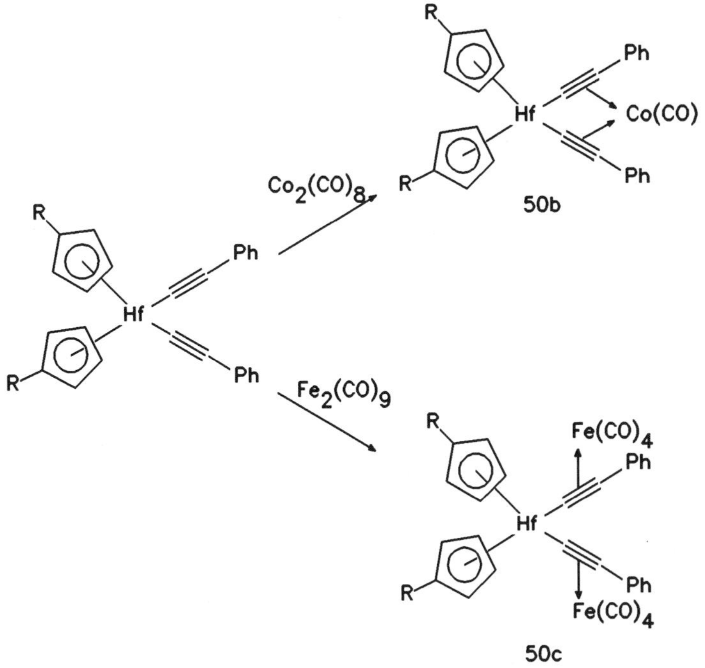

(27)

(28)

Lang and co-workers (*165,166*) have also investigated the ability of (η^5-$C_5H_4SiMe_3)_2Hf(C{\equiv}CPh)_2$ to act as an organometallic chelate toward different metal salts or complexes with labile ligands. The organometallic diyne reacts with $Ni(CO)_4$ or $Co_2(CO)_8$ to afford in high yields $L_2Hf\{\mu\text{-}\eta^1{:}\eta^2\text{-}C{\equiv}CPh\}M(CO)$ (**50**) [M = Ni (**50a**), Co (**50b**)] and L_2Hf-$(C{\equiv}CPh)\{\mu\text{-}(\eta^2{:}\eta^{2'}\text{-}C{\equiv}CPh)Co_2(CO)_6\}$ (L = η^5-$C_5H_4SiMe_3$). With the use of more starting material $Co_2(CO)_8$, $L_2Hf\{\mu\text{-}(\eta^2\text{-}C{\equiv}CPh)Co_2(CO)_6\}_2$ was the major product [Eq. (27)] (*165*). However, reaction of the Hf precursor with $Fe_2(CO)_9$ affords $L_2Hf\{\mu\text{-}(\eta^2\text{-}C{\equiv}CPh)Fe(CO)_4\}_2$ (**50c**), a compound in which both of the phenylethynyl ligands are side-on η^2-coordinated to $Fe(CO)_4$ fragments [Eq. (28)].

The two σ-acetylide ligands of **A** in $L_nM\{(\mu\text{-}\eta^1{:}\eta^2\text{-}C{\equiv}CR)_2M'L_n\}$ can open up, resulting in an increase in the C–M–C angle, which allows the $M'L_n$ fragment to move closer to the M center (**B**). As $M'L_n$ moves closer to M, the acetylene ligands will become increasingly asymmetrically coupled with the formation of a M–M' metal bond, ultimately giving rise to the structural type **C**. The reaction between $Cp_2Ti(C{\equiv}C'Bu)_2$ and *cis*-$Pt(C_6F_5)_2(THF)_2$ affords the bimetallic complex $Cp_2Ti(\mu\text{-}\eta^1\text{-}C{\equiv}C'Bu)_2Pt(C_6F_5)_2$ (**51**), which has been shown by X-ray crystallography (Table X) to contain two asymmetric $\mu\text{-}\eta^1$-alkynyl ligands and to show evidence of direct bonding interaction between the metal centers (**C**) (*167*).

The bimetallic complexes $L_2Pt(\mu\text{-}\eta^1{:}\eta^2\text{-}C{\equiv}CR)_2CuCl$ (**52**) [L_2 = dppe, R = Ph (**52a**), $'Bu$ (**52b**); L = PMe_2Ph, R = $'Bu$ (**52c**)] and $[(dppe)Pt(\mu\text{-}\eta^1{:}\eta^2\text{-}C{\equiv}CPh)_2Cu(NCMe)]BF_4$ (**52d**) were synthesized from the *cis*-bis(alkynyl)diphosphineplatinum precursors with 0.3 equiv of CuCl and $[Cu(NCMe)_4]BF_4$, respectively (*168*). Recrystallization in acetone-acetonitrile of $[(dppe)Pt(\mu\text{-}\eta^1{:}\eta^2\text{-}C{\equiv}CPh)_2Cu(NCMe)]BF_4$ afforded the trimetallic complex $[\{(dppe)Pt(C{\equiv}CPh)_2\}_2Cu]BF_4$ (**53**), which was confirmed by an X-ray structure determination. Treatment of anionic

TABLE X

SELECTED X-RAY DATA[a] FOR BIS σ,π-BRIDGED ACETYLENE MOIETIES IN
BIMETALLIC COMPLEXES

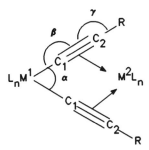

Complex		Bond distances (Å)					Bond angles (°)			
No.	M^1 M^2	M^1–M^2	M^1–C_1	M^2–C_1	M^2–C_2	C_1–C_2	α	β	γ	Ref.
46a	Ti Fe	3.26	2.07	2.23	2.20	1.20	85.4	171	163	(160)
46b	Ti Co	2.819	2.071	1.981	2.000	1.260	98.2	160.4	152.1	(161)
48d	Ti Ag	3.16	2.113	2.31	2.42	1.208	93.8	171	170	(163)
47	Ti Cu	2.95	—	—	—	—	91.0	167	163	(160)
50a	Hf Ni	2.968	2.170	2.043	2.026	1.250	87.0	160.7	153.0	(165)
50b	Hf Co	2.933	2.20	2.004	1.994	1.266	86.2	159.9	149.7	(166)
51	Pt Ti	2.831	2.016	2.248	—	1.218	103.8	159.8	172	(167)
52c	Pt Cu	3.129	2.03	2.16	2.28	1.22	84.6	186	166	(168)
55k	Pt Pd	3.049	2.008	2.284	2.353	1.219	81.8	167.6	166.5	(172)

[a] Standard deviations omitted.

polynuclear platinum–silver derivatives $\{(C_6F_5)_2Pt_2(C\equiv CR)_2\}_2Ag_2$ (169) with PR_3 affords the alkynyl-bridged complexes $(Bu_4N)[\{Pt(C_6F_5)_2$-$(\mu\text{-}\eta^1{:}\eta^2\text{-}C\equiv CR)_2\}AgL]$ (52) [R = Ph, L = PPh_3 (52e), PEt_3 (52f); R = ^tBu, L = PPh_3 (52g) PEt_3 (52h)] (170). The reaction of the tetranuclear precursor with dppe yielded $\{Pt(C_6F_5)_2(\mu\text{-}C\equiv CR)_2Ag\}_2(\mu\text{-dppe})$ (R = Ph, ^tBu) wherein the dppe ligand bridges two identical Pt units. The structure of 52e was established by single-crystal X-ray diffraction studies. NMR data (^1H, ^{19}F, ^{31}P) indicate that all the complexes exhibit dynamic behavior in solution.

54

55

56

Terminal transition metal acetylides $L_nMC\equiv CR$, $trans$-Pt(C$\equiv$CR)$_2$-L$_2$, cis-Pt(C$\equiv$C^{t}Bu)$_2$L$_2$, Pt(C$\equiv$CR)$_2$(dppe), and Au(C$\equiv$CR)PPh$_3$ react with cis-Pt(C$_6$F$_5$)$_2$(CO)THF to give the μ-η^1:η^2-acetylide-bridged dinuclear zwitterionic complexes (CO)(C$_6$F$_5$)$_2$Pt$^-${μ-η^1:η^2-C$\equiv$CR}M$^+$L$_n$ (**54**) [$trans$-ML$_n$ = PtL$_2$(CCR), L = PPh$_3$, R = Ph (**54a**), ^{t}Bu (**54b**), L = PEt$_3$, R = Ph (**54c**), ^{t}Bu (**54d**); cis-ML$_n$ = PtL$_2$(CCR), R = ^{t}Bu, L = PPh$_3$ (**54e**), L = PEt$_3$ (**54f**); L = dppe R = ^{t}Bu (**54g**, SiMe$_3$ (**54h**); ML$_n$ = AuPPh$_3$, R = Ph (**54i**), ^{t}Bu, cis (**54j**), $trans$ (**54k**)] (*171*). In addition to **54e** the dinuclear complex cis-(PPh$_3$)$_2$Pt(μ-η^1:η^2-C$\equiv$C^{t}Bu)$_2$Pt(C$_6$F$_5$)$_2$ (**55a**) was also formed during the reaction. Complex **54a** rearranges to the dimer {Pt(PPh$_3$)(C$_6$F$_5$)(μ-σ/π-C$\equiv$CPh)}$_2$ (**56a**) in which each Pt is σ-bonded to one acetylide ligand and π-bonded to the other when refluxed in benzene for 1 hour. The structure of **54a** was confirmed by a crystal structure determination. Binuclear complexes of platinum with doubly bridging acetylide ligands L$_2$Pt(μ-η^1:η^2-C$\equiv$CR)$_2$M(C$_6$F$_5$)$_2$ (**55**) [M = Pt, R = Ph, L$_2$ = dppe (**55b**), COD (**55c**), L = PPh$_3$ (**55d**), R = ^{t}Bu, L$_2$ = dppe (**55e**), COD (**55f**), L = PPh$_3$ (**55g**); M = Pd, R = Ph, L = PPh$_3$ (**55h**), R = ^{t}Bu, L = PPh$_3$ (**55i**)] have been prepared by reacting cis-PtL$_2$(C$\equiv$CR)$_2$ with cis-M(C$_6$F$_5$)$_2$(THF)$_2$ (M = Pt, Pd) (*22*). The structure of (dppe)Pt(μ-η^1:η^2-C$\equiv$CPh)$_2$Pt(C$_6$F$_5$)$_2$ (**55a**) was established by X-ray diffraction methods.

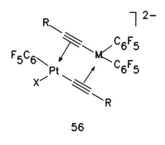

56	M	X	R
b	Pt	C$_6$F$_5$	Ph
c	Pt	C$_6$F$_5$	Bu^t
d	Pt	C$_2$R	Ph
e	Pt	C$_2$R	Bu^t
f	Pd	C$_6$F$_5$	Ph
g	Pd	C$_6$F$_5$	Bu^t

The reaction of [Bu$_4$N]$_2$[cis-Pt(C$_6$F$_5$)$_2$(C$\equiv$CSiMe$_3$)$_2$] with {Pt(C$_3$H$_5$)Cl}$_4$ in a 4:1 molar ratio and with {Pd(η^3-C$_3$H$_5$)Cl}$_2$ in a 2:1 molar ratio afforded cis-[{Pt(C$_6$F$_5$)$_2$(μ-η^1:η^2-C$\equiv$CSiMe$_3$)$_2$}M(η^3-C$_3$H$_5$)]$^-$ (**55**) [M = Pt (**55j**), Pd (**55k**)] (*172*). The dianionic complexes cis-[Pt(C$_6$F$_5$)$_2$-(C$\equiv$CR)$_2$]$^{2-}$ and [Pt(C$_6$F$_5$)(X)(C$\equiv$CR)$_2$]$^{2-}$ (R = Ph, ^{t}Bu) on the other hand, react with cis-M(C$_6$F$_5$)$_2$(THF)$_2$ to afford the bimetallic complexes [(C$_6$F$_5$)(X)Pt(μ-σ/π-C$\equiv$CR)$_2$M(C$_6$F$_5$)$_2$]$^{2-}$ (**56**) [M = Pt, X = C$_6$F$_5$, R = Ph (**56b**), R = ^{t}Bu (**56c**); M = Pt, X = C$\equiv$CR, R = Ph (**56d**), ^{t}Bu (**56e**); M = Pd, X = C$_6$F$_5$, R = Ph (**56f**), ^{t}Bu (**56g**)], where one of the bridging σ,π-acetylide ligands migrates, thereby changing coordination modes between the two metal atoms (*22*). This conversion can be represented by:

$$\{Pt(\eta^1\text{-}C_2R), M(\eta^2\text{-}C_2R)\} \rightarrow \{Pt(\eta^2\text{-}C_2R), M(\eta^1\text{-}C_2R)\}$$

The structure of one of the products $[PMePh_3]_2[(C_6F_5)_2Pt(\mu\text{-}C{\equiv}CPh)_2\text{-}Pt(C_6F_5)_2]$ (**55b**) was confirmed by an X-ray structure analysis (*22*). It seems reasonable to assume that the migration (**A**→**D**) proceeds via a $\mu\text{-}\eta^1{:}\eta^1\text{-}C{\equiv}CR$ intermediate (**C**). In fact, the structural data of **51** show that although the acetylene ligands were σ-bonded to the Ti center initially, the bridging ligands are tilted in such a way that the σ-bonding orbitals point more to the platinum atom in the final product (*167*).

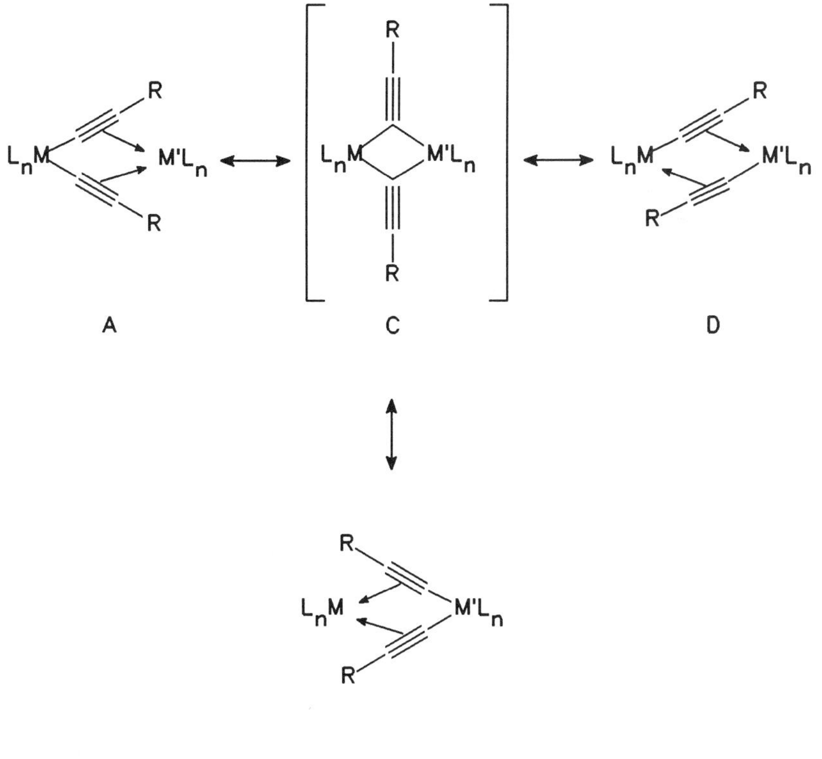

The neutral binuclear double-bridged acetylide complexes $\{Pt(\mu\text{-}\sigma/\pi\text{-}C{\equiv}CR)(C_6F_5)(PPh_3)\}_2$ (**56**) [R = Ph (**56h**), ^{t}Bu (**56i**), $SiMe_3$ (**56j**)] have been synthesized from *trans*-$Pt(C{\equiv}CR)_2(PPh_3)_2$ and *cis*-$Pt(C_6F_5)_2(THF)_2$ (*173*). The crystal structure of **56h** shows that two identical $Pt(C_6F_5)(PPh_3)(C{\equiv}CPh)$ units are joined together through η^2-bonding of M—$C{\equiv}CPh$ moieties. This again establishes a site preference for ligands and involves a rearrangement of groups between two metal centers. The π-coordination modes can be displaced by ligands L (L = PPh$_3$, pyridine) to give bridge-cleaved mononuclear products.

The reaction of Ir(COD)Cl dimer with bis(trimethylsilyl)acetylene gives the bimetallic complex Ir$_2$(COD)$_2$(μ-η^1:η^2-C≡CSiMe$_3$)$_2$ (57), in which each Ir is σ-bonded to one alkynyl ligand and π-bonded to the other (174). The synthesis of the analogous Rh complex requires the addition of methyllithium. The dinuclear compounds are cleaved by trimethylphosphine to form mononuclear complexes with σ-alkynyl ligands. Complex 57 reacts with an excess of COD by cleaving the dimer and adding over the alkyne ligand to give an unusual bicyclic ligand with allyl- and alkene-bonding functions, Ir(COD)(2-4-η^3:9-η^2-bicyclo[3.3.1]non-2-en-4-yl-9-trimethylsilylmethylene).

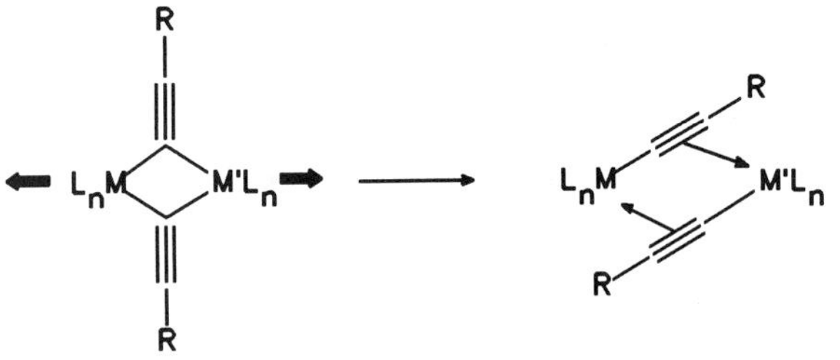

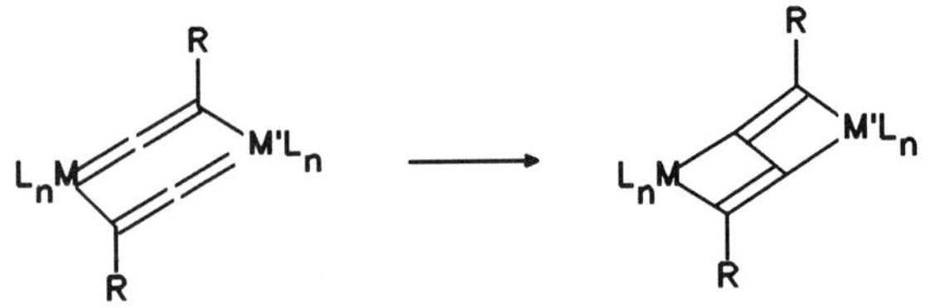

Zirconocene, "Cp$_2$Zr," was generated in situ and treated with 2 equiv of trimethylsilylacetylene to afford, in very low yield, the dimer {ZrCp$_2$(μ-η^1:η^2-C≡CSiMe$_3$)}$_2$ (58a) (175). The corresponding titanium analog {Cp$_2$Ti(C≡CSiMe$_3$)}$_2$ (58b) was prepared earlier by Hawthorne and co-workers (176) through the addition of an ether slurry of NaC≡CSiMe$_3$ to titanocene dichloride. Structural analysis of 58a and 58b reveal that the two metal atoms and the four acetylene carbon atoms lie in a plane and that the metal–carbon σ and π bond distances of the two acetylene ligands do not differ significantly. The structure is well represented by structural type D.

Erker and co-workers (143,177) reacted M(R'Cp)$_2$(C≡CR)$_2$ (M = Zr,

Hf; R = Ph, Me; R' = H, Me, 'Bu) with zirconocenes (R''Cp)$_2$Zr (R'' = H, Me, 'Bu), generated *in situ* from the corresponding (η^4-butadiene)zircono-cene precursors to give, in yields of around 60%, binuclear σ,π-alky-nyl-bridged complexes (R'Cp)$_2$M(μ-η^1:η^2-C≡CR)(μ-η^2:η^1-C≡CR)Zr-(R''Cp)$_2$ (59) [M = Zr, R' = R'' = H, R = Me (59a), Ph (59b), R' = R'' = Me, R = Me (59c), Ph (59d), R' = R'' = 'Bu, R = Me (59e), Ph (59f); M = Hf, R' = R'' = H, R = Me (59g), Ph (59h)]. Again, by combining metal fragments with different alkyl substituents, it was possible to show from the spectroscopic and structural data of the products that during the course of the reaction a σ-acetylide migration between the two metal centers occurred, which again demonstrates that in bimetallic systems acetylide ligands may have a site preference with respect to σ- and π-coordination modes. Although the complex {Zr(Cp')(μ-C≡CPh)}$_2$ (59d) displays essentially the same structural parameters as 58a with respect to the bridging units, the Zr–C bond distances are significantly shorter than the corresponding distances of mononuclear bis(acetylide) complexes of zirconium (*143*). Bond length values indicate considerable π interaction between the metal centers and adjacent organic π system across the connecting σ bond, which justifies contributions from structure types **D** and **E**.

Molecular orbital studies (*178*) of bis(μ-alkynyl)-bridging complexes of the type L$_2$M(μ-η^1:η^2-C≡CR)$_2$ML$_2$ (L$_2$M = Cp$_2$Zr, Cp$_2$Ti) (**D**) reveal that the in-plane π and π* orbitals of the bridging acetylenes find bonding partners among them. Bending the bridging ligands (**B**) enhances such interactions, lengthens the carbon–carbon triple bond distances, and may ultimately result in weak bonding interactions between the two σ-bonded carbon atoms of the two acetylene ligands. In the titanium complex the M–C and C$^\alpha$–C$'^\alpha$ nonbonding distances are shorter compared to the corresponding separation in a zirconium analog. It is predicted that this separation may be small for Ti and that the bonding interaction between the carbon atoms may go all the way to form C$^\alpha$–C$'^\alpha$ coupled products (**F**). The resulting binuclear complex L$_2$Ti(μ-RC$_4$R)TiL$_2$ with R = H has a bridging HC$_4$H ligand which can be described as a zigzag butadiyne li-gand. In fact, it is known that (phenylethynyl)titanocene spontaneously dimerizes under conditions of preparation by C–C coupling (*179*). The structure type **F** was established for the dinuclear product **60** by X-ray analysis (Table XI), which showed a C–C coupling distance of 1.485 Å and Ti–Ti distance of 4.227 Å (*180*). Interestingly, no C–C coupling was observed for **58b**, where the corresponding C–C distance was 2.706 Å and the Ti–Ti distance 3.550 Å (*176*).

The reaction of the RCp$_2$TiCl dimer (R = SiMe$_3$) with LiC≡CPh results in the coupling of two acetylide ligands to give a bimetallic complex

TABLE XI

SELECTED X-RAY DATA[a] FOR BIS σ,π-BRIDGED ACETYLENE MOIETIES IN BIMETALLIC COMPLEXES

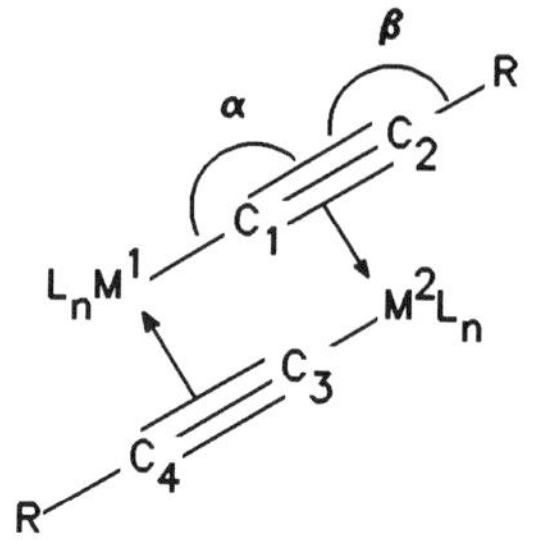

Complex			Bond distances (Å)					Bond angles (°)		
No.	M¹	M²	M¹–M²	M¹–C₁ M²–C₃	M²–C₁ M¹–C₃	M²–C₂ M¹–C₄	C₁–C₂ C₃–C₄	α	β	Ref.
55b	Pt	Pt	3.43	1.978	2.263	2.267	1.230	170.4	152.6	(22)
				2.023	2.369	2.304	1.219	173.2	154	
57	Ir	Ir	3.339	1.98	2.23	2.18	1.23	176	146	(174)
				2.01	2.24	2.18	1.21	175	146	
58a	Zr	Zr	3.522	2.191	2.426	2.407	1.249	—	—	(175)
				2.191	2.420	2.399	1.260	172.7	142.5	
59b	Zr	Zr	3.506	2.188	2.431	2.407	1.261	172.3	213.2	(143)
60ᵇ	Ti	Ti	4.227	2.153	2.325	2.083	1.325	155.8	128	(180)
62ᶜ	Ti	Ti	—	2.069	2.393	2.318	1.244	176.4	141.5	(182)

[a] Standard deviations omitted.
[b] C₁–C₃ distance 1.485 Å.
[c] C₁–C₃ distance 2.726 Å.

containing a bridging 1,4-diphenyl-1,3-butadiene ligand (181). Two dinuclear intermediates were isolated from the reaction, the first, containing two μ-η^1-C≡CPh (C) units, formed at low temperature after 2 hours and converted to the second over 16 hours, which contains two μ-η^1:η^2-C≡CPh (D) ligands. The final product, [RCp₂Ti(μ-PhC₄Ph)TiRCp₂] (61) (F), was isolated after stirring at room temperature.

The reaction of {CpZr(μ-Cl)}₂(μ-η^5:η^5-C₁₀H₈), where the two Zr centers are held closely together by the π-coordination to fulvalene and two bridged chloro ligands, with 2 equiv of LiC≡CPh give {CpZr(μ-C≡CPh)}₂(μ-η^2:η^2-C₁₀H₈) (181). Spectroscopic data indicate that both acetylene bridging ligands are σ-coordinated to the one and π-coordinated to the other metal center. The bridging ligands display definite alkylidene character, which corresponds to structure **E**.

Whereas the reaction of PhC≡C-C≡CPh with coordinatively unsaturated "Cp₂Ti" affords **60**, the analogous Me₃SiC≡C-C≡SiMe₃ gives {Cp₂Ti(η^1:η^2-C≡CSiMe₃)}₂ (**62**) (**D** and **E**) with cleavage of the original butadiyne ligand (*182*). The complex **62** can be obtained from the reduction of titanocene dichloride by Mg in the presence of 1,4-bis(trimethylsilyl)butadiyne or from the reductive elimination of Me₃SiC≡C-C≡CSiMe₃ from Cp₂Ti(C≡CSiMe₃)₂ and subsequent reaction with unreacted Cp₂Ti(C≡CSiMe₃)₂. Coupling of two phenylethynyl ligands has also been confirmed in a structural study of (Cp*₂Sm)₂{(μ-η^2:η^2-PhC₄Ph)} (*183*).

Homoleptic σ-bonded tetrakis(acetylide) complexes [Pt(C≡CR)₄]²⁻ react with *cis*-Pt(C₆F₅)₂(THF)₂ in a 1:1 ratio to give dinuclear complexes [NBu₄]₂[{Pt(C₆F₅)(C≡CR)(μ-σ/π-C≡CR)}₂] (**63**) [R = ᵗBu (**63a**), SiMe₃ (**63b**)] and in a 1:2 ratio to give the trinuclear complexes [(C₆F₅)₂Pt(μ-σ/π-C≡CR)₂Pt(μ-η^1:η^2-C≡CR)Pt(C₆F₅)₂]²⁻ (**64**) [R = Ph (**64a**), ᵗBu (**64b**), SiMe₃ (**64c**)] (*184*). The crystal structure of **64a** (Fig. 2) reveals many interesting features with respect to the central Pt(C≡CPh)₄ unit and rearrangement processes. The expected acetylide migrations occurred only on the one side of the molecule, leaving two adjacent Pt fragments formally negatively charged. One is tempted to predict that this complex under thermal conditions should, after further rearrangement, which also involves the other side of the molecule, afford a more symmetrically coordinated trimetallic product, with both the outermost Pt fragments negatively charged.

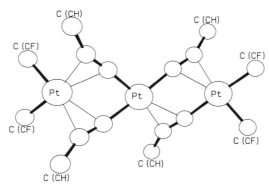

FIG. 2.   X-Ray structure of [(C₆F₅)₂Pt(μ-σ/π-C≡CPh)₂Pt(μ-η^1:η^2-C≡CPh)Pt(C₆F₅)₂]²⁻ (**64a**) [Forniés *et al.* (*184*)].

D. σ-Carbene, Carbyne/π-Alkene, Allyl, and Alkyne Complexes

1. σ-Carbene/π-Alkene

The coordination of a second transition metal fragment to the π-electron density of a metal–carbene bond is represented by the bonding type **A**. In fact, it was the recognition of the potential of such an interaction that lead to the well-known discovery by Stone and co-workers (89,185) that molecules containing unsaturated metal–carbon bonds (Lewis bases) can additively react with complexes containing vacant coordination sites (Lewis acids). The resulting product is best formulated with a three-membered dimetallopropane ring-structure, the discussion of which falls outside the scope of this article (8). The reactive fragment $Cp_2Ti=CH_2$ reacts with a variety of late transition metal complexes containing bridging halide ligands, $\{ML_n(\mu\text{-Cl})\}_2$, to yield early–late binuclear complexes (**B**) containing $\mu\text{-}CH_2$ and $\mu\text{-}Cl$ ligands (186,187). Owing to the bridging ligands the metal–metal bond distances are sufficiently short to assume that some direct metal–metal interaction is present.

A

B

The deprotonation of acidic hydrogen atoms in the α position to the carbene carbon atom of many Fischer carbene complexes affords the very reactive and versatile carbene anions $[L_nM\{=C(OR)CH_2\}]^-$ (188–190). Macomber et al. (191,192) have demonstrated that these carbene anions react effectively with organometallic substrates to give bimetallic complexes of chromium and tungsten with σ,σ-bonded bis(carbene) bridges. This bis(carbene) complex $(CO)_5W=C(NMe_2)CH_2CH_2CH(SiPh_3)C(OMe)=W(CO)_5$ underwent a clean rearrangement whereby the silyl group migrated to the carbene carbon to give $(CO)_5W\{=C(NMe_2)CH_2CH_2CH=C(OMe)SiPh_3\}$ (193). It was proposed that this novel rearrangement is facilitated by the metal center through a 1,3-shift of the

SiPh$_3$ group to the tungsten, yielding $(CO)_5W\{=C(NMe_2)CH_2CH_2$ $CH=C(OMe)\}W\{SiPh_3\}(CO)_5$, followed by a reductive elimination supported by a σ,π-bridged bimetallic intermediate $(CO)_5W\{\mu\text{-}\eta^1\text{:}\eta^2\text{-}$ $C(NMe_2)CH_2CH_2CH=C(OMe)SiPh_3\}W(CO)_5$.

One of several complexes formed by photochemically reacting $CpRe(CO)_3$ and $PhC\equiv CH$ in THF, was the binuclear complex $Cp\text{-}$ $(CO)_2Re\{\mu\text{-}\eta^1\text{:}\eta^2\text{-}(=C=C(Ph)C(Ph)=CH_2)\}ReCp(CO)_2$ (65) (129). The formation of 65 involves a photoinduced coupling of two acetylene units. The structure of 65 reveals that the vinyl substituent of a vinylidene complex is η^2-bonded to a second rhenium center.

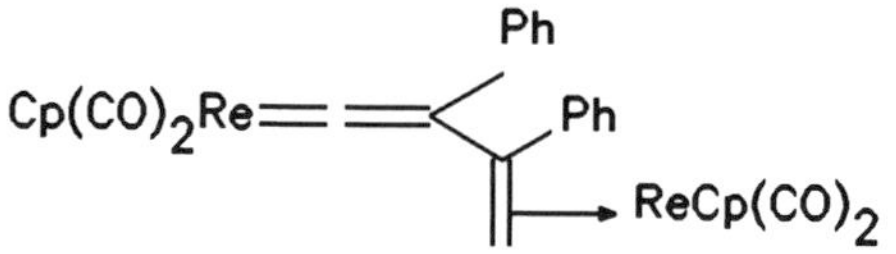

65

Although very few examples of bimetallic complexes having carbene–alkene or carbene–alkyne σ,π-bridging ligands have been reported, their existence as possible intermediates has been postulated (194). Electronic and geometric properties of metal fragments and bridging bifunctional ligands could promote intermolecular coordination (B, C), which presents an alternative to intramolecular interaction (A).

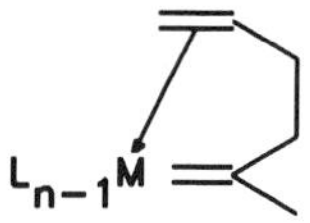

A 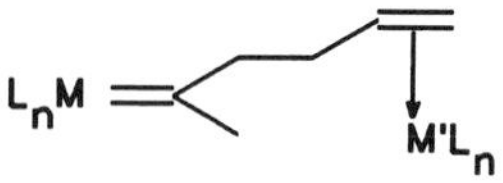   B

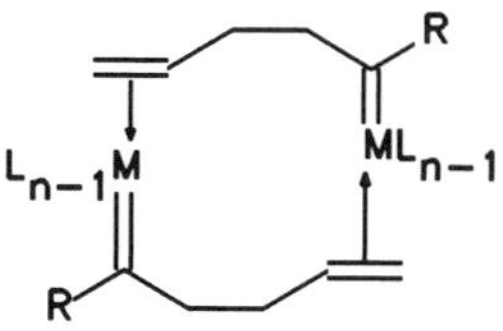

C

The reaction of the chromium and tungsten carbene anions $[(CO)_5\text{-}$ $M=C(OMe)CH_2\}]^-$ (M = W, Cr) with coordinated alkenes has been studied by Beck and co-workers (195,196), as well as by Geoffroy and co-workers (197). The reaction of the carbene anions with cationic complexes having a coordinated ethylene ligand leads to the formation of

σ,σ-bridged bimetallic carbene–alkyl complexes, $(CO)_5M\{=C(OMe)$-$CH_2CHR'CHR''\}M'L_n$ [$M'L_n$ = Re(CO)$_5$, M = W, Cr, R' = R'' = H) (195,196); $M'L_n$ = FeCp(CO)$_2$, M = Cr, R' = Me, R'' = Me, H, R' = Ph, R'' = Me (197)]. The carbene anions react with the cationic allyl complex $[Cp(CO)(NO)Mo(\eta^3$-$C_3H_5)]^+$ to give σ,π-bridged bimetallic carbene–alkene complexes, $[(CO)_5M\{=C(OMe)CH_2CH_2CH=CH_2$-$\eta^2$-$\}MoCp$-(CO)(NO)] (66) [M = Cr (66a), W (66b)] [Eq. (29)] (195).

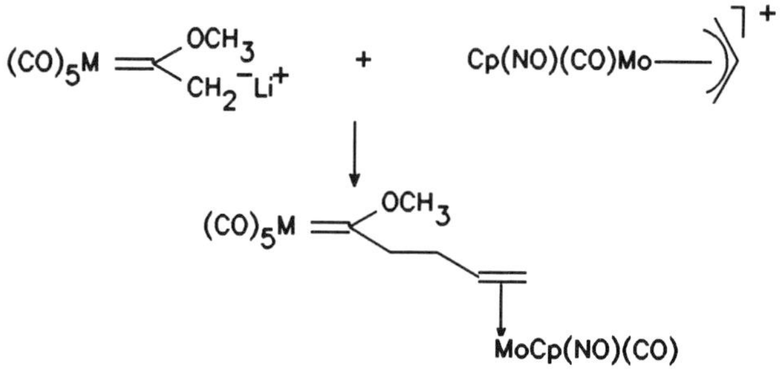

$$(CO)_5M = \!\!<^{OCH_3}_{CH_2^-Li^+} \quad + \quad Cp(NO)(CO)Mo\!-\!\!\big\rangle\big\rangle^{\Big]^+} \qquad (29)$$

$$\downarrow$$

$$(CO)_5M = \!\!<^{OCH_3}$$

MoCp(NO)(CO)

66

Addition of the carbene anions $[(CO)_5M\{=C(OMe)CH_2\}]^-$, to the cationic tropylium complexes, $[M'(CO)_3(\eta^7$-$C_7H_7)]^+$, gave $(CO)_5M\{=C(OMe)CH_2$-C_7H_7-η^6-$\}M'(CO)_3$ (67) [M = Cr, M'' = Cr (67a), Mo (67b); M = W, M' = Cr (67c), Mo (67d)] (196). Aumann and Runge (198) have studied the reaction of the carbene complexes $M\{=C(OEt)Me\}(CO)_5$ with the uncoordinated tropylium ion, $C_7H_7^+BF_4^-$, in the presence of the base NEt$_3$. Thermolysis of the product, the 2-cycloheptatrienyl–ethylidene complex, afforded as the major reaction product 4,5-homotropilidene and as a by-product from a side reaction $(CO)_5Cr\{=C(OEt)$-CH_2-C_7H_7-η^7-$\}Cr(CO)_3$ (67e). A crystal structure determination of 67e revealed a W=C bond distance which is typically found for tungsten carbene complexes (199) in which the carbene moiety is located on the exo position of the C$_7$H$_7$ ligand.

The bimetallic complex $(CO)_5Cr\{=C(OMe)CH_2$-C_6H_7-η^4-$\}Fe(CO)_3$ (68a) was synthesized by reacting the anionic carbene precursor with the cationic cyclohexadienyl complex $[Fe(\eta^5$-cyclohexadienyl)(CO)$_3]^+$ (197). Complex 68a is chiral, and a structure determination showed that both enantiomers are present in the unit cell. Deprotonation of a second acidic proton from the position α to the carbene carbon of 68a by n-BuLi and subsequent treatment with another equivalent of the iron substrate afforded the trimetallic σ,π,π complex $(CO)_5Cr=C(OMe)CH\{C_6H_7$-$\eta^4$-$Fe(CO)_3\}_2$ (68b). The complex 68b has three chiral centers and can as a

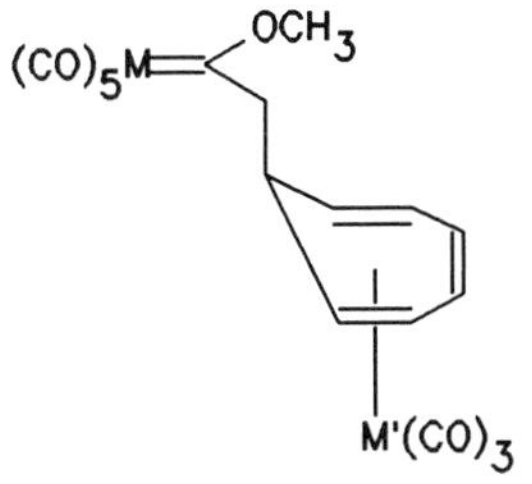

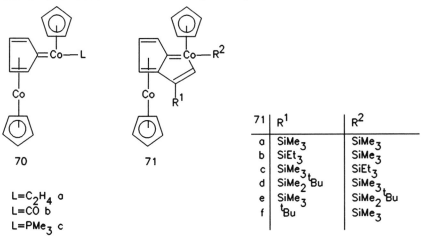

<table>
<thead>
<tr><th colspan="3">67</th></tr>
<tr><th></th><th>M</th><th>M'</th></tr>
</thead>
<tbody>
<tr><td>a</td><td>Cr</td><td>Cr</td></tr>
<tr><td>b</td><td>Cr</td><td>Mo</td></tr>
<tr><td>c</td><td>W</td><td>Cr</td></tr>
<tr><td>d</td><td>W</td><td>Mo</td></tr>
</tbody>
</table>

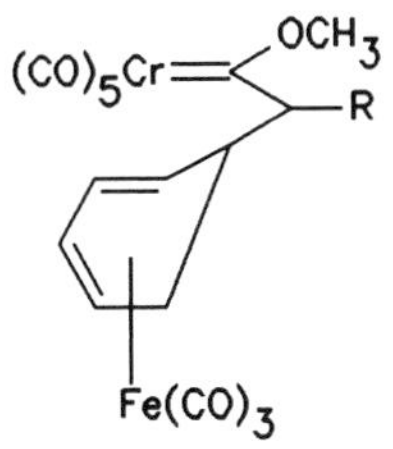

R=H a

R=$(\eta^4$-$C_6H_7)$Fe(CO)$_3$ b

result exist as a pair of enantiomers or as two meso forms. Two of these were isolated and separately characterized by X-ray crystallography. As predicted from the method of synthesis, the structure displays the chromium carbene moiety to be situated on the exo faces of the coordinated cyclohexadiene rings.

Deprotonation of W{=C(NEt$_2$)Me}(CO)$_5$ and subsequent treatment with carbon disulfide generates *in situ* the anion [W{=C(NEt$_2$)CH$_2$CS$_2$}(CO)$_5$]$^-$, which reacts with the cationic allyl complex [MoCp(CO)(NO){η^3-C$_3$H$_5$}]$^+$ to give (CO)$_3$W{=C(NEt$_2$)CH$_2$C(S)SCH$_2$CH=CH$_2$-η^2-}MoCp(CO)(NO) (**69**) (*200*). The ^{13}C NMR spectrum of **69** shows a duplication of resonances, indicating the presence of diastereoisomers in solution and a nonstereoselective reaction pathway.

70   71

L=C$_2$H$_4$ a
L=CO b
L=PMe$_3$ c

<table>
<thead>
<tr><th>71</th><th>R^1</th><th>R^2</th></tr>
</thead>
<tbody>
<tr><td>a</td><td>SiMe$_3$</td><td>SiMe$_3$</td></tr>
<tr><td>b</td><td>SiEt$_3$</td><td>SiMe$_3$</td></tr>
<tr><td>c</td><td>SiMe$_3$</td><td>SiEt$_3$</td></tr>
<tr><td>d</td><td>SiMe$_2$^{t}Bu</td><td>SiMe$_3$</td></tr>
<tr><td>e</td><td>SiMe$_3$</td><td>SiMe$_2$^{t}Bu</td></tr>
<tr><td>f</td><td>^{t}Bu</td><td>SiMe$_3$</td></tr>
</tbody>
</table>

The ethylene ligand in $CoCp\{\eta^2\text{-}C_2H_4\}_2$ readily undergoes substitution and reacts at 50°C with $CoCp(\eta^4\text{-}C_5H_6)$, whereby a Co fragment displaces the activated hydrogens of the η^4-cyclopentadiene ligand to give $CpCo\{\mu\text{-}C_5H_4\}CoCp(\eta^2\text{-}C_2H_4)$ (**70a**) (*201*). The ethylene ligand in **70a** is easily substituted by ligands L to give $CpCo\{\mu\text{-}C_5H_4\}CoCp(L)$ [L = CO (**70b**), PMe_3 (**70c**)]. Two structures are possible for the σ,π-bridging ligand. The bridging ligand may be of the cyclopentadienylidene type and have η^4-cyclopentadiene and η^1-carbene bonding functions, $\eta^1\text{:}\eta^4\text{-}C_5H_4$, or of a zwitterion cyclopentadienyl type and have η^5-cyclopentadienyl and η^1-alkyl bonding functions. According to the X-ray structure analysis and structural data of the bridging ligand, complex **70c** is better represented by a structure that incorporates a bridging diolefin–carbene ligand. The reaction of **70a** with silylalkynes is unusual and gives σ,π-bimetallic complexes **71a–f** with a cobalt bicyclic ring system as ligand (*202*). Alkynes with different silyl substituents give nonseparable mixtures of the stereoisomers in which each of the silyl substituents is bonded to the cobalt center. Characteristic are the resonances in the ^1H NMR spectrum of the cobalt–enyl hydrogen (δ 9.9–11.1 ppm) and the broad signal at δ values 30–40 ppm for the cobalt-coordinated silyl group in the ^{29}Si NMR spectrum. Insight into the bonding was provided from a crystal structure determination of an analogous compound **71g**. In **71g** two trimethylsilyl groups are present, but the Cp ligand of the π-bonded Co is replaced by an $\eta^5\text{-}C_5Me_4Et$ ligand. The carbacyclic five-membered ring of the cobaltapentalene ligand is bonded through all five carbon atoms.

2. σ-Carbene/π-Allyl

Erker and co-workers (*13,206–224*) have introduced a new synthetic route for the synthesis of Ficher-type metallocene oxocarbene complexes. This method, whereby a terminal carbonyl ligand is converted to a carbene by a second, early transition metal center, involves C–C coupling of a coordinated alkene with an η^2-coordinated terminal carbonyl at an early transition metal center [Eq. (30)]. Review articles (*13,207–209*) have appeared in the literature that extensively cover this area of chemistry, so only a short summary concerning bimetallic complexes with σ-carbene and π-allyl bridging ligands is presented.

$$ \tag{30} $$

TABLE XII

SELECTED DATA OF METALLACYCLIC ZIRCONIUMOXO CARBENE[a] COMPLEXES

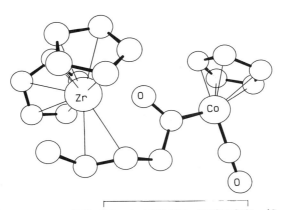

	Bond distances (Å)					^{13}C NMR (ppm),	
M^1L_n	Zr–C_1	Zr–C_2	Zr–C_3	M^1–C_x	C_x–O	C_x	Ref.
VCp(CO)$_3$	2.443	2.514	2.653	2.102	1.270	—	(210)
Cr(CO)$_5$	2.420	2.491	2.624	2.063	1.254	363.7	(212)
W(CO)$_5^b$	2.445	2.535	2.667	2.212	1.267	342.2	(213)
	2.421	2.497	2.621	2.198	1.262		
CoCp(CO)	2.423	2.492	2.614	1.815	1.287	311.9	(217)
RhCp(CO)	2.434	2.514	2.640	1.925	1.278	300.8	(218)

[a] C_x represents the carbene carbon atom.
[b] Two orientations for the allyl moiety afford crystallographically independent units.

In the compounds the carbene ligand of M' is attached to a second transition metal (M) by coordination through the oxygen and allyl functions of the carbene substituents. Early transition metal metallocenes with coordinated butadiene, $MCp_2(\eta^4\text{-}C_4H_6)$ (M = Zr, Hf, Th), react with metal carbonyls $L_nM'CO$ to give $Cp_2M(\mu\text{-}\eta^3\text{:}\eta^1\text{-}CH_2CH=CH\text{-}$

FIG. 3. X-Ray structure of $Cp_2Zr\{\mu\text{-}\eta^3\text{:}\eta^1\text{-}CH_2CH=CHCH_2C(O)\}=\{CoCp(CO)\}$ (72) [Eker et al. (217)].

$CH_2C={M'L_n}O)$ (**72**) [M = Zr, M' = Zr, Hf, L_n = Cp(CO) (*13*); M'L_n = VCp(CO)$_3$, NbCp(CO)$_3$ (*13,210,211*); Cr(CO)$_5$, Mo(CO)$_5$, W(CO)$_5$ (*212–215*); Fe(CO)$_3$L, L = CO, PPh$_3$ (*216*); Co(XCp)(CO), X = H, Cl (*217*); RhCp(CO) (*218*); *cis*- and *trans*-Pt(C$_6$F$_5$)$_2$L, L = CO, P^{i}Pr$_3$ (*219*); M = Hf, M'L_n = VCp(CO)$_3$ (*220*); M = Th, M'L_n = Cr(CO)$_5$ (*221*)]. The most important structural features of these compounds (*210,212–214*) are an asymmetrically Zr-bonded η^3-allyl group, shorter zirconium–carbon bond lengths for the terminal bonded allyl carbon, a very short carbon–oxygen distance, and a relatively long carbon–metal bond distance for the carbene ligand (Table XII, Fig. 3). Although significant acylmetal complex character is present in the majority of the examples studied, very short M–C(carbene) distances were recorded for the cobalt (*217*) and rhodium compounds (*218*). The dialkylboryl complex HfCp$_2${η^4-*trans*-(9-BBN)-butadiene} (9-BBN is borabicyclo[3.3.1]nonane), generated by low temperature photolysis, reacts regioselectively with C–C coupling at the less substituted (9-BBN)butadiene terminus to yield a 4:1 mixture of the isomeric (π-allylmetaloxy)carbene and seven-membered (σ-allylmetaloxy)-carbene complexes (*222*). In the case of the hafnium metallocycle with niobium and vanadium carbenes, an equilibrium is established in solution between the π-allyl and seven-membered σ-allyl forms (*220*). The complexes **72** react with ketones and aldehydes by insertion into the metal–allyl bond to give extended nine-membered metallocyclic zirconoxycarbene complexes (*223,224*). The C–C double bond of the nine-membered ring is located directly in front of the metal atom (Zr, Hf) and is found in one of two isomeric forms. Although the metal is formally a 16-electron species, the rigid nature of the metallocyclic framework places the double bond outside the range for π-bonding interaction (*220*).

3. σ-Carbene/π-Alkyne

As far as we are aware only isolated examples of stable, fully character-ized bimetallic complexes which could be classified having σ-carbene and π-alkyne bridging ligands and without metal-metal bonds have been re-corded. Gladysz and co-workers (*203,204*) developed synthetic routes to synthesize σ,σ-bimetallic complexes of the general formula $L_nMC_xM'L_n$ (*x* = 2, 3). These methods were based on the deprotonation of ter-minal acetylide complexes and their metallation with organometallic substrates or reaction with metal carbonyls. The reaction of Re-Cp*(PPh$_3$)(NO)(C≡CLi) with dirheniumdecacarbonyl and subsequent al-kylation with Me$_3$OBF$_4$ afforded the trimetallic Fischer carbene complex (**X**). Alkoxy abstraction by the electrophilic BF$_3$ resulted in cleavage of the Re–Re metal bond and formation of a novel trimetallic complex in which the C_3 chain is anchored by σ-bonded transition metal fragments on each end and spanned by a third Re fragment which is π-bonded [Eq. (31)] (205). The structure of $[Cp^*(Ph_3P)(NO)Re(\mu-\eta^1:\eta^1:\eta^3-CCC\{Re(CO)_5\})]Re(CO)_4]BF_4$ (73) was confirmed by X-ray crystallography. It is possible to draw many resonance forms for the bridging C_3 ligand and the descriptions $[Re]\!=\!C\!-\!C\!\equiv\!C\!-\![Re]$ and $[Re]\!=\!C\!=\!C\!=\!C\!-\![Re]$ were presented for the chain. However, the pattern of Re–C and C–C bond lengths suggests that a structural formulation of $Cp^*(Ph_3P)(NO)Re\!=\!C\!-\!C\!\equiv\!C\!-\!Re(CO)_5$ dominates.

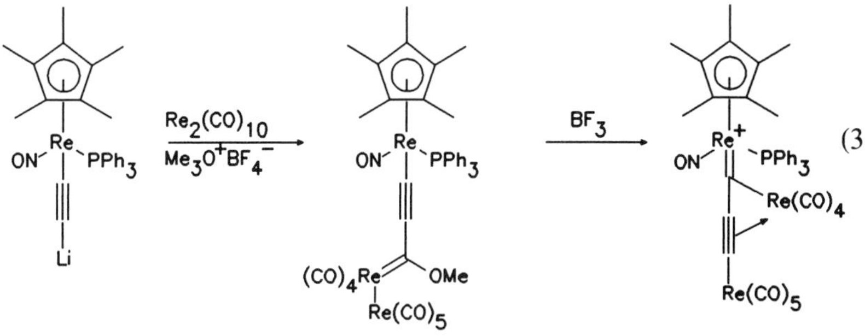

$$(31)$$

An interesting carbene–alkyne intermediate was postulated for the conversion of carbene complexes to chrysene derivatives, which again illustrates the potential of σ,π-activation of bridging ligands by metal centers [Eq. (32)] (225). The thermal decomposition of the carbene complex Cr-

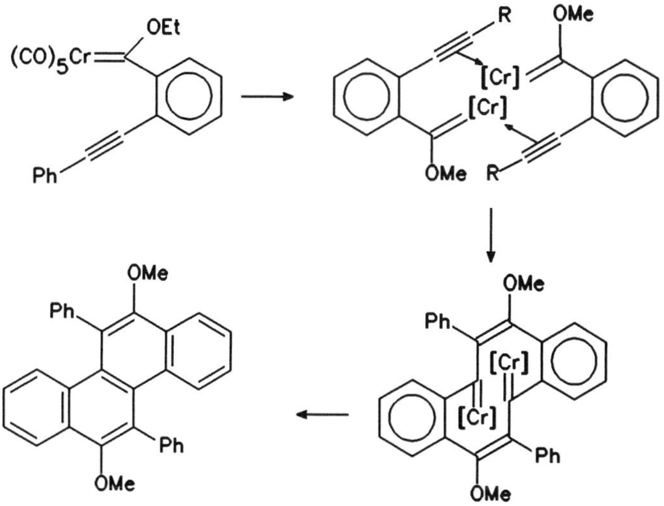

$$(32)$$

$\{C(OMe)C_6H_4\text{-}o\text{-}C\equiv CPh\}(CO)_5$ leads directly to the formation of a chrysene derivative via the formal dimerization of the carbene ligand. A plausible explanation for the formation of the final product involves a doubly alkyne-bridged dinuclear complex, alkyne insertions into metal–carbene bonds, and coupling of the carbene carbons.

4. σ-Carbyne/π-Alkene

Utilization of the electron-rich metal–carbyne bond to function as a π-ligand toward coordinatively unsaturated metal complexes has widely been employed by Stone and co-workers (226,227) and represents a versatile route for the synthesis of bimetallic and trimetallic complexes with metal–metal bonds. Bimetallic carbido-bridged complexes [M]=C=[M]

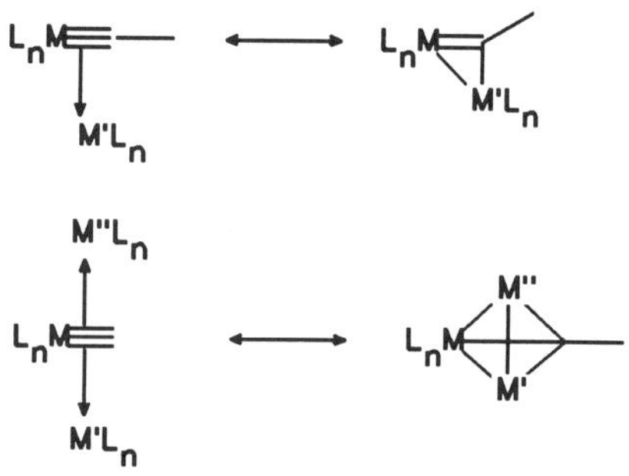

(228,229) or metallated carbyne complexes [M]≡C—[M] (230–232) and their chemistry are growing in interest. Templeton and co-workers (233) prepared the μ-carbide complex $Tp(CO)_2Mo\equiv CFeCp(CO)_2$ [Tp is tris(3,5-dimethylpyrazolyl)borate] by reacting the chlorocarbyne complex $Tp(CO)_2Mo\equiv CCl$ with $K[FeCp(CO)_2]$ and studying the chemical reactivity of the bimetallic complex. The complex was protonated with fluoroboric acid to give a cationic μ-methyne product. Spectroscopic data are representative of a product displaying agostic character for the proton and a small barrier against the oscillation of the metal fragments.

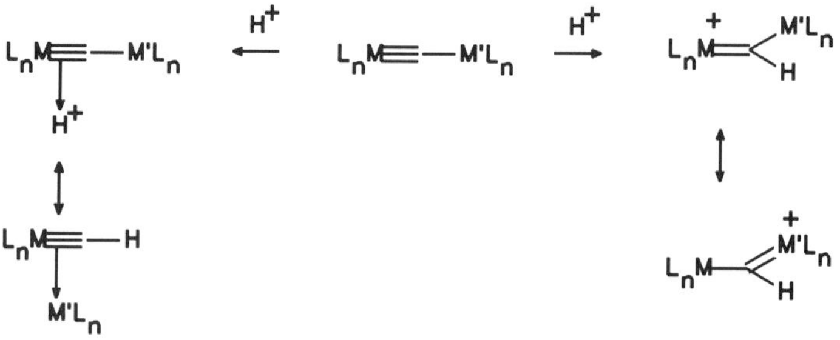

The addition of the anionic thiocarbyne complex $Tp(CO)_2M\equiv CS^-$ (M = Mo, W) (234) to cationic organometallic Lewis acids $L_nM^+BF_4^-$, $[ML_n(\eta^2\text{-}C_2H_4)]^+$, and $[ML_n(\eta^2\text{-}C_2H_2)]^+$ affords the σ,σ-bimetallic complexes $Tp(CO)_2M\{\equiv CS\text{-}\}ML_n$, $Tp(CO)_2M\{\equiv CSCH_2CH_2-\}ML_n$, and $Tp\text{-}(CO)_2M\{\equiv CSCH=CH-\}ML_n$, respectively (18,195). The complexes $Tp(CO)_2M\equiv CS^-$ (M = Mo, W) react with the cationic allyl complexes $[M'L_n\{\eta^3\text{-}C_3H_5\}]^+$ to give bimetallic complexes with σ-carbyne and π-alkene bridges, $Tp(CO)_2M\{\equiv CSCH_2CH=CH_2\text{-}\eta^2\text{-}\}M'L_n$ (**74**) [M = Mo, $M'L_n = Fe(CO)_4$ (**74a**), $MoCp(CO)(NO)$ (**74b**); M = W, M' = $Fe(CO)_4$ (**74c**), $MoCp(CO)(NO)$ (**74d**)] (195). Whereas the thiocarbyne precursors attack the cyclohexadienyl ring in $[Fe\{\eta^5\text{-}C_6H_7\}(CO)_3]^+$ to give the corresponding $\eta^1:\eta^4$-carbyne–diolefin complex $Tp(CO)_2M\{\equiv CS-C_6H_7\text{-}\eta^4\}Fe\text{-}(CO)_3$ (**75**) [M = Mo (**75a**), W (**75b**)] attack on $[Mo\{\eta^7\text{-}C_7H_7\}(CO)_3]BF_4$ is not on the ring but on the metal center to give $Tp(CO)_2M\{\equiv CS\}Mo\{\eta^7\text{-}C_7H_7\}(CO)_2$.

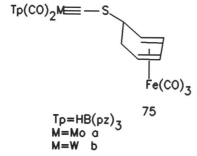

$Tp(CO)_2M\equiv{-}S$

$Fe(CO)_3$

$Tp=HB(pz)_3$ **75**
M=Mo a
M=W   b

E. σ-Alkyl, Carbene, and Carbyne/π-Arene Complexes

Bimetallic complexes where the σ-bonded metal center is bonded to an arene ring carbon atom, and is thus in direct contact with the π system of the arene ring via metal d orbitals, are distinguished from those where the σ-metal fragment forms part of a nonconjugated side chain of the ring.

1. π-Cyclopentadienyl Complexes

a. *σ-Alkyl Complexes.* σ,π-Bonded bridging cyclopentadienyl complexes fall into two classes of compounds, those where the cyclopentadienyl ligand is the only link between the two metal centers **A** and those where the Cp bridge is supported by other bridging ligands **C**. A special case of $L_nM\{\eta^5:\eta^1\text{-}C_5H_4\}M'L_n$ (**A**) has the ferrocenyl fragment σ-bonded to the second metal fragment, $FcM'L_n$ [Fc is $Fe(\eta^5\text{-}C_5H_5)(\eta^5\text{-}C_5H_4)$] (**B**). Similarly, the linking of a σ,π-bridged cyclopentadienyl bimetallic complex $L_nM\{(\mu\text{-}X)_m\}\{\eta^5:\eta^1\text{-}C_5H_4\}M'L_n$ **(C)** by a second cyclopentadienyl ligand ($X_m = C_5H_4$) in a σ,π fashion **(D)** represents merely a subgroup of compounds with supportive bridging ligands.

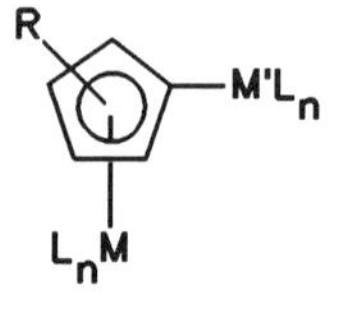

Bimetallic complexes of bonding type **A** were synthesized and studied extensively for the cyclopentadienyl complexes of especially Fe and Mn by Russian scientists between 1970 and 1980 (*235*). A facile method of synthesizing these compounds is via the reaction of the lithiated cyclo-pentadienyl complex $LiCpML_n$ with a metal halide, $M'L_nX$. In this way, many examples of bimetallic complexes with a wide range of σ-bonded transition metal fragments have been prepared. For example, $\{\mu\text{-}\eta^1:\eta^5\text{-}C_5H_4\{M'L_n\}\}Mn(CO)_3$ **(76)** [$M'L_n$ = Fp = $CpFe(CO)_2$ **(76a)** (*236*), $CpW(CO)_3$, $Re(CO)_5$, $Mn(CO)_5$, $Pt(PPh_3)Cl$, $CpNi(PPh_3)$, $AuPPh_3$, Ag, Cu (*235*)]. Decarbonylation of the acyl complex $(\eta^1:\eta^5\text{-}C_5H_4C(O)\{Fp\})Mn\text{-}(CO)_3$ **(77)** provides another route to $(\eta^1:\eta^5\text{-}C_5H_4\{Fp\})Mn(CO)_3$. The complex $(\eta^1:\eta^5\text{-}C_5H_4\{Fp\})Fe(CO)_2R$ **(78a)** (R = Me) was also synthesized from the metallation reaction with n-BuLi and FpI (*237*). Complex **78a** under-went further metallation at the unsubstituted Cp-ring and was sub-sequently treated with Me_3SiCl or FpI to give $(\eta^1:\eta^5\text{-}C_5H_4\{Fe(\eta^5\text{-}C_5H_4R')\}(CO)_2\}Fe(CO)_2R$ **(78)** [R' = $SiMe_3$, R = Me **(78b)**, CH_2Ph **(78c)**; R' = H, R = CH_2Ph **(78d)**] (*238*). The trimetallic complex $Cp(CO)_2Fe\{\mu\text{-}\eta^1:\eta^5\text{-}C_5H_4\}(CO)_2Fe\{\mu\text{-}\eta^1:\eta^5\text{-}C_5H_4\}Mn(CO)_3$ **(79a)** was synthesized simi-larly from $Cp(CO)_2Fe\{\mu\text{-}\eta^1:\eta^5\text{-}C_5H_4\}Mn(CO)_3$ **(76a)**. X-Ray structure de-termination of **76a** and **79a** confirms that the metals are bonded to bridging Cp rings in a σ,π fashion, that the rings are asymmetrical with the ipso-

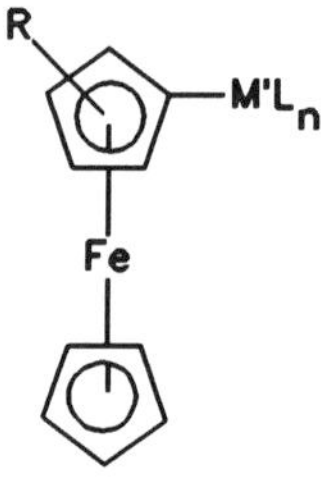

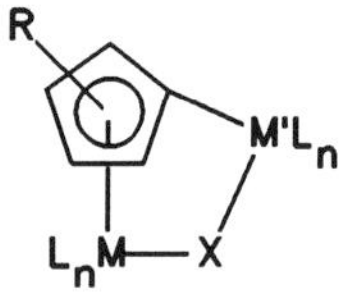

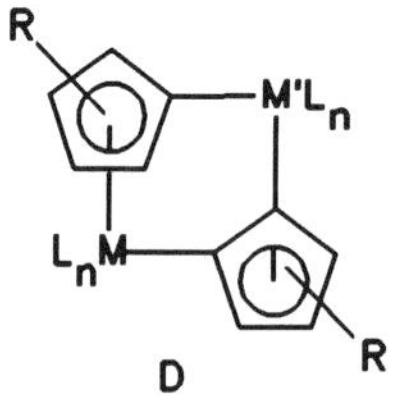

carbon atom noticeably further from the metal than the other ring carbons, and that the σ-bonded metal atom is bent away from the π-coordi-
nated metal toward the open side of the ring. The complex **76a** was
metallated and treated with C_3F_7I to give **79b** (R = I) (*239*). The photolysis
of **78d** in the presence of PPh_3 gave $(\eta^1:\eta^5\text{-}C_5H_4\{FeCp(CO)\text{-}}$
$(PPh_3)\})Fe(CO)_2(CH_2Ph)$ **(78c)** (*240*).
    The perhalogenated cyclopentadienyl complex $Mn(\eta^5\text{-}C_5Cl_4Br)(CO)_3$
reacts with n-BuLi and subsequently with $Mn(CO)_5Br$ or $Au(PPh_3)Cl$ to
give $\{\eta^1:\eta^5\text{-}C_5Cl_4(M'L_n)\}Mn(CO)_3$ **(80)** $[M'L_n = Mn(CO)_5$ **(80a)**, AuPPh$_3$
(80b)] (*241*). Reactions of **80a** with PPh_3 give as the major reaction product cis-$(\eta^1:\eta^5\text{-}C_5Cl_4\{Mn(CO)_4(PPh_3)\})Mn(CO)_3$ **(80c)** and with n-BuLi and
$Au(PPh_3)Cl$, **80b**. The X-ray data of the crystal structure determinations
of **80a,b** show no or very weak interaction between the metal centers.
    Dinuclear complexes of iridium containing σ-cyclopentadienyl and hy-
drido bridges, $[CpM(\mu\text{-}H)_2\{\mu\text{-}\eta^5:\eta^1\text{-}C_5H_4\}IrHL_2]^+$ **(81)** [M = W, L =
PEt_3 **(81a)**, PMe_2Ph **(81b)**, PPh_2Me **(81c)**, PPh_3 **(81d)**, PCy_3 **(81e)**; M =
Mo, L = $PMePh_2$ **(81f)**, PPh_3 **(81g)**] were synthesized from cationic complexes of Ir with labile ligands of the type $[IrH_2L_2(solvent)_2]^+$ (solvent =
acetone, ethanol, L = tertiary phosphines) and hydrido complexes
which act as Lewis bases, Cp_2MH_2 (M = W, Mo) (*242–245*). One
hydrido bridge of **81d** can be opened by the addition of pyridine (py) or
dppe to give $[(H)(Ph_3P)_2(py)Ir\{\eta^1:\eta^5\text{-}C_5H_4\}\{\mu\text{-}H\}WCp(H)]PF_6$ **(82a)** and
$[(H)(Ph_3P)(dppe)Ir(\eta^1:\eta^5\text{-}C_5H_4)\{\mu\text{-}H\}WCp(H)]PF_6$ **(82b)**, respectively
(*245*). The crystal structures of **81c**, **81f**, and **82b** have been determined by
using X-ray diffraction methods, and of note is the decrease in Ir–M
interaction in going from Mo to W (Ir-W distance 2.706 Å, and Ir–Mo
distance 2.641 Å) (*244,245*). In **82b** the Ir–W distance of 3.077 Å is significantly longer than the corresponding distance in **81c**, which has two hydrido bridges (*245*).
    A binuclear zirconium(IV) complex with σ,π-cyclopentadienyl and
phosphine bridges, $Cp_2(H)Zr\{\eta^1:\eta^5\text{-}C_5H_4\}\{\mu\text{-}P(Ph_2)CH_2\}Zr\{CH_2PPh_2\}Cp$
(83), has been prepared as the minor product from $Cp_2Zr(CH_2PPh_2)Cl$ by
reduction with sodium amalgam (*246,247*). Both the free and coordinated

84a

phosphorus environments are identifiable in the ^{31}P NMR spectrum of **83**. Heating of **83** leads to the formation of a bis(σ,π-cyclopentadienyl)-bridged dimer with a metal–metal bond, $\{ZrCp(PPh_2Me)(\eta^1:\eta^5\text{-}C_5H_4)\}_2$. Reduction of Cp_2MCl_2 with magnesium in the presence of 1 equiv of PMe_3 affords the dimeric $\eta^1:\eta^5\text{-}C_5H_4$-bridged complexes $\{MCp(PMe_3)(\eta^1:\eta^5\text{-}C_5H_4)\}_2$ **(84)** [M = Ti **(84a)**, Zr **(84b)**, Hf **(84c)** in very high yields (*248*). The X-ray crystal structure of **84a** reveals two bridging Cp ligands as well as the two phosphine ligands which are cis to one another. Both of the bridgehead carbon atoms of the bridging Cp ligands are involved in a three-center bond with the two Ti atoms thereby disrupting the aromatic character of the Cp rings. A nonbonding Ti–Ti separation of 3.223 Å was observed.

$$\text{(33)}$$

84b

Under conditions of controlled oxidation of Zr(II)–bis(phosphine) complexes, it was shown that the oxidation proceeds over a tetranuclear species to afford the dimeric Zr(III) complex, $\{Cp(R_3P)Zr(\mu\text{-}\eta^1:\eta^5\text{-}C_5H_4)\}_2$ **(85)** [$PR_3 = PPhMe_2$ **(85a)**, PPh_2Me **(85b)**] (*249*). The oxidation of **85b** with I_2 afforded the dinuclear compound which has a Zr(III)–Zr(III) metal–metal bond, two bridging iodo ligands, and an $\eta^5:\eta^5$-fulvalene bridging ligand. The σ,π-coordinated cyclopentadienyl ligands couple to afford a final product with a π,π-coordinated fulvalene ligand, $Cp_2Zr_2\{\mu\text{-}I\}_2\{\mu\text{-}(\eta^5:\eta^5\text{-}C_5H_4\text{—}C_5H_4)\}$ [Eq. (33)]. The reaction of the Zr(II) compound $Cp_2Zr(PMe_3)_2$ and the Zr(IV) complex Cp_2ZrCl_2 leads to the corresponding bimetallic compound, but with bridging chloro ligands (*250*). The Zr–Zr distance in the solid structure of the molecule was 3.233 Å, and the plane for the Zr_2Cl_2 core was symmetrically folded along the Zr–Zr vector. In an experiment to establish the origin of the Cp ligands which combine to afford the fulvalene ligand, bis(methylcyclopentadienyl)zirconium dichloride was used as reaction partner with the phosphine precursor. The result clearly indicated that the fulvalene ligand originated only from the two Cp rings of the $Cp_2Zr(PMe_3)_2$ molecule and that the chloro ligands came from the zirconocene dichloride (*251*).

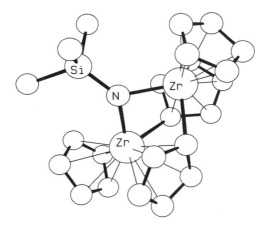

FIG. 4.   X-Ray structure of [{CpZr(μ-σ/π-C_5H_4)}$_2${μ-NSiMe$_3$}] (**86**) [Wielstra *et al.* (*252*)].

The oxidation of {Cp(Me$_3$P)Zr(η^1 : η^5-C$_5$H$_4$)}$_2$ (**84b**) by Me$_3$SiN$_3$ led to the formation of the dinuclear imido-bridged complex Zr$_2$Cp$_2$(μ-σ/π-C$_5$H$_4$)$_2${μ-NSiMe$_3$} (**86**), the crystal structure of which was determined (Fig. 4) (*252*). Complex **86** reacts with the mild oxidant PhSSPh via cleavage of the σ bond of one of the two bridging σ,π-cyclopentadienyl ligands, forming an asymmetric dinuclear complex Cp(PhS)Zr{μ-NSiMe$_3$}{μ-(η^1 : η^5-C$_5$H$_4$)}Zr(C$_5$H$_4$SPh)Cp (**87**). Complex **86** is surprisingly stable, and, notwithstanding the fact that the nonbonded C–C separation of the two bridgehead carbon atoms of the σ,π-cyclopentadienyl ligands is only 2.893 Å, reductive elimination with the formation of fulvalene could not be achieved by either thermolysis or photolysis. Oxidation of **84b** with several reagents (R$_3$PX$_2$, R = *n*-Bu, Me; X = Cl, Br, I; PhSSPh; Me$_2$PPMe$_2$) yielded the reductive coupling of the two η^1 : η^5-C$_5$H$_4$ rings (*253*).

The alkyne–Zr(II) complex Cp$_2$(THF)Zr{η^2-Me$_3$SiC≡CSiMe$_3$} readily loses the THF ligand and is transformed by the transfer of a hydrogen to the alkyne ligand, affording a dinuclear complex with a σ-alkenyl ligand, Cp{Me$_3$SiCH=C(SiMe$_3$)Zr{(σ/π-C$_5$H$_4$)$_2$}Zr{C(SiMe$_3$)C=CHSiMe$_3$}Cp (**88**) (*254*). The NMR data and X-ray crystal structure determination of **88** reveal agostic interactions of the alkenyl C–H bonds and the Zr centers. Complex **88** is reactive toward CO$_2$ and water, giving the dimeric zirconafuranone metallocycle and zirconoxane complex, respectively (*255*).

The compound {MoCp$_2$(H)Li}$_2$ reacts with N$_2$O to afford yellow dimers {MoCp(H)(η^1 : η^5-C$_5$H$_4$)}$_2$ (**89**) [*cis* (**89a**) and *trans* (**89b**)] (*256*). Complexes **89a,b** react with I$_2$ or MeI, yielding {MoCp(I)(η^1 : η^5-C$_5$H$_4$)}$_2$ (**86c**) or after irradiation in toluene and subsequent treatment with MeBr or MeI the corresponding bromo or iodo dimer **86d** (256,257). Reaction of the iodo dimer with Na(OEt)/NH$_4$PF$_6$ affords the cationic complex with a bridging ethoxylato ligand, [CpMo{μ-OEt}{(σ/π-C$_4$H$_4$)$_2$}MoCp]PF$_6$ **(90)**. The compounds **89a,b** rearrange thermally, affording Cp(I)Mo{μ-η^5:η^5-C$_5$H$_4$—C$_5$H$_4$}Mo(I)Cp which has a metal–metal bond and a π,π-fulvalene bridge. Photolysis of Cp$_2$WH$_2$ in ether or thermal decomposition of Cp$_2$W(Me)H in cyclohexane gives isomers of CpRW{(μ-σ/π-C$_5$H$_4$)$_2$}WCp(H) **(91)** [R = H **(91a)**, Me **(91b)**] (256). The bridging σ,π-cyclopentadienyl brings the two metal centers in close proximity and determines the separation between the two metal atoms (258,259). The existence or absence of metal–metal bonds in bimetallic complexes is not always unambiguous and is often based solely on bond length data. Therefore, the inclusion of dative covalent bonds between metals by authors to validate the 18-electron rule is common.

$$92 \tag{34}$$

The bridging η^5 : η^5-fulvalene ligand also fixes two metals in sufficiently close proximity to enable metal–metal interaction while allowing complexes to retain dinuclear status after bond fission. Irradiation of Ru$_2$(CO)$_4${η^5 : η^5-fulvalene} (Ru–Ru distance 2.821 Å) in the absence of an added ligand leads to rapid formation of the thermally unstable complex {Ru(CO)$_2${μ-η^1 : η^5-C$_5$H$_4$}}$_2$ **(84d)** (Ru–Ru distance 3.456 Å) which reverts back to the precursor at room temperature in THF (260). The binuclear compound Cp$_2$(CO)Zr{η^1 : η^5-C$_5$H$_4$}Ru(CO)$_2$ inserts ethylene in the σ bond of the bridged η^1 : η^5-cyclopentadienyl ligand while retaining the Ru–Zr metal–metal bond (261). Treatment of the reaction product with 'BuOH leads to metal–metal bond cleavage and formation of (CO)$_2$(H)Ru{η^5 : η^1-C$_5$H$_4$CH$_2$CH$_2$}Zr{O'Bu}Cp$_2$ **(92)** [Eq. (34)]. The reaction of titanocene dichloride with 2 equiv of Mn(η^5-C$_5$H$_4$Li)(CO)$_3$ afforded Cp$_2$Ti{μ-η^1 : η^5-C$_5$H$_4$Mn(CO)$_3$}$_2$ **(93)** (262). The X-ray structure determination of **93** was recorded.

Ferrocene-containing metal complexes have been reported over the years, and the topic has been reviewed (263). Again, the preferred method to prepare such complexes is the utilization of ferrocenyllithium with metal halides. Examples of bimetallic ferrocenyl compounds are known where the σ-bonded metals are in both high and low formal oxidation states.

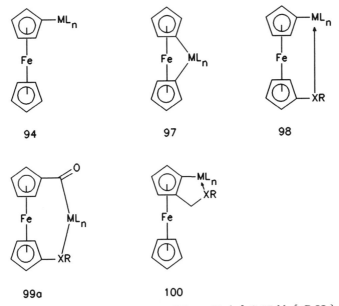

The σ,π complexes, FcML$_n$ (94) [Fc = Fe$\{\eta^5$-C$_5$H$_5\}\{\eta^5$-C$_5$H$_4\}$, ML$_n$ = Ti$\{$NR$_2\}_3$ (R = Me (94a), Et (94b)) (264); ZrCp$_2$Cl (94c) (265); WCp(NO)$_2$ (94d), WCp(O)$_2$ (94e) (266); Mn(CO)$_5$ (94f) (267); FeCp(CO)$_2$ (94g) (268); RuCp(CO)L (94h) (L = CO, CN^{t}Bu, PPh$_3$, PMe$_3$); RuCp*(CO)L (94i) (L = CO, CN^{t}Bu, PPh$_3$, PMe$_3$); Ru(C$_6$Me$_6$)(CO)X (94j) (X = Cl, p-tolyl, n-butyl) (269); Ru(PMe$_3$)$_3$(CO)Cl (94k). Fc = $\{\eta^5$-C$_5$H$_4$Cl$\}\{\eta^5$-C$_5$H$_3$Cl$\}$ (94l), ML$_n$ = FeCp(CO)$_2$ (94m), Mn(CO)$_5$ (94n), Ir(PPh$_3$)$_2$(CO) (94o), AuPPh$_3$ (94p) (270)] have been prepared directly or indirectly from a salt elimination reaction of FcLi and ML$_n$X or FcCl and Na[ML$_n$].

The σ-ferrocenyl compound Cp(NO)$_2$W$\{$Fc$\}$ (94d) was oxidized by air in a dichloromethane solution to give Cp(O)$_2$W$\{$Fc$\}$ (94e) (266). The complex Cp(CO)$_3$W$\{$Fc$\}$ (94q), which is isoelectronic with 94d, was obtained from the corresponding ferrocenoyl complex (η^1 : η^5-C$_5$H$_4$C(O)$\{$WCp-(CO)$_3\}$)FeCp (95a) after thermal or photochemical-induced decarbonylation (268). The complex Cp(CO)$_2$FeC(O)Fc (95b), which was prepared from the metal carbonylate anion and the halo-substituted ferrocene, was decarbonylated similarly. The stability of the σ bond in 94f was tested in different solvents and under conditions suited for hydroformylation reactions, and the products, Fc—CHO, Fc—COOMe and Fc—CH$_2$OH, were characterized spectroscopically (267). Whereas a carbonyl in 94f was displaced by PPh$_3$ to give the corresponding ferrocenyl complex Mn(CO)$_4$(PPh$_3$)Fc (94r), *tert*-butylisonitrile ligands displace more than one carbonyl and yield the ferrocenoyl complexes *fac*-Mn(CO)$_3$-

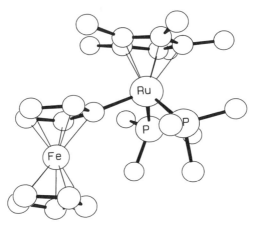

FIG. 5.   X-Ray structure of Cp*(Me₃P)₂RuFc (**94s**) [Lehmkuhl et al. (271)].

$(CN^tBu)_2\{C(O)Fc\}$ (**95c**) and cis-Mn(CO)$_4$(CN^{t}Bu){C(O)Fc} (**95d**). The complex **94f** was oxidized by nitrosyl–tetrafluoroborate and iodine to [Mn(CO)$_5$Fc]$^+$ (**96a**) and could be reduced back to **94f** by aqueous thiosulfate solutions. The complexes **94h** (CO), **94i** (CO), and **94j** (Cl) were oxidized by AgBF$_4$ to [RuL$_3$(CO)L$'$Fc]$^+$ (**96**) [L$'$ = CO, L$_3$ = Cp (**96b**), Cp* (**96c**); L$'$ = Cl, L$_3$ = C$_6$Me$_6$ (**96d**)]. The Ru complexes with Cp and Cp* ligands were also characterized by cyclovoltammetry (269). By refluxing CpRu(PMe$_3$)$_2$CH$_2$^{t}Bu and ferrocene in THF for 2 days the bimetallic complex Cp*(PMe$_3$)$_2$Ru{Fc} (**94s**) was formed (Fig. 5) (271).

The doubly metallated 1,1′-dilithioferrocene reacted with metallocene dichlorides at room temperature by the displacement of both chloro ligands affording the [1]ferrocenophanes $\{RC_5H_4\}_2M\{(\eta^1 : \eta^5\text{-}C_5H_4)_2Fe\}$ (**97**) [M = Ti, R = H (**97a**), Bu (**97b**); M = Zr, R = H (**97c**), ^{t}Bu (**97d**), M = Hf, R = H (**97e**), ^{t}Bu (**97f**)] (272). The X-ray crystal structure determination of **97d** displays a Zr atom which is pseudotetrahedrally surrounded by two σ-bonded and two π-bonded cyclopentadienyl ligands with the Zr acting as the link between the two Cp rings of the ferrocene. As a result of this, the rings are in an eclipsed conformation and are tilted toward the Zr atom with an angle of 6.0° and a Zr–Fe distance of 2.962 Å.

The [1]ferrocenophanes (1,1′-ferrocenediyl)phenylphosphine and -phenylarsine react with phenyllithium by opening the P or As bridge to give 1′-lithio-1-(diphenylphosphino)ferrocene and 1′-lithio-1-(diphenylarsino)ferrocene, respectively (273,274). The reaction of Fe(η^5-C$_5$H$_4$Li)(η^5-C$_5$H$_4$PPh$_2$) with M(CO)$_6$ followed by Me$_3$OBF$_4$ affords the zwitterions Fe(η^5-C$_5$H$_4$\{M$^-$(CO)$_5$\})\{η^5-C$_5$H$_4$P$^+$Ph$_2$Me\} (**94**) [M = Cr (**94t**), W (**94u**)] and carbene-bonded ferrocenophanes (vide infra, **109q,r**) (275).

94 s,t

Reaction of the metallated phosphine precursor and $CpFe(CO)_2I$ afforded $1'$-iodo-1-(diphenylphosphino)ferrocene, the [2]ferrocenophane $Cp(CO)Fe\{\eta^1: \eta^5\text{-}C_5H_4\}Fe(\eta^5\text{-}C_5H_4PPh_2)$ (98), and the CO-inserted [3]ferrocenophane $Cp(CO)Fe\{\eta^1: \eta^5\text{-}C(O)C_5H_4\}Fe(\eta^5\text{-}C_5H_4PPh_2)$ (99a) (276). The arsine analog gave only the σ-bound derivative $Cp(CO)_2Fe\{\eta^1: \eta^5\text{-}C_5H_4\}Fe(\eta^5\text{-}C_5H_4AsPh_2)$, and no evidence for the formation of complexes analogous to 98 or 99a was obtained. The tilt of the cyclopentadienyl rings away from the parallel plane toward the center of the molecule is 8.6° in 98 and 6.0° in 97d, compared to the corresponding value of 26.7° in the [1]ferrocenophane containing the PPh link (272,274,276).

Dimethylaminomethylferrocene reacts with Na_2PdCl_4 in the presence of bases giving optically active planar chiral 2-dimethylaminomethylferrocenylpalladium chloride (277). The cyclopalladated product was isolated as a dimer and used in the synthesis of a σ,π-bimetallic monoglycinate complex (278) which reacts further with a second equivalent of sodium glycinate to afford a chiral anionic hexacoordinated complex of palladium (279). The dimer is cleaved by ligands to give neutral bimetallic complexes $(\eta^5\text{-}C_5H_3CH_2N(Me)_2\{Pd(L)Cl\})FeCp$ (100) [L = PPh_3 (100a), $AsPh_3$ (100b), py (100c), $P(OR)_3$, R = Ph (100d), Et (100e), Me (100f)] and the cationic complex $[(\eta^5\text{-}C_5H_3CH_2N(Me)_2\{Pd(dppe)\})FeCp]PF_6$ (100g) (280). Oxidation of the iron center gives an Fe(III)–Pd(II) compound which exhibits a near-infrared band for heterometal electron transfer. Reaction of N,N-dimethylferrocenecarbothioamide with lithium tetrachloropalladate in methanol resulted in cyclopalladation in dimeric Fe–Pd complexes which were cleaved by ligands to give cis and trans isomers of $(\eta^5\text{-}C_5H_3C(NMe_2)S\{Pd(L)X\})FeCp$ (101) [X = Cl or I, L = $AsPPh_3$ (101a), PPh_3 (101b), P(p-tolyl)$_3$ (101c), PBu_3 (101d), py (101e), p-^{t}Bupy (101f)] (281).

Phenylazoferrocene undergoes nickelation and palladation exclusively on the phenyl group, giving $CpFe\{\eta^5\text{-}C_5H_4N{=}NC_6H_4\text{-}o\text{-}MCp\}$ (102) [M = Ni (102a), Pd (102b)] (282). The homoannular metallation products, $(\eta^5\text{-}(2\text{-}C_5H_3C(O)Me)\{M(CO)_4\})FeCp$ (100) [M = Mn (100h), Re (100i)] were

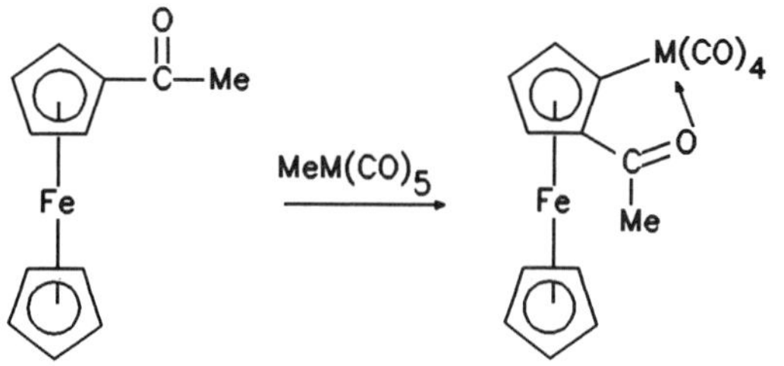

(35)

100 h,i

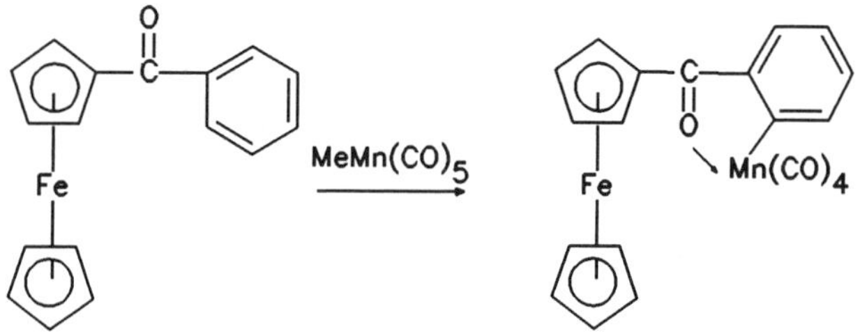

(36)

101g

obtained from the reaction of $M(CO)_5Me$ and acetylferrocene [Eq. (35)] (283). The competition of metallation by a phenyl substituent versus the cyclopentadienyl ring of a ferrocenyl group was explored in the reaction of benzoylferrocene with $Mn(CO)_5Me$. The sole product of the reaction was tetracarbonyl(2-ferrocenylcarbonylphenyl)manganese **101g**, indicating that metallation occurred only on the phenyl ring [Eq. (36)].

A cationic ruthenium complex containing the 6-ferrocenyl-2,2′-bipyridine unit as a ligand, $[\{Fc—C_6H_3—C_5H_4N\}Ru(bipy)_2]^+$ **(101h)** (bipy is 2,2′-bipyridyl), was prepared and characterized spectroscopically (284).

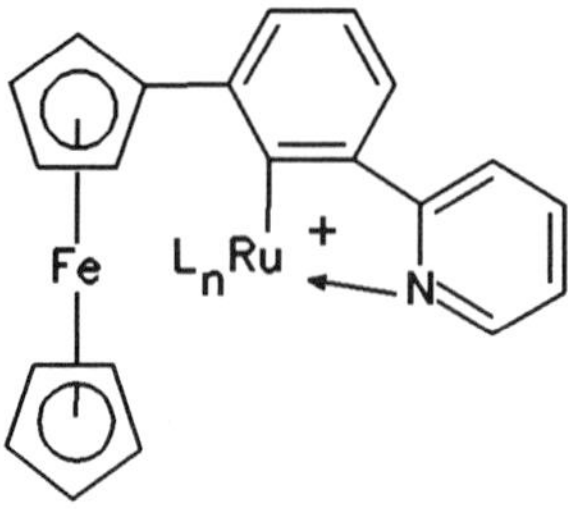

101h

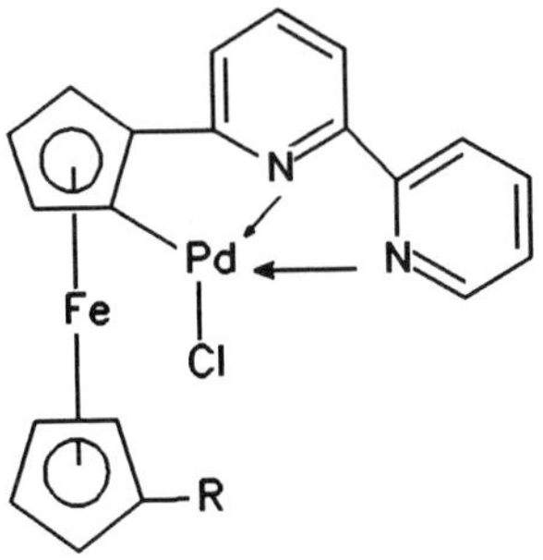

102

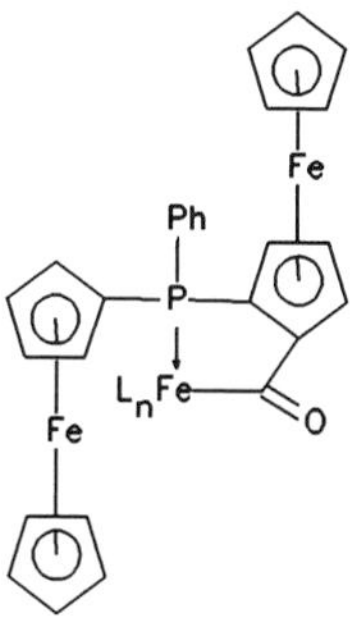

103

The reaction of Pd(COD)Cl$_2$ with ferrocenylbipyridine ligands leads to the removal of a proton α to the bipyridyl substituent and the palladation of this position (*284*). The bimetallic complexes (η^5-RC$_5$H$_4$)Fe{(η^5 : η^1-C$_5$H$_3$—C$_5$H$_3$N—C$_5$H$_4$N)PdCl} (102) [R = H (102a), 2,2′-bipyrid-6-yl (100b)] crystallized as solvates, and a crystal structure determination of 102b was recorded. The ruthenocenyl analog 102c of 102a was synthesized similarly. When ferrocenyl- and ruthenocenyl-1,10-phenanthroline precursors were used no proton abstraction or metallation occurred, and the palladium was coordinated to the N and N′ atoms of the 1,10-phenanthrolin-2-yl substituent. Deprotonation of CpFe{η^5-C$_5$H$_4$—P(Ph)C$_5$H$_5$} and subsequent reaction with 2 equivalents of FpI afforded the trimetallic complex CpFe{η^5-C$_5$H$_4$P(Ph)FeCp(CO)C(O)C$_5$H$_3$-η^5}FeCp (103), in which two ferrocenyl moieties are linked by a phosphorus unit which forms part of a metallocycle containing the third Fe center as an

FeCp(CO) fragment (*285*). A crystal structure determination of **103** confirmed the molecular formulation.

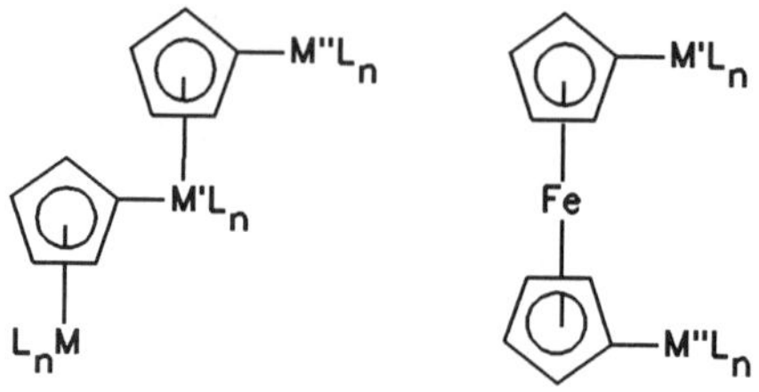

E   F

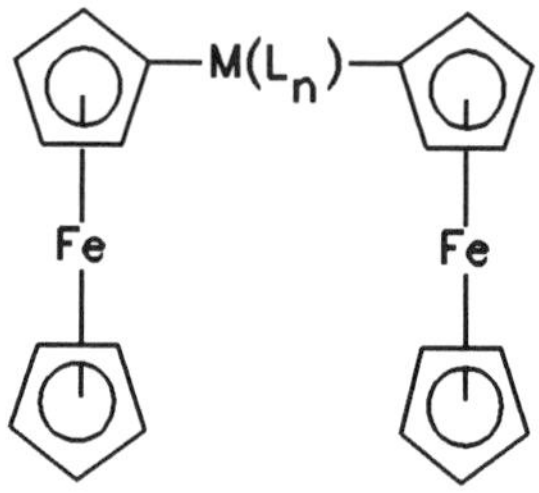

104

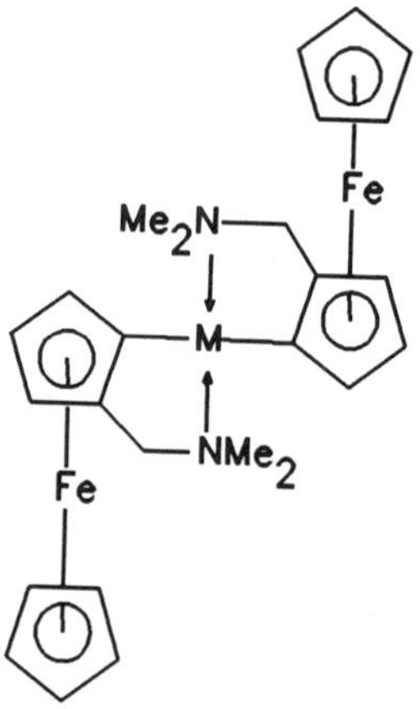

105

Complexes where two Fc substituents are σ-bonded to a central transition metal fragment form part of a general class of trimetallic compounds **E**. On the other hand, two metal fragments, one on each of the cyclopentadienyl rings of ferrocene, represent another type of arrangement (**F**) of metal fragments in trimetallic complexes. The ferrocenyl compounds

{Cpfe(η^5-C$_5$H$_4$)}$_2$MCp$_2$ (104) [M = Ti (104a), Zr (104b), Hf (104c)] were synthesized by treatment of dicyclopentadienylmetal dichloride with excess ferrocenyllithium (286). These air-sensitive crystalline solids were purified by sublimation and characterized spectroscopically. Titanocene dichloride reacts with 2 equiv of 1,1'-dichloro-2-lithioferrocene, and as a consequence of the chiral nature created by the substituents of the 1,1'-Cl$_2$Fc group, the trimetallic product {Cl$_2$Fc}$_2$TiCp$_2$ (104d) exists in dl and meso forms (271). By extraction of the product mixture with hexane/dichloromethane (9:1), it was possible to leach out the more soluble dl-isomer. Along the same route the trimetallic complexes of zinc and cadmium with 2-(dimethylaminomethyl)ferrocenyl as chelating ligands, {Cpfe(η^5: η^1-C$_5$H$_4$CH$_2$NMe$_2$-C,N)}$_2$M (105) [M = Zn (105a), Cd (105b)], were prepared (287). Paramagnetic (RCp)$_2$V units have been bridged by ferrocene in a perpendicular fashion (E) (288). The reaction of dilithiated ferrocene with vanadocene chloride under the elimination of lithium chloride afforded (η^5-RC$_5$H$_4$)$_2$V{η^1:η^5-C$_5$H$_4$}Fe{η^5:η^1-C$_5$H$_4$}V(η^5-RC$_5$H$_4$) (106) [R = H (106a), Me 106b)]. The ferrocene bridge is distorted, and the structure can be described as having a significant contribution of η^4-coordinated Cp rings. A fairly long V–C σ bond of 2.17 Å was observed in the solid state for 106b (Table XIII). As far as we are aware, no examples of the structure type F where two different metal fragments are directly σ-bonded to the two cyclopentadienyl rings of ferrocene have been recorded.

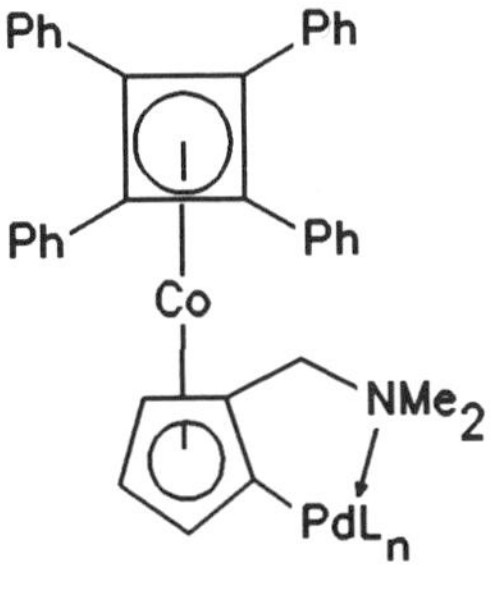

107

The complex [{η^5-(dimethylaminoethyl)cyclopentadienyl}(η^4-tetraphenylcyclobutadiene)cobalt(I) reacts with dilithium tetrachloropalladate(II) in the presence of sodium acetate to give an ortho-palladated chloro-bridged dimer (289). The dimer can be cleaved by PPh$_3$ or Tlacac to give (η^4-C$_4$Ph$_4$)Co{η^5:η^1-C$_5$H$_3$CH$_2$N(Me$_2$)}PdLX (107) [L = PPh$_3$, X = Cl (107a); LX = acac (107b)].

TABLE XIII

SELECTED DATA FOR BIMETALLIC COMPLEXES CONTAINING FERROCENYL FRAGMENTS

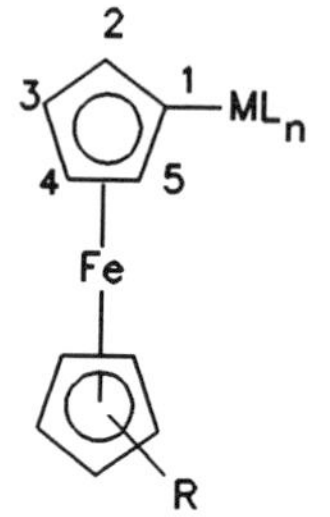

Complex			Bond distances (Å)			
No.	ML_n	R	$M-C_1$	$Fe-C_1$	$FeC(C_2-C_5)$	Ref.
94c	$ZrCp_2Cl$	—	2.314	2.143	2.039	(265)
94d	$WCp(NO)_2$	—	2.164	2.059	2.036	(266)
94s	$RuCp^*(PMe_3)_2$	—	2.123	2.197	2.045	(271)
97d[a]	$Zr(^tBuCp)_2$		2.284	2.029	2.048	(272)
94t	$Cr(CO)_5$	PPh_2Me	2.149	—	—	(275)
98	$FeCp(CP)$	PPh_2	2.002	2.049	2.039	(276)
106b	VCp'_2	VCp'_2	2.171	2.119	2.037	(288)

[a] The Zr–Fe distance is 2.962 Å.

b. *σ-Carbene Complexes.* Fischer-type carbene complexes with bridging ligands which have π-cyclopentadienyl and σ-carbene bonding functionalities have been synthesized and studied. The complexes $L_nM\{\eta^1:\eta^5-\{=C(OR)C_5H_4\}\}M'L_n$ (**108**) [ML_n = $W(CO)_5$, $M'L_n$ = $(\eta^5-C_5H_4)Mn(CO)_3$, R = Et (**108a**) (290); ML_n = $CpMn(CO)_3$, M' = $(\eta^5-C_5H_4)Mn(CO)_3$, R = Me (**108b**), $M'L_n$ = Fc, R = Et (**108c**)] (291) were obtained in good yields from reactions of lithiated cyclopentadienyl complexes and appropriate metal carbonyls.

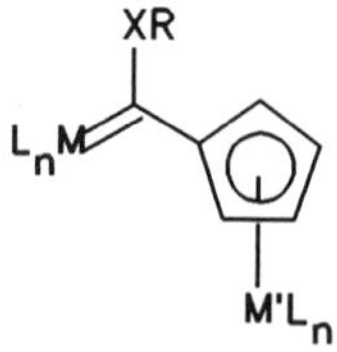

108 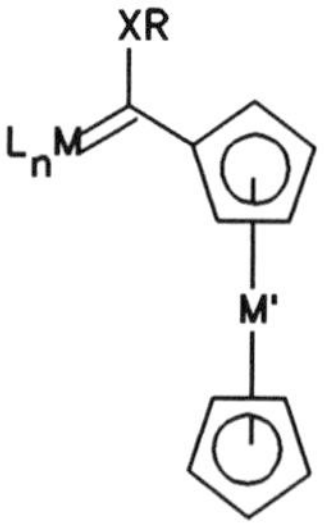   109

A series of ferrocenyl carbene complexes of chromium, tungsten, and manganese $L_nMC(Fc)XR$ **(109)** $[ML_n$ = $Cr(CO)_5$, XR = $O^-NMe_4^+$ **(109a)**, OMe **(109b)**, OEt **(109c)**, NH_2 **(109d)**, NMe_2 **(109e)**, NC_4H_8 **(109f)**; ML_n = $W(CO)_5$, XR = $O^-NMe_4^+$ **(109g)**, OMe **(109h)**, OEt **(109i)**, NH_2 **(109j)**, NMe_2 **(109k)**, NC_4H_8 **(109l)**; ML_n = $Mn(MeCp)(CO)_2$, XR = OMe **(109m)**] was prepared to study the electronic effects of the ferrocenyl substituent in the carbene ligand (*292*). The ferrocenyl substituent acts as an electron donor by means of resonance interaction with the carbene carbon atom. The lithiated metallocenes ruthenocene and 1,1'-dimethylferrocene react with the group VIB metal hexacarbonyls to yield bimetallic acylato complexes which, after subsequent alkylation with Et_3OBF_4, gave the corresponding carbene complexes $(CO)_5M-\{=C(OEt)(\eta^5-C_5H_3R)M'(\eta^5-C_5H_4R)\}$ **(109)** [M' = Ru, R = H, M = Cr **(109n)**, Mo **(109o)**, W **(109p)**; M' = Fe, R = Me, M = Cr **(109q)**, Mo **(109r)**, W **(109s)**] (*293*). The solid state structure of **109p** displays a ruthenocenyl substituent with coplanar cyclopentadienyl rings which are in a staggered conformation relative to one another, and located between two *cis*-carbonyls of the $W(CO)_5$ fragment. The W–C(carbene) distance of 2.23 Å is longer than typical values recorded for alkyl substituents on carbene ligands of tungsten complexes (*199*).

The cationic bimetallic carbyne complex $[Cp'(CO)_2Mn\{\equiv CFc\}]^+$ (*vide infra*) reacts with LiXPh by attack on the carbyne carbon atom to afford the neutral carbene complexes $MeCp(CO)_2Mn\{=C(Fc)XPh\}$ **(108)** (X = S **(108d)**, Se **(108e)**, Te **(108f)**] (*294*) and with $Na[Co(CO)_4]$ to yield $MeCp(CO)_2Mn\{=C(Fc)Co(CO)_4\}$ **(108g)** (*295*). The corresponding reaction with $K[Mn(CO)_5]$ leads to the formation of a trimetallic complex in which two Mn fragments are joined by a metal–metal bond and the ferrocenyl substituent forms part of a bridging ketene ligand.

110   109 q,r

Chiral binuclear carbene complexes of the Fischer type were synthesized utilizing (R,R)-1-(1-dimethylaminoethyl)-2-lithioferrocene (*296*). The lithiated precursor reacts with tungsten hexacarbonyl to give a bimetallic acylate, $CpFe\{\eta^5-2-C_5H_3\{C(Me)HNMe_2\}C(OLi)W(CO)_5\}$ **(109t)**, which can be converted to a chiral amino–carbene complex, $CpFe\{\eta^5\text{-}2\text{-}C_5H_3\{C(Me)(OH)H\}C(NMe_2)\text{=}W(CO)_5\}$ (**109u**) in the presence of a proton donor like *tert*-BuCl. Whereas diastereoisomeric furanoid carbene complexes, $CpFe\{\eta^5\text{-}2\text{-}C_5H_3C(Me)HOC\text{=}W(CO)_5\}$ (**110**), were produced at low temperature in the presence of a strong alkylating agent, the formation of the bimetallic carbene $CpFe\{\eta^5\text{-}2\text{-}C_5H_3(CH\text{=}CH_2)\{C(OEt)\text{=}W(CO)_5\}$ (**109v**) was favored at higher temperatures. In addition to the zwitterions **94t** and **96u**, the isomeric carbene-based [3]ferrocenophanes $(CO)_4M\{C(OMe)C_5H_4FeC_5H_4PPh_2\}$ (**109**) [M = Cr (**109q**), W (**109r**)] were isolated as stable, red crystals in yields ranging from 20 to 40% from the reaction of the [1]ferrocenophane, 1,1′-ferrocenediyl-phenylphosphine and PhLi with $M(CO)_6$ (*275*).

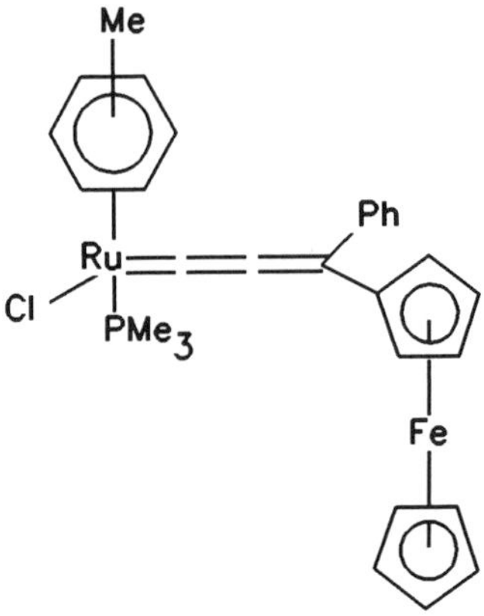

111

To study the influence of an electron-donating substituent on the allenylidene moiety in ruthenium complexes, the reaction of $\{\eta^6\text{-}C_6Me_6\}Ru(Cl)_2(PMe_3)$ and $FcC(Ph)(OH)C\equiv CH$ was studied (*297*). The product, a stable violet phenylallenylidene complex of ruthenium, $\{\eta^6\text{-}C_6Me_6\}(PMe_3)(Cl)Ru\{\text{=}C\text{=}C\text{=}C(Ph)Fc\}$ (**111**), was isolated. The compound $Li[W\{C(NMe_2)CHSiMe_3\}(CO)_5]$, which was formed by the deprotonation reaction of $W\{C(NMe_2)CH_2SiMe_3\}$ and *n*-BuLi, reacted with $CpFe(\eta^5\text{-}C_5H_4CHO)$ to yield $(E)\text{-}[(\eta^1\!:\!\eta^5\text{-}C_5H_4CH\text{=}CHC(NMe_2)W(CO)_5)FeCp]$ (**112**) (*298*). Although some disorder was observed at the carbon atom of the double bond adjacent to the ferrocenyl substituent, the stereochemistry is clearly of the *E* configuration, placing the carbene fragment and the ferrocenyl group on opposite sides of the double bond.

    c. *σ-Carbyne Complexes.*   Bimetallic complexes with bridging ligands which contain σ-carbyne and π-cyclopentadienyl links are generally prepared from the corresponding bimetallic carbene complexes and suitable Lewis acids. The complex **108a** reacts with Al_2Br_6 in dichloromethane at low temperatures to give $W(\equiv C\{\eta^5\text{-}C_5H_4Mn(CO)_3\})(CO)_4Br$ (**113a**) (*290*) and **109c** with BCl_3 in pentane to give $Cr\{\equiv CFc\}(CO)_4Cl$ (**113b**) (*299*). Ferrocenyl carbene complexes **109c,i** react with Al_2X_6 affording the *trans*-carbyne complexes $M\{\equiv CFc\}(CO)_4X$ (**113**) [M = Cr, X = Br (**113c**); M = Mo, X = Br (**113d**); M = W, X = Cl (**113e**), Br (**113f**)] (*299,300*). The complexes **113c** and **113f** are converted to the corresponding iodocarbyne complexes **113g** and **113h**, respectively, by LiI. The X-ray structure determination of **113c** reveals cyclopentadienyl rings which are coplanar and almost in an eclipsed configuration relative to one another. Other structural data are in good agreement with data recorded for mononuclear carbyne complexes (*301*).

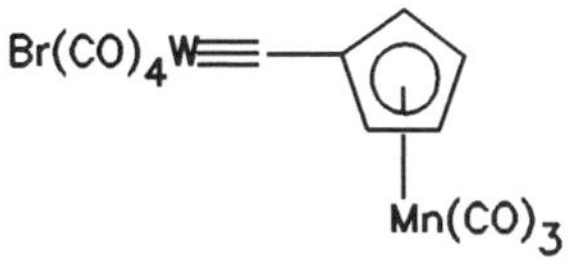

$Br(CO)_4W\equiv$

$Mn(CO)_3$

113a

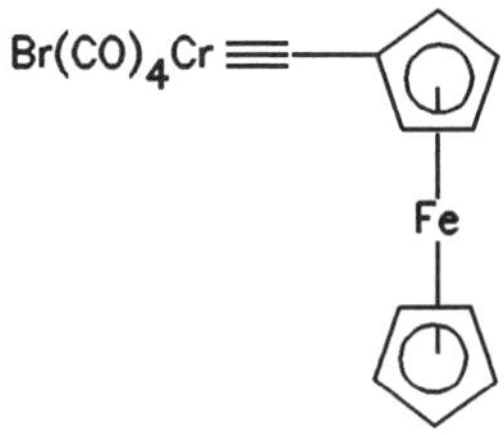

$Br(CO)_4Cr\equiv$

Fe

113c

Ferrocenyl- and ruthenocenyl carbyne complexes $W\{\equiv CC_5H_3$ $RM(C_5H_4R)\}(CO)_4X$ (**113**) [M = Ru, R = H, X = Cl (**113g**), Br (**113h**); M = Fe, X = Br, R = Me (**113i**)] were synthesized from the corresponding bimetallic carbene complexes **109p** and **109s** and MX_3 (M = B, X = Cl, Br; M = Al, X = Cl, Br) (*293*). The reaction of $Mn(\eta^5\text{-}C_5H_4Li)(CO)_3$ with $M(CO)_6$, $(CF_3CO)_2O$, and bidentate ligands afforded the cymantrenylmethylidyne complexes $M\{\equiv C\text{—}C_5H_4Mn(CO)_3\}(LL)(O_2CCF_3)\text{-}$ $(CO)_2$ (**114**) [M = W, LL = N,N,N',N'-tetramethylethylenediamine (TMEDA) (**114a**); M = Mo, LL = 2,2'-bipyridyl (bipy) (**114b**)] (*302*). Treatment of the molybdenum complex **114b** with $NaC_5H_5 \cdot MeO\text{-}$ C_2H_4OMe or $K[HB(pz)_3]$ (pz is pyrazol-1-yl) provides the complexes $Mo\{\equiv C\text{—}C_5H_4Mn(CO)_3\}(LLX)(CO)_2$ (**114**) [LLX = $\eta^5\text{-}C_5H_5$ (**114c**), $HB(pz)_3$ (**114d**)].

2. π-Benzene Complexes

Although the route utilizing lithiated cyclopentadienyl complexes and organometallic complexes with halogen ligands in the synthesis of σ,π-heterobimetallic complexes with bridging cyclopentadienyl ligands was established long ago, the same procedure, as a general method for the synthesis of the corresponding bimetallic complexes with bridging benzene ligands, has only relatively recently been exploited (*303–305*). The alternative method starting with π-haloarene complexes and reacting them with anionic organometallic substrates has also been used, but with limited success (*306*). Whereas the first method works well if the metallation step is successful, the main drawback of the second method relates to the strength of nucleophilic character of the anionic metal substrate. By comparison, the nucleophilic character in the first case is localized on the metallated carbon, whereas the negative charge on the anionic complex in the second case is delocalized over the entire molecule (*307*). The ease of preparation and availability of (arene)Cr(CO)$_3$ as a starting material, especially from refluxing Cr(CO)$_6$ in dibutyl ether in the presence of the appropriate arene (*308*) or by utilizing Cr(CO)$_3$(NH$_3$)$_3$ (*309*) as a source of Cr(CO)$_3$, has resulted in the promotion of arene–chromium chemistry above the study of the chemistry of many other analogous transition metal arene–complexes.

(x=0,1··)

a. *σ-Alkyl Complexes.* Phenyl-bis(η^5-cyclopentadienyl)tungsten hydride reacts with Cr(CO)$_3$(NH$_3$)$_3$ to form benzene-bridged chromium complexes of tungsten, W{(η^1:η^6-C$_6$H$_5$)Cr(CO)$_3$}Cp$_2$H (**115a**), which were converted with halomethanes to the corresponding halogen-substituted compounds, W{(η^1:η^6-C$_6$H$_5$)Cr(CO)$_3$}Cp$_2$X (**115**) [X = Cl (**115b**), Br (**115c**), I (**115d**)] (*310*). The highly nucleophilic metal carbonyl anions [CpFe(CO)$_2$]$^-$ (Fp$^-$) and [Cp*Fe(CO)$_2$]$^-$ (Fp*$^-$) react with para- and meta-substituted haloarenes (η^6-C$_6$H$_4$RX)Cr(CO)$_3$ to form (η^6-*p*-C$_6$H$_4$R{Fp})Cr(CO)$_3$ (**115**) [R = CF$_3$, (**115e**), CO$_2$Me (**115f**), Cl (**115g**), F (**115h**), Me (**115i**)], (η^6-*m*-C$_6$H$_4$R{Fp})Cr(CO)$_3$ (**115**) [R = CF$_3$ (**115j**), MeO (**115k**), Me (**115l**)], and (η^6-C$_6$H$_5${Fp*})Cr(CO)$_3$ (**115m**) (*306*). A variety of less reactive metal nucleophiles CpMo(CO)$_3^-$, CpNi(CO)$^-$, Mn(CO)$_5^-$, and Co(CO)$_4^-$ failed to react. Spectroscopic data reveal that the Fp substituent is electron-donating and that a low barrier of rotation about the σ bond of the arene to iron fragment exists. The molecular structure of **115g** indicates that the cyclopentadienyl is on the open side of the arene ring and therefore anti with respect to the Cr(CO)$_3$ group. The Cr–Fe separation of 3.943 Å falls in the nonbonding range. The tetracarbonyl ferrate anion, Na$_2$[Fe(CO)$_4$], reacts with (η^6-C$_6$H$_5$Cl)Cr(CO)$_3$ to produce the anionic bimetallic complex $Na[(\eta^6-C_6H_5\{Fe(CO)_4\})Cr(CO)_3]$ (**116a**) (*304*). Spectroscopic data suggest that the negative charge is located on the iron and that the bimetallic complex adopts an η^6-arene–alkyl structure as opposed to the alternative η^5-cyclohexadienyl–carbene structure. The reaction of the lithiated arene $(\eta^6-C_6H_5Li)Cr(CO)_3$ and iron pentacarbonyl, on the other hand, afforded the bimetallic acylate $(\eta^6-o-C_6H_4ClC(O)\{Fe(CO)_4\})Cr(CO)_3$ (**117a**) which reacts through C–C bond formation with electrophiles to produce $\{\eta^6-C_6H_4(Cl)C(O)R\}Cr(CO)_3$. The carbonylates $M(CO)_5^{2-}$ (M = Cr,W) react with $(\eta^6-C_6H_5F)Cr(CO)_3$ affording $[Et_4N][(\eta^6-C_6H_5\{M-(CO)_5\})Cr(CO)_3]$ (**116**) [M = Cr (**116b**), W (**116c**)] (*311*). Spectroscopic data suggest that the molecules adopt the classic η^6-arene structure, which was confirmed by an X-ray crystal structure determination of **116c**.

115   117

Hunter and co-workers (*24,312–315*) synthesized a series of phenylene-bridged polymetallic σ,π complexes to serve as models for the production of linear-chain polymers with conducting properties. One-electron reversible oxidation processes were recorded at a stationary platinum-bead electrode for $(\eta^6-1,4-C_6H_4X\{Fp\})Cr(CO)_3$ (**117**) [X = H (**117a**), Cl (**117b**), Me (**117c**), OMe (**117d**)]. Treatment with 1 equiv of $AgPF_6$ leads to the formation of the thermally stable but air-sensitive blue paramagnetic complexes $(\eta^5-p-C_6H_4X\{Fp\})Cr(CO)_3]^{\cdot+}PF_6^-$ (*312*). Complexes **117a–d**, as well as the complexes $[(\eta^6-C_6H_5\{Fp^\#\})Cr(CO)_3]$ (**117**) [Fp$^\#$ = (indenyl)Fe(CO)_2 (**117e**), Fp$^\#$ = Fp' = $(\eta^5-C_5H_4Me)Fe(CO)_2$ (**117f**)], were synthesized by the reaction of NaFp$^\#$ and the corresponding fluoroarene complex (*24,312*). Electrochemical data confirm observations made from the spectroscopic data that the Fp substituent is a strong electron-donating group (*313*). The σ-bonded Fp–acyl complex, $p-C_6H_5$—$C_6H_5C(O)Fp$ when refluxed in dibutylether with $Cr(CO)_6$ gave $(\eta^6-1,4-C_6H_4Ph\{Fp\})Cr(CO)_3$ (**117g**) (*314*). From the spectroscopic data, it was clear that the $Cr(CO)_3$ group was π-bonded to the more electron-rich and sterically crowded benzene ring containing the Fp substituent.

The advantage of using lithiated arene complexes of chromium as starting materials to synthesize σ,π-bimetallic complexes of chromium and manganese was clearly demonstrated by reacting $(\eta^6-o-C_6H_4XLi)Cr(CO)_3$ (X = H, F) with $Mn(CO)_5Br$ (*305*). The products, $(\eta^6-C_6H_4XC(O)\{Mn-$

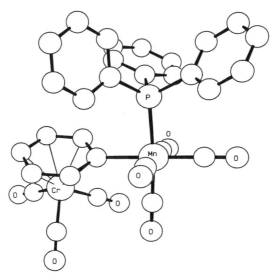

FIG. 6. X-Ray structure of cis-[(η¹:η⁶-C₆H₅{Mn(CO)₄PPh₃})Cr(CO)₃] (117i) [Lotz et al. (305)].

(CO)₅})Cr(CO)₃ (118) [X = H (118a), F (118b)] have σ,π-bridging benzoyl ligands. When the reaction was performed in the presence of P(OMe)₃, a complex with a bridging benzoyl ligand, cis-(η⁶-C₆H₅C(O){Mn(CO)₄-P(OMe)₃})Cr(CO)₃ (118c) as well as a complex with a bridging phenyl ligand, cis-(η⁶-C₆H₅{Mn(CO)₄P(OMe)₃})Cr(CO)₃ (117h), were isolated. Refluxing 118c in hexane converted it quantitatively to 117h (315). However, when the reaction was performed in the presence of PPh₃ only cis-(η⁶-C₆H₅{Mn(CO)₄PPh₃})Cr(CO)₃ (117i) was isolated. The more basic phosphine ligand inhibits acyl formation. Bridging benzoyl complexes of the type fac-(η⁶-C₆H₅C(O){Mn(LL)(CO)₃})Cr(CO)₃ (118) [LL = dppe (118d), tmeda (118e)] were obtained on performing the reactions in the presence of LL (305). Whenever σ,π-benzoyl-bridged compounds were synthesized, π,π-benzophenone-bridged bis(chromium) complexes were also formed. The solid state structure of 117i (Fig. 6) reveals a highly distorted phosphine ligand located on the open side of the benzene ligand, away from the Cr(CO)₃ group. The disorder caused by the bulky phosphine ligand is reflected in an exceptionally long Mn–C σ bond distance (2.103 Å). Interestingly, the structure of 118c displays the phosphite on the same side of the Cr(CO)₃ group and interlocked between the carbonyl ligands (315).

Metallation of $(\eta^6$-C$_6$H$_5$R)Cr(CO)$_3$ with n-butyllithium and subsequent reaction with 1 equiv of titanocene dichloride at low temperatures afforded $(\eta^1$:η^6-C$_6$H$_4$R{TiCp$_2$Cl})Cr(CO)$_3$ (**117**) [R = H (**117j**), o-F (**117k**), p-Me (**117l**), m-Me (**117m**), o-OMe (**117n**), m-NMe$_2$ (**117o**)], in high yields (*316,317*). The nature of substituents on the arene ring directs the positions of metallation and determines the distribution of isomers obtained. Using the same procedure, but replacing Cp$_2$TiCl$_2$ by Au(PPh$_3$)Cl, the σ,π-benzene bridged-bimetallic complexes $(\eta^1$:η^6-C$_6$H$_4$R{AuPPh$_3$})-Cr(CO)$_3$(**117**) [R = H (**117p**), o-F (**117q**)] were synthesized (*316*). Spectroscopic data reveal that both σ-bonded metal fragments increase the electron density on the arene ring in comparison to the unsubstituted arene complex. The transfer of charge onto the arene ring by Fp (d^6), Mn(CO)$_4$L (d^6), and AuPPh$_3$ (d^{10}) can also be explained by π-resonance effects, whereas an increase in the electron density by TiCp$_2$Cl (d^0) was ascribed to inductive effects of metal fragments resulting in a highly polarized σ bond (*24,25,316*).

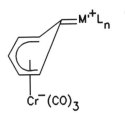

Structural studies of **117** (Table XIV) reveal small but significant distortions in the arene ring as a result of the σ substituent being a metal fragment. Hunter *et al.* (*24*) observed that Fp substituents, and the ipso-carbon atom to which they were attached, were bent away from the Cr(CO)$_3$ group. It was shown that the magnitude and direction of distortions were independent of the steric bulk of the substituent and that deviations originated primarily from electronic effects. A model based on π-resonance effects was developed to explain the direction of distortions (*24,318*). However, little if any multiple bonding interaction between the ipso-carbon atom and σ-bonded titanium (d^0) was observed in the X-ray analysis of **117j,k,n** (*25,316*). Arguments based on inductive charge effects at the ipso-carbon atom by the electropositive Ti substituent satisfactorily explain the observed arene distortions. The endocyclic bond angle α will decrease, which is accompanied by the lengthening of the adjacent bonds. As a result of this, ring strain develops, which will lead to a movement out of the plane of the arene ring and away from the Cr(CO)$_3$ group by the ipso-carbon atom (*25,319*). In fact, when this effect was enhanced by introducing electron-withdrawing substituents (R) in the

TABLE XIV

SELECTED X-RAY DATA[a] FOR σ,π-BRIDGED ARENE LIGANDS IN BIMETALLIC COMPLEXES

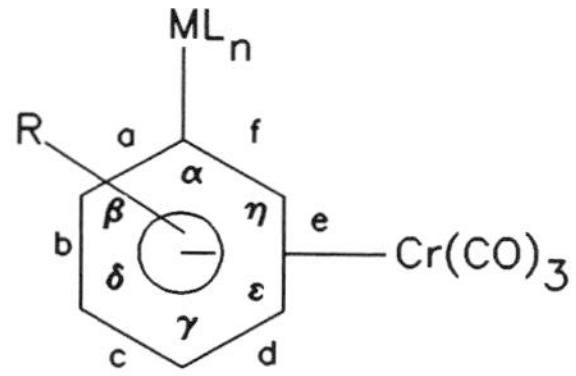

117

No.	ML$_n$	R	M–C$_1$	a	b	c	d	e	f	Cr–C$_1$	Cr–C$_n$	α	β	δ	γ	ε	η	Ref.
116c	W(CO)$_5$$^-$	H	2.217	1.46	1.36	1.46	1.41	1.32	1.51	2.357	2.259	—	—	—	—	—	—	(311)
115g	FeCp(CO)$_2$	p-Cl	1.990	1.414	1.413	1.379	1.388	1.404	1.417	—	—	—	—	—	—	—	—	(306)
117f	Fe(Cp')(CO)$_2$	H	1.998	1.423	1.406	1.398	1.374	1.419	1.401	2.294	2.216	116.2	121.0	120.9	118.9	120.8	121.6	(24)
117c	FeCp(CO)$_2$	p-Me	1.991	1.422	1.402	1.400	1.402	1.420	1.399	2.286	2.217	115.9	122.1	121.3	117.6	120.8	122.2	(24)
117e	Fe(indenyl)(CO)$_2$	H	1.987	1.408	1.414	1.385	1.413	1.387	1.434	2.300	2.226	116.6	121.5	120.5	119.2	120.5	121.6	(24)
117i	Mn(CO)$_4$PPh$_3$	H	2.103	1.422	1.406	1.405	1.395	1.408	1.415	2.308	2.212	114.6	123.0	119.5	119.8	119.2	123.7	(305)
117j	TiCp$_2$Cl	H	2.228	1.418	1.405	1.394	1.413	1.395	1.439	2.307	2.226	115.9	121.9	120.8	119.3	122.2	119.9	(25)
117k	TiCp$_2$Cl	o-F	2.217	1.396	1.393	1.381	1.395	1.390	1.431	2.333	2.200	111.3	126.3	120.7	115.7	122.8	122.9	(316)
117n	TiCp$_2$Cl	o-OMe	2.240	1.401	1.404	1.387	1.387	1.400	1.443	2.319	2.225	113.4	124.2	119.6	120.1	123.5	119.2	(317)
117o	TiCp$_2$Cl	m-NMe$_2$	2.236	1.392	1.413	1.379	1.425	1.430	1.417	2.300	2.254	117.5	121.5	119.6	122.4	123.1	115.4	(317)
117l	TiCp$_2$Cl	p-Me	2.213	1.405	1.399	1.390	1.395	1.409	1.441	2.278	2.208	114.6	123.5	120.2	117.6	119.5	122.6	(316)

[a] Standard deviations omitted.

[b] Cr–C$_n$ represents averaged value, $n = 2$ to $n = 6$.

ortho position of the bridging benzene ligand, the α angle decreased with a corresponding increase in the electronegativity of the R substituent. The coordination of chromiumtricarbonyl to chlorobenzene and subsequent carbonylation by palladium complexes to esters, aldehydes, amides, and α-oxo-amides is an example of bimetallic activation in homogeneous catalysis (*320*). The oxidative addition of coordinated chlorobenzene to zero-valent palladium phosphine complexes gave σ,π-bimetallic complexes $(\eta^6\text{-}C_6H_5\{Pd(L_2)Cl\})Cr(CO)_3$ (**118**) [L = PPh$_3$ (**118a**), PEt$_2$Ph (**118b**), PPh$_2$Et (**118c**)] which are monomers. The dimers formed during the reactions with PPh$_3$ can be cleaved by the addition of the monodentate phosphines (L = PPh$_3$, PPh$_2$Et, PPhEt$_2$) or bidentate phosphines (L$_2$ = dppe or dpph), giving *cis*-$(\eta^6\text{-}C_6H_5\{Pd(L_2)Cl\})Cr(CO)_3$ (**118**) [L$_2$ = dppe (**118d**), dpph (**118e**), dpph is bis(diphenylphosphine)hexane]. These aryl–palladium bimetallic compounds react with CO under atmospheric pressure by CO insertion, yielding $(\eta^6\text{-}C_6H_5C(O)\{Pd(L_2)Cl\})Cr(CO)_3$ (**119**) [L = PPh$_3$ (**119a**), PPhEt$_2$ (**119b**), L$_2$ = dpph (**119c**)]. The insertion of CO for the monomeric bimetallic complex with L$_2$ being dppe was unsuccessful, but the corresponding product *cis*-$(\eta^6\text{-}C_6H_5C(O)\{Pd(dppe)Cl\})Cr(CO)_3$ (**119d**) was nevertheless obtained after cleavage by dppe of the bis(acyl) dimer of triphenylphosphine. Cationic π-arene complexes of iron, Fe(η^6-*p*-C$_6$H$_4$RCl)Cp, reacted with ML$_n^-$ to give [(η^6-*p*-C$_6$H$_4$R{ML$_n$})FeCp]$^+$ (**120**) [R = H, ML$_n$ = Re(CO)$_5$ (**120a**), CpW(CO)$_3$ (**120b**); R = Me, ML$_n$ = Re(CO)$_5$ (**120c**); R = Cl, ML$_n$ = Re(CO)$_5$ (**120d**)] (*114*). The very reactive (CO)$_5$MFBF$_3$ reacts with Re(CO)$_5$Ph by eliminating two carbonyl ligands to afford the cationic σ,π-bridged benzene bimetallic complex [(η^6-C$_6$H$_5${Re(CO)$_5$})M(CO)$_3$]BF$_4$ (**120**) [M = Mn (**120e**), Re (**120f**)].

121a   121b

Diphenylacetylene reacts regiospecifically with bis(cyclopentadienyl)(*o*-fluorobenzyne)titanium, which was prepared by thermal pyrolysis of Cp$_2$Ti(*o*-fluorophenyl)$_2$, to give the corresponding titanole. Treatment of the titanium metallocycle with Cr(CO)$_3$(NCMe)$_3$ afforded the

σ,π-bimetallic product **121a** (*321*). A structure determination of **121a** indicated a reaction mechanism of strong steric discrimination during the acetylene insertion process. The reaction of 3 equiv of *o*-fluorolithiobenzene with CpFe(CO)$_2$I results in an intramolecular cyclization to give a metallocyclic product, which was characterized after derivatization with Cr(CO)$_3$(NCMe)$_3$, giving (η^6-C$_6$H$_4$-*o*-{Fe(Cp)(CO)}{C(O)C$_6$H$_4$-2'-C$_6$H$_4$-2''-F})Cr(CO)$_3$ (**121b**) (*322*).

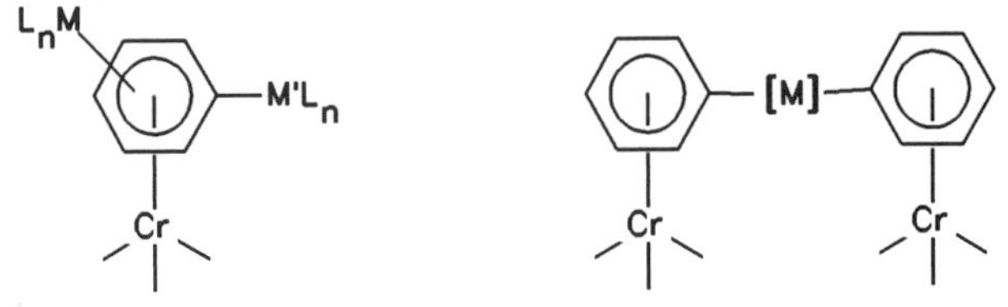

$$\sigma,\sigma',\pi \qquad \sigma,\pi \qquad \sigma',\pi'$$

The acyls 1,3- and 1,4-C$_6$H$_4${C(O)Fp$^\#$}$_2$ are decarbonylated under conditions of reflux in dibutylether, and the resulting bis(iron–aryl) complexes, in the presence of an excess Cr(CO)$_6$, give (η^6-1,*n*-C$_6$H$_4${Fp$^\#$}$_2$)Cr(CO)$_3$ (**122**) [Fp$^\#$ = Fp = CpFe(CO)$_2$, *n* = 3 (**122a**), *n* = 4 (**122b**); Fp$^\#$ = Fp' = (η^5-C$_5$H$_4$Me)Fe(CO)$_2$, *n* = 3 (**122c**), *n* = 4 (**122d**)] (*323,324*). The analogous compounds (η^6-1,4-C$_6$H$_4${Fp}$_2$)M(CO)$_3$ (**122**) [M = Mo (**122e**), W (**122f**)] were prepared similarly. Noteworthy is that when an excess of NaFp is added to solutions of (η^6-1,4-C$_6$H$_4$Cl{Fp})Cr(CO)$_3$, no evidence for the formation of **122a** was observed (*313*). The observed chemical shifts of the ipso-carbon atoms for the 1,4-substituted trimetallic compounds show an additional shielding in the ^{13}C NMR spectra, which implies that the Fp centers in these complexes are involved in an Fe–arene–Fe conjugative interaction. High electron density on the arene ring is supported by IR data, which show a substantial transfer of electron density from the iron-containing fragment to Cr(CO)$_3$ on complexation. Refluxing the bis(Fp) σ-arene complexes in the pres-

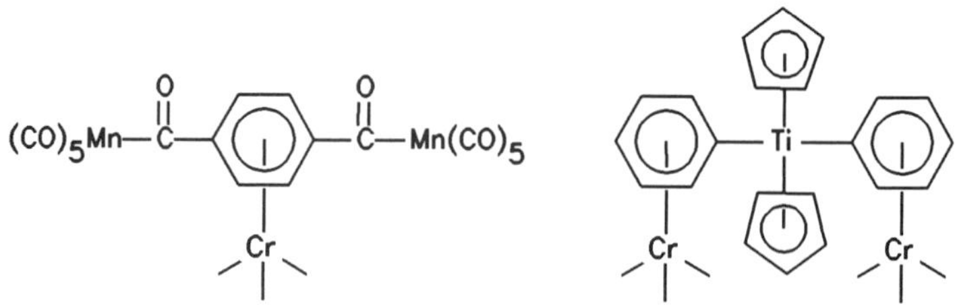

123   124

ence of $Cr(CO)_6$ afforded $(\eta^6\text{-}2,6\text{-}C_{10}H_6\{Fp\}_2)Cr(CO)_3$ (**122g**) and $(\eta^6\text{-}4,4'\text{-}$ $C_{12}H_8\{Fp\}_2)Cr(CO)_3$ (**122h**). Cyclic voltammetry and spectroscopic data indicate that the $Cr(CO)_3$ center is more electron-rich in the following order: **122g** > **122h** > **117g** (*314*).

Treatment of (1,4-dibromobenzene)$Cr(CO)_3$ with 2 equiv of n-BuLi followed by 2 equiv of $[Mn(CO)_5Br]$ afforded the σ,σ,π-bridged compound $(\eta^6\text{-}C_6H_4\{C(O)Mn(CO)_5\}_2)Cr(CO)_3$ (**123**) (*305*). Two equivalents of lithiated arenes react with titanocene dichloride to give the trimetallic complexes $Cp_2Ti\{(\eta^6\text{-}C_6H_4R)Cr(CO)_3\}_2$ (**124**) [R = H (**124a**), p-Me (**124b**)] (*316*). Irrespective of whether 1 or 2 equiv of the lithiated fluorobenzene derivative was used, only one chloro ligand was displaced from titanocene dichloride to give **117k** and the π,π-bimetallic compound $(\eta^6\text{:}\eta^6\text{-}$ $C_6H_5\text{-}C_6H_4F\text{-}o\text{-}F)\{Cr(CO)_3\}_2$.

Lithiated $(\eta^6\text{-}C_6Me_5CH_2Li)Cr(CO)_3$ reacted with Cp_2TiCl_2 to give the air-sensitive product $(\eta^6\text{-}C_6Me_5CH_2\{Ti(Cp_2)Cl\})Cr(CO)_3$ (**125**) (*316*). Benzyl chloride when reacted with $Na[Co(CO)_4]$ affords an equilibrium mixture of the σ-bonded η^1-benzyl and π-bonded η^3-benzyl complexes, which undergoes carbonyl insertion to give η^1-phenylacetylcobalt tetracarbonyls (*325–327*). The reaction of $(\eta^6\text{-}C_6H_5CH_2Cl)Cr(CO)_3$, on the other hand, gave the σ,π-benzyl-bridged complex $(\eta^6\text{-}C_6H_5CH_2\{Co(CO)_4\})Cr(CO)_3$ (**126a**) in pure form. The X-ray structure of **125** shows the benzyl group in the axial position and a relatively long Co–C(alkyl) bond length of 2.13 Å (*325,328*). The reaction of $(\eta^6\text{-}C_6H_5CH_2Cl)Cr(CO)_3$ or $\{\eta^6\text{-}1,4\text{-}C_6H_4\text{-}$ $(CH_2Cl)_2\}Cr(CO)_3$ with 1 or 2 equiv of $[Co(DH)_2(py)Cl]$ (DH is the monoanion of dimethylglyoxime) gave the bimetallic and trimetallic compounds $(\eta^6\text{-}C_6H_5CH_2\{Co(DH)_2(py)\})Cr(CO)_3$ (**126b**) or $(\eta^6\text{-}1,4\text{-}C_6H_4\{CH_2Co(DH)_2\text{-}$ $(py)\}_2)Cr(CO)_3$ (**126c**) (*329*). The Co–C(alkyl) bond distance of **126b** is, as expected for a Co(III) ion in an octahedral ligand environment, significantly shorter than the corresponding distance in **126a**. The interaction of $Mn(CO)_5FBF_3$ with $CpFe(CO)_2\{CH_2Ph\}$ affords the complex $[(\eta^6\text{-}C_6H_5\text{-}$ $CH_2\{Fp\})Mn(CO)_3]BF_4$ (**127**) (*330*).

128	ML_n	R
a	$Re(CO)_5$	H
b	$FeCp(CO)_2$	H
c	$Re(CO)_5$	Me

128

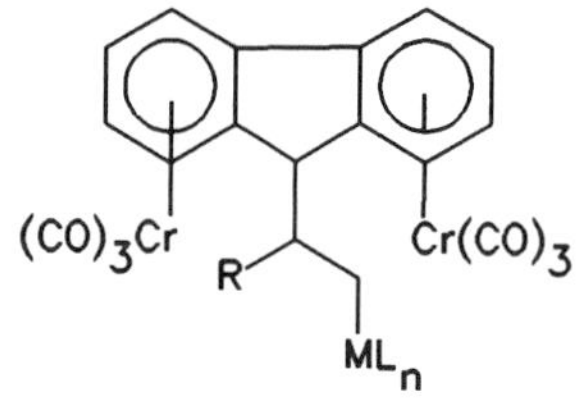

129	ML_n	R
a	$Re(CO)_5$	H
b	$FeCp(CO)_2$	H
c	$WCp(CO)_3$	H
d	$Re(CO)_5$	Me

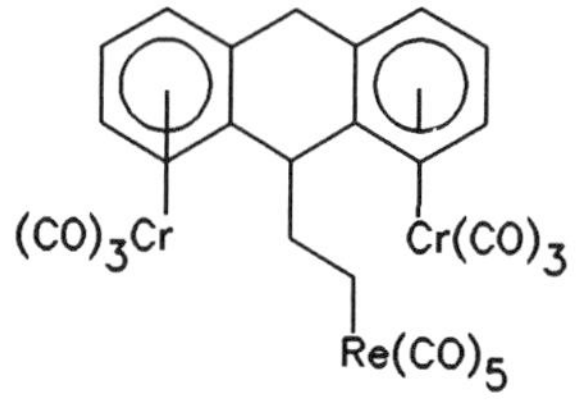

129

130

The addition of the anions μ-(diphenylmethane)-, μ-fluorene-, and μ-(9,10-dihydroanthracene)-bis(tricarbonylchromium) to cationic complexes with π-bonded olefin ligands, $[L_nM(\eta^2\text{-}C_2H_3R)]^+$ (R = H, Me), provided a synthesis for the σ,π,π-trimetallic complexes 128, 129, and 130 (331).

b. σ-Carbene and σ-Carbyne Complexes.   Addition of the organometallic alkylating agent $Fe(\eta^5\text{-}C_6Me_5CH_2)Cp$, which contains an exocyclic methylene group, to the methoxycarbene complex $[Fp^*\{C(OMe)H\}]PF_6$ effects electrophilic condensation to yield the bimetallic complex $(\eta^6\text{-}C_6Me_5CH_2C(OMe)H)\{Fp^*\})FeCp$ (131a) (332). The formation of a dicationic dinuclear carbene intermediate $[\{\eta^6\text{-}(\eta^6\text{-}C_6Me_5CH_2C(H)\{Fp^*\})\text{-}FeCp\}]^{2+}$ (131b) was postulated after methoxide abstraction from reacting 131a with $CF_3SO_3SiMe_3$ at low temperatures. Treatment of 131c with HBF_4–OEt_2 in the presence of PMe_3 quantitatively afforded the stable σ,π-vinylarene diiron complex $(E)\text{-}[(\eta^6\text{-}C_6Me_5CH{=}CH\{Fp^*\})FeCp]^+$ (131c).

The high nucleophilicity of a metallated arene carbon atom was demonstrated by Fischer et al. (333) and utilized in the synthesis of bimetallic complexes with a σ-carbene and π-arene bridging ligand (B). They reacted $(\eta^6\text{-}C_6H_5Li)Cr(CO)_3$ with $M(CO)_6$, which after subsequent alkylation with Et_3OBF_4 gave the carbene complexes $(\eta^6\text{-}C_6H_5C(OEt){=}\{M(CO)_5\})Cr(CO)_3$ (132) [M = Cr (132a), Mo (132b), W (132c)]. Spectroscopic data are consistent with a higher electron density on the arene ring compared with the mononuclear arene complex. The carbene complexes **132** react with borontrihalides to give bimetallic complexes with a σ-carbyne and π-arene bridging ligand (**C**).

The compounds $(\eta^6\text{-}C_6H_5C\equiv\{M(CO)_4X\})Cr(CO)_3$ (**133**) [M = Cr, X = Cl (**133a**), Br (**133b**); M = W, X = Cl (**133c**), Br (**133d**)] were synthesized along this route. Bimetallic carbyne complexes $(\eta^6\text{-}o\text{-}C_6H_4(OMe)C\equiv\{M(O_2CCF_3)(CO)_2L_2\})Cr(CO)_3$ (**133**) [M = Mo, L_2 = tmen (**133e**), bipy (**133f**); M = W, L_2 = tmen (**133g**), bipy (**133h**), tmen is tetramethylethylenediamine] were synthesized from the metallated anisole precursor, suspensions of the hexacarbonyls, and subsequent treatment with (CF$_3$-CO)$_2$O followed by the addition of an equivalent amount of the bidentate ligand (*334*). Reaction of **133** with NaCp or K{HB(pz)$_3$} gave similar carbyne compounds, $(\eta^6\text{-}C_6H_4(OMe)C\equiv\{M(CO)_2L\})Cr(CO)_3$ (**133**) [M = Mo, L = Cp (**133i**), HB(pz)$_3$ (**133j**); M = W, L = Cp (**k**)].

III

BRIDGING LIGANDS WITH PARTICIPATING HETEROATOMS

A. Ligands with Carbon–Heteroatom π-Bonds

Complexes with σ-coordinated heteroatoms and π-coordinated hydrocarbons have been synthesized and studied extensively, but they are not included in this article. On the other hand, complexes with σ-bonded carbon atoms and π-bonded CX groups (X = heteroatom), with the exception of species with metal–metal bonds and η^1:η^2-coordinated carbonyl or isocyanide ligands, have been very rarely reported in the literature. In most cases heteroatoms preferentially coordinate by lone pair electrons of the heteroatom and not through the adjacent unsaturated bond.

134   **135**

Shaw and co-workers (*152*) observed broad and poorly defined resonances in the ^{31}P NMR spectrum of (CO)$_3$Mo(μ-dppm)$_2$Rh(CO)Br, which suggested that two or more species were in rapid dynamic equilibrium. Low-temperature studies revealed well-defined resonances of two independent heterobimetallic species. A possible explanation is that the two metal centers, which are held in close proximity by the bridging dppm ligands, promote the scrambling of carbonyls between the metal centers. Comparisons can be drawn with bimetallic activation of acetylides and acetylide migrations (Section II,C). The η^2-coordination of the terminal carbonyl ligand was assigned on the basis of carbonyl absorptions recorded in the IR spectra of $[(CO)_2Mo\{\mu\text{-dppm}\}_2\{\mu\text{-}(\eta^1:\eta^2\text{-CO})\}M(CO)L]^+$ (**134**) [M = Rh, L = NCMe (**134a**), CN^{t}Bu (**134b**); M = Ir, L = NCMe (**134c**), NCPh (**134d**)] (*152*). The combination of early transition metals (Zr, Ti) with late transition metals (Mo, W) via a single bridging monophosphido ligand was utilized by Stephan and co-workers (*335–337*) to study the cooperative activation of carbon oxide substrates. The presence of an open coordination site on the Zr or Ti permits Lewis acid activation and "side on" coordination of the carbonyl ligand bound to Mo or W in $Cp(CO)M(\mu\text{-PR}_2)(\mu\text{-}\eta^1:\eta^2\text{-CO})M'Cp_2$ (**135**) [M' = Zr M = Mo, R = Cy (**135a**), Ph (**135b**), M = W, R = Ph (**135c**) (*335*); M' = Ti, R = Et, M = Mo (**135d**), M = W (**135e**) (*336,337*); M = Mo, R = Et, M' = Zr (**135f**) (*337*)]. Crystal structure determinations of **135b,d,e,f** reveal M–M bond distances in the range of 3.21–3.25 Å, M–P–M angles of 82–79°, and M'–O distances which are shorter than the M'–C bond distances for the η^2-coordinated carbonyl ligand.

The activation of carbon monoxide in early/late heterobimetallic complexes has been the subject of numerous studies, especially because of possible applications in carbon oxide reduction chemistry (*17*). Examples of such compounds where the $\mu\text{-}\eta^1:\eta^2$-CO ligand is supported by a second bridging ligand are $Cp(CO)Mo\{\mu\text{-MeC}(O)\text{-}C,O\}\{\mu\text{-}\eta^1:\eta^2\text{-CO}\}ZrCp_2$ (**135g**) (*338,339*), $CpCo(\mu\text{-CO})\{\mu\text{-}\eta^1:\eta^2\text{-CO}\}ZrCp_2$ (**135h**) (*340*), $Cp(CO)\text{-}W\{\mu\text{-CR}\}\{\mu\text{-}\eta^1:\eta^2\text{-CO}\}TiCp_2$ (**135i**) (*341,342*), and $Cp(CO)W\{\mu\text{-CR}CH_2\}\{\mu\text{-}\eta^1:\eta^2\text{-CO}\}TiCp_2$ (**135j**) (*343*).

The bimetallic fulvalene complex $CpZr(\mu\text{-}\eta^5:\eta^5\text{-}C_5H_4\text{—}C_5H_4)(\mu\text{-Cl})_2ZrCp$ reacts with *tert*-butyl isocyanide to give $Cp(Cl)Zr\{\mu\text{-}C_{10}H_8\}\{\mu\text{-}\eta^1:\eta^2\text{-}C\equiv N^t Bu\}ZrCp(Cl)$ (**136**) (*344,345*). During a sequence of insertion

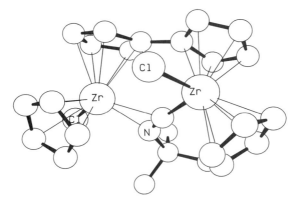

FIG. 7. X-Ray structure of $Cp(Cl)Zr\{\mu\text{-}C_{10}H_8\}\{\mu\text{-}\eta^1:\eta^2\text{-}C{\equiv}NBu^t\}ZrCp(Cl)$ (136) [Herrmann *et al.* (*344*)].

reactions and oxidation reactions of the dimeric Zr(III) precursor, it was shown that the fulvalene bracket between the two zirconium atoms remains intact. The structure of **136** in the solid state (Fig. 7) reveals a bridging (σ and π) coordinated *tert*-butyl isocyanide ligand and an intermetallic distance which suggests that the isocyanide was inserted into a Zr–Zr bond during the reaction. The optimal orientation for the σ,π-bridging isocyanide ligand is achieved by rotation of 22.6° of the two Zr fragments around the C–C bond connecting the two cyclopentadienyl rings of fulvalene. The insertion of isocyanides into Zr–C bonds was observed in the reaction of zirconium and ruthenium dimetalloalkenes and *tert*-butyl isocyanide to give $Cp(Me_3P)_2Ru\{\mu\text{-}\eta^1:\eta^2\text{-}(CH{=}CRCN^t\text{-}Bu)\}ZrCp_2Cl$ (**137**) [R = H (**137a**), Me (**137b**)] (*346*). The product initially formed has the nitrogen of the η^2-iminoacyl ligand on the outside position of the three ligation sites of the Cp_2Zr moiety and rearranges to the thermodynamically favored isomer with the nitrogen on the inside. Spectroscopic and crystallographic data for **137a** indicated a significant contribution from a zwitterionic resonance form, which is facilitated by the disparate electronic properties of the two metal fragments. Contrary to the case for **137a**, the metals and organic bridging group of **137b** are twisted out of planarity to minimize interaction between the methyl and *tert*-butyl groups. The reaction of carbon monoxide with $Cp(PMe_3)_2Ru\{\mu\text{-}CH{=}CR\}ZrCp_2Cl$ affords the analogous η^2-acyl complexes $Cp(Me_3P)_2Ru\{\mu\text{-}\eta^1:\eta^2\text{-}(CH{=}CRC(O)\}ZrCp_2Cl$ (**138**) [R = H (**138a**), ^{t}Bu (**138b**)] (*347*). The spectroscopic data and structural data of **137a** were interpreted to indicate that the zwitterionic forms are of greater importance than the neutral forms. The cyclic bimetallic titanoxycarbene complex $Cp^*Ti\{\mu\text{-}$

$CH_2CH_2C(O)\}W(CO)_5$ inserts *tert*-butyl isocyanide to give $(CO)_5W\{\mu-\eta^1:\eta^2-C(O)CH_2CH_2C\equiv NC'Bu\}TiCp*_2(CN'Bu)$ (**138**) (*348*).

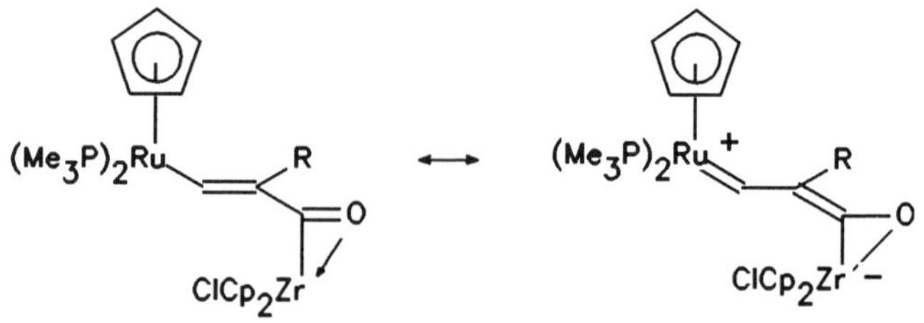

138

The coordination chemistry of compounds containing phosphorus–carbon multiple bonds, such as phosphaalkynes, phosphaalkenes, phosphaallenes, phosphaalkenyls, and phosphaallyls, has been studied extensively (*349,350*). The chemistry is dominated by the donor properties of the phosphorus atom, and, as far as we are aware, no examples of bimetallic or trimetallic compounds with bridging ligands which link metal centers by a carbon σ bond and a $C=P$ or $C\equiv P$ π bond have been recorded. Once the lone pair on the phosphorus atom is involved in bonding, the unsaturated bond becomes a possible site for further coordination.

B. Pseudoaromatic Heterocyclic π-Ligands

Five-membered and six-membered monoheterocycles, with the heteroatom part of the π conjugation in the ring, have planar structures. One important feature of these compounds is the degree of aromaticity in the ring. Heterocycles are often classified as being π-excessive or π-deficient (*351*) or in another approach as having hard or soft donor atoms according to the hard and soft acids and bases (HSAB) concept (*352*). It is these properties that will determine the coordinating properties of the heterocycles. The chemistry of heterocyclic ligands which are π-coordinated to transition metals in sandwich and half-sandwich compounds is well documented (*353,354*). They are potentially the best candidates to act as σ,π-bridges (σ-carbon) in bimetallic and trimetallic complexes. However, until relatively recently, very few examples of π-bonded pseudoaromatic heterocycles, which are σ-bonded to a second transition metal fragment through a ring carbon atom and without a metal–metal bond, have been recorded.

In the cobalt and rhodium sandwich complexes $MCp\{\eta^5-CRCR-BR'CHR''BR\}$ the acidic proton of the diborole ring was removed and

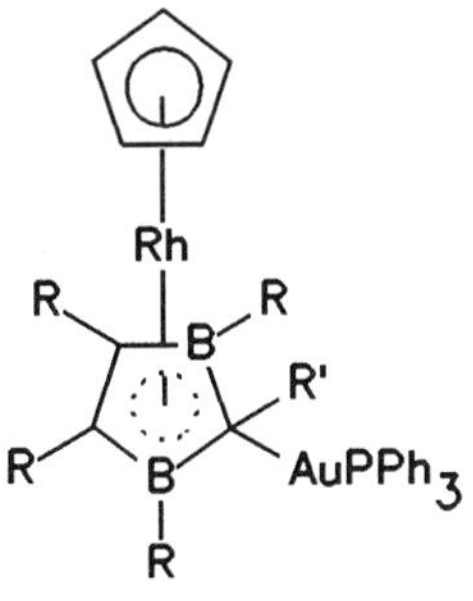

139

replaced by the isolable $AuPPh_3$ group to give the bimetallic complexes $(\eta^5\text{-BR}'CRCRBR'CR''\{AuPPh_3\})MCp$ (**139**) [M = Co, R = Me, R' = Me, R'' = H (**139a**); M = Rh, R = R' = Et, R'' = Me (**139b**)] (*355*). Spectroscopic data and X-ray structure determination of **139a** suggest that the ipso-carbon atom of the ring is pentacoordinated with a Au–C–M three-center two-electron interaction.

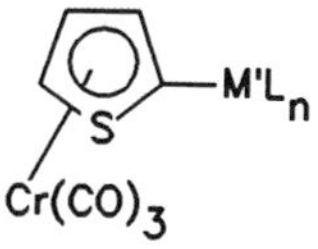

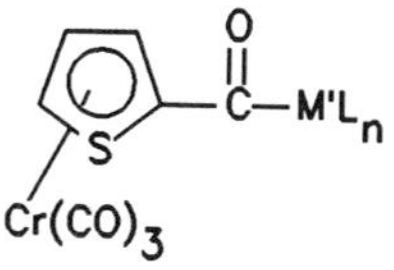

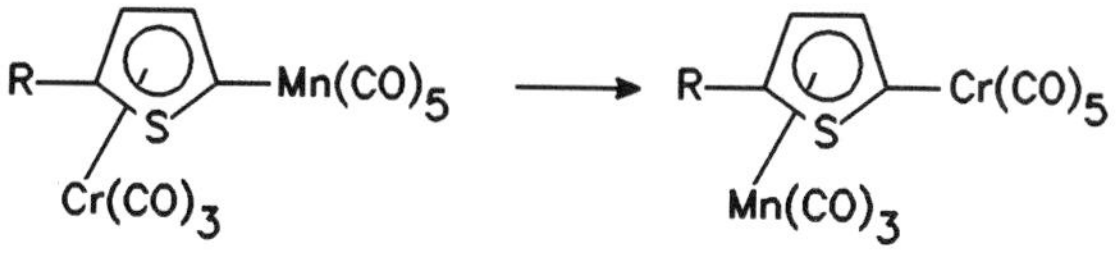

$$(37)$$

140 **141**

The activation of thiophene by transition metals has been the topic of many investigations (*356,357*). A solution of $(\eta^5\text{-2-LiC}_4H_2RX)Cr(CO)_3$ (X = S, Se) reacts with 1 equiv $Mn(CO)_5X$ (X = Br, Cl) to give the bimetallic complex $(\eta^5\text{-XCRCHCHC}\{Mn(CO)_5\})Cr(CO)_3$ (**140**) (X = S, R = H (**140a**), Me (**140b**); X = Se, R = H (**140c**)) (*23,358*). Complexes **140** are thermally unstable and irreversibly convert to an isomeric form in which the transition metals have exchanged coordination sites to give the remarkably stable bimetallic complexes $(\eta^5\text{-XCRCHCHC}\{Cr(CO)_5\})$-$Mn(CO)_3$ [Eq. (37)] (**141**). The process of metal exchange for the thienyl (tn) ligand can be represented by

$$(\eta^1\text{-tnMn}, \eta^5\text{-tnCr}) \rightarrow (\eta^1\text{-tnCr}, \eta^5\text{-tnMn}).$$

Spectroscopic data and structural data for **140a** reveal a typically π-coor-

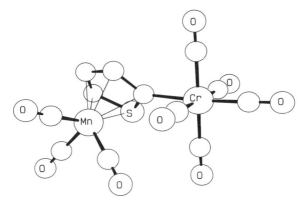

FIG. 8. X-Ray structure of (η^5-$\overline{\text{SCHCHCHC}}${Cr(CO)$_5$})Mn(CO)$_3$ (**141a**) [Waldbach et al. (23)].

dinated thiophene ligand with a substituent in the 2 position. In contrast, the corresponding data for **141a** (Fig. 8) reveal a thiophene ring with allylic character and a σ-bonded chromium fragment with carbene character (23).

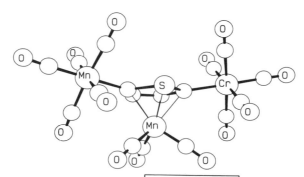

When the reaction was performed under a CO atmosphere the major product of the reaction was a carbonyl-inserted bimetallic compound (η^5-

FIG. 9. X-Ray structure of (η^5-2,5-$\overline{\text{SC}}${Mn(CO)$_5$}$\overline{\text{CHCHC}}${Cr(CO)$_5$})Mn(CO)$_3$ (**143**) [Waldbach et al. (359)].

TABLE XV

Selected X-ray Data for σ,π-Bridged Thiophene Moieties in Bimetallic and Trimetallic Complexes

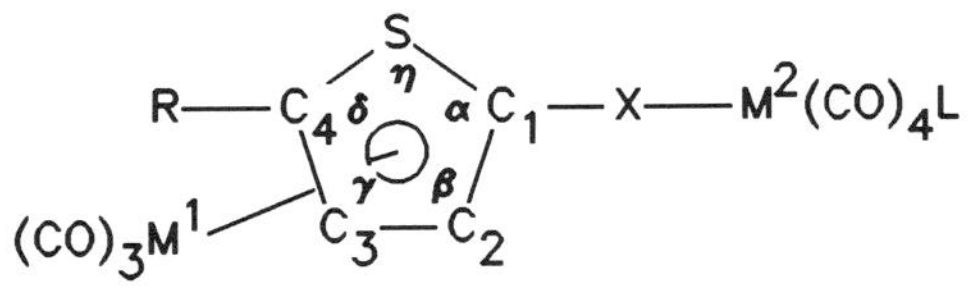

	Complex				Bond distances (Å)					Bond angles (°)					
No.	M¹	M²L	X	R	C_1–C_2	C_2–C_3	C_3–C_4	C_1–S	C_4–S	α	β	γ	δ	η	Ref.
A[a]	—	—	—	—	1.370	1.424	1.370	1.714	1.714	115.3	112.5	112.5	115.3	92.2	(357)
140a	Cr	MnCO	—	H	1.390	1.428	1.371	1.764	1.723	107.6	115.5	111.8	111.7	93.3	(23)
141a	Mn	CrCO	—	H	1.485	1.472	1.399	1.768	1.729	109.7	109.4	115.3	111.0	94.5	(23)
143	Mn	CrCO	—	Mn(CO)₅	1.421	1.425	1.385	1.768	1.763	101.9	117.6	116.6	103.9	99.9	(359)
144	Cr	MnCO	CO	Mn(CO)₅	1.388	1.427	1.391	1.768	1.767	109.1	113.3	115.7	107.9	93.8	(359)
142	Cr	MnCO	CO	H	1.373	1.424	1.375	1.748	1.735	114.3	110.6	112.5	113.2	88.5	(358)

[a] **A** represents thiophene.

$\overline{\text{S}\text{CRCHCHCC(O)}}\{Mn(CO)_5\})Cr(CO)_3$ **(142)** [R = H **(142a)**, Me **(142b)**] *(23,358)*. The reaction of the dilithiated thiophene complex of chromium and bromopentacarbonylmanganese afforded the σ,σ,π-trimetallic compound $(\eta^5\text{-}2,5\text{-}SC_4H_2\{Mn(CO)_5\}_2)Cr(CO)_3$ **(142)** *(359)*. The complex **142** will irreversibly participate in a metal exchange process, which is solvent and temperature dependent, to give the thermodynamically favored trimetallic product $(\eta^5\text{-}2,5\text{-}\overline{\text{SC}\{Mn(CO)_5\}CHCHC\{Cr(CO)_5\}})Mn(CO)_3$ **(143)**. Under a CO atmosphere a carbonyl was inserted into one of the Mn–C(thiopene) σ bonds, and no metal exchange was observed to give $(\eta^5\text{-}2,5\text{-}\overline{\text{SC}\{C(O)Mn(CO)_5\}CHCHC\{Mn(CO)_5\}})Cr(CO)_3$ **(144)**. The solid state structure of **143** (Fig. 9) reveals a σ-bonded chromium fragment with some carbene character in the σ bond and a manganese fragment which is twisted out of the plane of the thiophene ring and away from the Mn(CO)$_3$ group. The structural data of **144** on the other hand, are consistent with the delocalization of electron density from the thiophene ring onto the inserted carbonyl which lies in the same plane (Table XV).

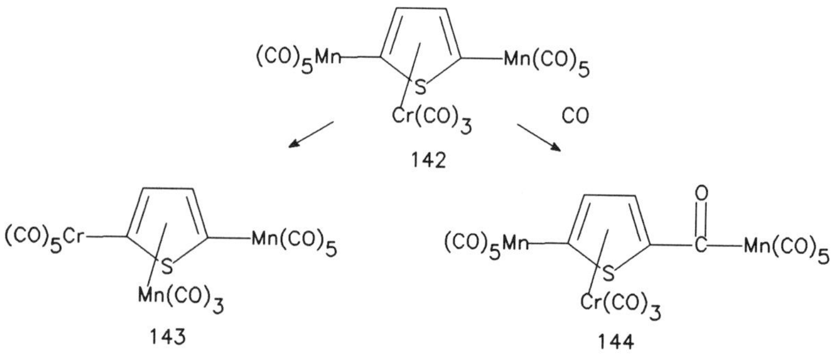

The condensation of $Cr(CO)_6$ with $FeCp(CO)_2\{2\text{-}\overline{\text{CSCHCHCH}}\}$ and $FeCp(CO)_2\{2\text{-benzofuran}\}$ is facilitated by the iron fragment and gave the bimetallic complexes **145** and **146** *(360)*.

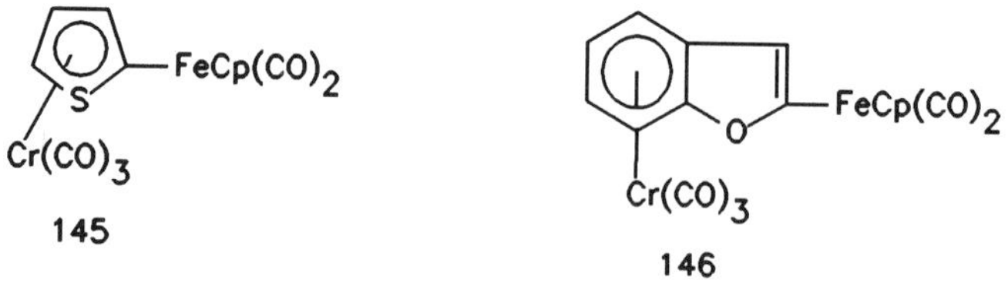

IV

CONCLUDING REMARKS

Bimetallic and trimetallic complexes, having an organic substrate as a bridging ligand by σ- and π-attachments constitute an area of research only starting to develop. Many aspects have been neglected and need to be addressed by introducing new synthetic approaches. We can expect that future work in this area of organometallic chemistry will lead to new discoveries with fundamental implications and practical applications. We are well aware that this article is far from complete, but instead have rather opted to illustrate that σ,π-bimetallic chemistry creates new challenges and opportunities in developing unique reaction pathways resulting from the activation of organic substrates by more than one metal center. We have distinguished between two types of activation of a bridging ligand by two or more metal centers. The σ-bonded metal fragment may be in contact with the π-bonded metal center through π-conjugation over the bridging ligand or may act as two independent activators of a nonconjugated system. In the latter case, although bridging ligand activation is of a more local nature, this may be used advantageously in organic synthesis by exploiting electronic and steric properties of metal fragments. In the first case, π-resonance effects of the σ-bonded metal fragment affect or are affected by the π-bonded metal fragment. This brings the electronic properties of the two metal centers in direct contact with one another with potentially interesting consequences. σ,π-Bridging ligands of dimetallic complexes may scramble or fluxionate between the two metal centers or may even have a preferred mode of coordination for a particular metal center (Sections II,C and III,B). Thus, it is not unlikely that under certain conditions a ligand will migrate from one metal center to the other. When more coordination positions are involved, it may even appear that the metals are exchanging positions (Section III,B) or selecting a preferred site with respect to the ligand environment. The idea or concept of a preferred location or site for a ligand or metal fragment in bimetallic and trimetallic complexes is established. Studies that will elucidate the mechanisms of processes of this kind will stimulate further research in this area.

We have shown that metal activation of σ,π-coordinated ligands in bimetallic complexes may facilitate carbon-carbon bond formation. However, only a few examples of C–C coupling reactions of organic substrates have been recorded (Sections II,C–E), and this aspect of bimetallic activation has not yet been fully exploited. Furthermore, activation of a bridging ligand in both the σ-frame and π-frame by different transition metals, and in a cooperative manner, enables the researcher to study the chemistry of a bridging ligand under conditions of extreme activation.

ACKNOWLEDGMENTS

The perspective drawings were produced using Ortep[361] from data obtained from the Cambridge Crystallographic Data File[362]. The authors are grateful to Elizabeth van Rooyen, Marilé Landman, Yvette Terblans, and Annatjie Kok for assistance in the preparation of this article. Simon Lotz wishes to thank the following for their kindness in supplying information and material prior to publication: Proff. M. Akita (Yokohama), W. Beck (München), J. Forniés (Zaragoza), J. A. Gladysz (Salt Lake City), A. D. Hunter (Youngstown), H. Lang (Heidelberg), C. M. Lukehart (Nashville), J. R. Moss (Cape Town), and A. Wojcicki (Columbus).

REFERENCES

(1) Gladfelter, W. L.; Geoffroy, G. L. *Adv. Organomet. Chem.* **1980**, *18*, 207.
(2) Vahrenkamp, H. *Adv. Organomet. Chem.* **1983**, *22*, 169.
(3) Deeming, A. J. *Adv. Organomet. Chem.* **1986**, *26*, 1.
(4) Salter, I. D. *Adv. Organomet. Chem.* **1989**, *29*, 249.
(5) Farrugia, L. J. *Adv. Organomet. Chem.* **1990**, *31*, 301.
(6) Holton, J.; Lappert, M. F.; Pearce, R.; Yarrow, P. I. W. *Chem. Rev.* **1983**, *83*, 135.
(7) Nesmeyanov, A. N.; Rybinskaya, M. I.; Rybin, L. V.; Kaganovich, V. S. *J. Organomet. Chem.* **1973**, *47*, 1.
(8) Herrmann, W. A. *Adv. Organomet. Chem.* **1982**, *20*, 159.
(9) Casey, C. P.; Audett, J. D. *Chem. Rev.* **1986**, *86*, 339.
(10) Beck, W.; Sünkel, K. H. *Chem. Rev.* **1988**, *88*, 1405.
(11) Wadepohl, H. *Angew. Chem., Int. Ed. Engl.* **1992**, *31*, 247.
(12) Bullock, R. M.; Casey, C. P. *Acc. Chem. Res.* **1987**, *20*, 167.
(13) Erker, G. *Angew. Chem., Int. Ed. Engl.* **1989**, *28*, 397.
(14) He, X. D.; Maisonnat, A.; Dahan, F.; Poilblanc, R. *Organometallics* **1991**, *10*, 2443.
(15) Werner, H.; Treiber, M.; Nessel, A.; Lippert, F.; Betz, P.; Krüger, C. *Chem. Ber.* **1992**, *125*, 337.
(16) Kool, L. B.; Ogasa, M.; Rausch, M. D.; Rogers, R. D. *Organometallics* **1989**, *8*, 1785.
(17) Stephan, D. W. *Coord. Chem. Rev.* **1989**, *95*, 41.
(18) Beck, W.; Niemer, B.; Wieser, M. *Angew. Chem., Int. Ed. Engl.* **1993**, *32*, 923.
(19) Cotton, F. A. *Acc. Chem. Res.* **1968**, *1*, 257.
(20) Gell, K. I.; Williams, G. M.; Schwartz, J. *J. Chem. Soc., Chem. Commun.* **1980**, 550.
(21) Jackson, G. E.; Moss, J. R.; Scott, L. G. *S. Afr. J. Chem.* **1983**, *36*, 69.
(22) Forniés, J.; Gomez-Saso, M. A.; Lalinde, E.; Martinez, F.; Moreno, M. T. *Organometallics* **1992**, *11*, 2873.
(23) Waldbach, T. A.; van Rooyen, P. H.; Lotz, S. *Angew. Chem., Int. Ed. Engl.* **1993**, *32*, 710.
(24) Li, J.; Hunter, A. D.; McDonald, R.; Santarsiero, B. D.; Bott, S. G.; Atwood, J. L. *Organometallics* **1992**, *11*, 3050.
(25) Meyer, R.; Schindehutte, M.; van Rooyen, P. H.; Lotz, S. *Inorg. Chem.* **1994**, *33*, 3605.
(26) Hunter, A. D.; Szigety, A. B. *Organometallics* **1989**, *8*, 2670.
(27) Chukwu, R.; Hunter, A. D.; Santarsiero, B. D.; Bott, S. G.; Atwood, J. L.; Chassaignac, J. *Organometallics* **1992**, *11*, 589.

(28) Mort, J.; Pfinster, G. "Electronic Properties of Polymers"; Wiley: New York, 1982.
(29) Nalwa, H. S. *Appl. Organomet. Chem.* **1991**, *5*, 349.
(30) Cervantes, J.; Vincenti, S. P.; Kapoor, R. N.; Pannell, K. H. *Organometallics* **1989**, *8*, 744.
(31) Forsyth, C. M.; Nolan, S. P.; Stern, C. L.; Marks, T. J.; Rheingold, A. L. *Organometallics* **1993**, *12*, 3618.
(32) Carty, A. J. *Pure Appl. Chem.* **1982**, *54*, 113.
(33) Colbran, S. B.; Hanton, L. R.; Robinson, B. H.; Robinson, W. T.; Simpson, J. *J. Organomet. Chem.* **1987**, *330*, 415.
(34) Jordi, L.; Moreto, J. M.; Ricart, S.; Vinas, J. M.; Mejias, M.; Molins, E. *Organometallics* **1992**, *11*, 3507.
(35) Müller, A.; Jaegermann, W.; Enemark, J. H. *Coord. Chem. Rev.* **1982**, *46*, 245.
(36) Roof, L. C.; Kolis, J. W. *Chem. Rev.* **1993**, *93*, 1037.
(37) Weber, L. *Chem. Rev.* **1992**, *92*, 1839.
(38) Huttner, G.; Borm, J.; Zsolnai, L. *J. Organomet. Chem.* **1986**, *304*, 309.
(39) Caminade, A.-M.; Majoral, J.-P.; Mathieu, R. *Chem. Rev.* **1991**, *91*, 575.
(40) Scherer, O. J. *Angew. Chem. Int. Ed. Engl.* **1985**, *24*, 924.
(41) Scherer, O. J.; Brück, T.; Wolmershäuser, G. *Chem. Ber.* **1989**, *122*, 2049.
(42) Lindner, E.; Heckmann, M.; Fawzi, R.; Hiller, W. *Chem. Ber.* **1991**, *124*, 2171.
(43) Scherer, O. J.; Swarowsky, M.; Swarowsky, H.; Wolmershäuser, G. *Angew. Chem., Int. Ed. Engl.* **1988**, *27*, 694.
(44) Bleeke, J. R. *Acc. Chem. Res.* **1991**, *24*, 271.
(45) For references before 1981 the reader is referred to Wilkinson, G.; Stone, F. G. A. (Eds.), "Comprehensive Organometallic Chemistry"; Pergamon: Oxford, 1982; Vols. 1–9.
(46) Davies, S. G.; Green, M. L. H.; Mingos, D. M. P. *Tetrahedron* **1978**, *34*, 3047.
(47) Rashidi, M.; Puddephatt, R. J. *Organometallics* **1988**, *7*, 1636.
(48) Folting, K.; Huffman, J. C.; Lewis, L. N.; Caulton, K. G. *Inorg. Chem.* **1979**, *18*, 3483.
(49) Breimair, J.; Steimann, M.; Wagner, B.; Beck, W. *Chem. Ber.* **1990**, *123*, 7.
(50) Steil, P.; Beck, W.; Stone, F. G. A. *J. Organomet. Chem.* **1989**, *368*, 77.
(51) Tarazano, D. L.; Bodnar, T. W.; Cutler, A. R. *J. Organomet. Chem.* **1993**, *448*, 139.
(52) Nesmeyanov, A. N.; Rybinskaya, M. I.; Rybin, L. V.; Kaganovich, V. S.; Petrovskii, P. V. *J. Organomet. Chem.* **1971**, *31*, 257.
(53) Erker, G.; Kropp, K.; Atwood, J. L.; Hunter, W. E. *Organometallics*, **1983**, *2*, 1555.
(54) Horton, A. D.; Orpen, A. G. *Angew. Chem., Int. Ed. Engl.* **1992**, *31*, 876.
(55) Burch, R. R.; Muetterties, E. L.; Teller, R. G.; Williams, J. M. *J. Am. Chem. Soc.* **1982**, *104*, 4257.
(56) Burch, R. R.; Shusterman, A. J.; Muetterties, E. L.; Teller, R. G.; Williams, J. M. *J. Am. Chem. Soc.* **1983**, *105*, 3546.
(57) Fryzuk, M. D.; Jones, T.; Einstein, F. W. B. *Organometallics* **1984**, *3*, 185.
(58) Fryzuk, M. D.; Jones, T.; Einstein, F. W. B. *J. Chem. Soc., Chem. Commun.* **1984**, 1556.
(59) Keister, J. B.; Shapley, J. R. *J. Organomet. Chem.* **1975**, *85*, C29.
(60) Nubel, P. O.; Brown, T. L. *J. Am. Chem. Soc.* **1982**, *104*, 4955.
(61) Bonnet, J. J.; Mathieu, R.; Poilblanc, R.; Ibers, J. A. *J. Am. Chem. Soc.* **1979**, *101*, 7487.
(62) Akita, M.; Kondoh, A.; Kawahara, T.; Takagi, T.; Moro-oka, Y. *Organometallics* **1988**, *7*, 366.
(63) Bullock, R. M.; Lemke, F. R.; Szalda, D. J. *J. Am. Chem. Soc.* **1990**, *112*, 3244.

(64) Lemke, F. R.; Szalda, D. J.; Bullock, R. M. *J. Am. Chem. Soc.* **1991**, *113*, 8466.
(65) Raper, G.; McDonald, W. S. *J. Chem. Soc., Dalton Trans.* **1972**, 265.
(66) Hegedus, L. S.; Tamura, R. *Organometallics*, **1982**, *1*, 1188.
(67) Müller, H.-J.; Nagel, U.; Beck, W. *Organometallics* **1987**, *6*, 193.
(68) Hüffer, S.; Wieser, M.; Polborn, K.; Beck, W. *J.* personal communication, 1993.
(69) Wakefield, J. B.; Stryker, J. M. *J. Am. Chem. Soc.* **1991**, *113*, 7057.
(70) Booth, B. L.; Casey, S.; Haszeldine, R. N. *J. Organomet. Chem.* **1982**, *226*, 289.
(71) Dizikes, L. J.; Wojcicki, A. *J. Organomet. Chem.* **1977**, *137*, 79.
(72) King, R. B.; Bisnette, M. B. *J. Organomet. Chem.* **1967**, *7*, 311.
(73) Kerber, R. C.; Giering, W. P.; Bauch, T.; Waterman, P.; Chou, E.-H. *J. Organomet. Chem.* **1976**, *120*, C31.
(74) Laing, M.; Moss, J. R.; Johnson, J. *J. Chem. Soc., Chem. Commun.* **1977**, 656.
(75) Clive, D. L. J.; Kiel, W. A.; Menschen, S. M.; Wong, C. K. *J. Chem. Soc., Chem. Commun.* **1977**, 657.
(76) Kobayashi, M.; Wuest, J. D. *Organometallics* **1989**, *8*, 2843.
(77) Priester, W.; Rosenblum, M.; Samuels, S. B. *Synth. React. Inorg. Met.-Org. Chem.* **1981**, *11*, 525.
(78) Bodnar, T. W.; Cutler, A. R. *Organometallics* **1985**, *4*, 1558.
(79) Johnson, J. W.; Moss, J. R. *Polyhedron* **1985**, *4*, 563.
(80) Lennon, P. J.; Rosan, A.; Rosenblum, M.; Tancrede, J.; Waterman, P. *J. Am. Chem. Soc.* **1980**, *102*, 7033.
(81) Chang, T. C. T.; Foxman, B. M.; Rosenblum, M.; Stockman, C. *J. Am. Chem. Soc.* **1981**, *103*, 7361.
(82) Niemer, B.; Breimair, J.; Völkel, T.; Wagner, B.; Polborn, K.; Beck, W. *Chem. Ber.* **1991**, *124*, 2237.
(83) Lehmann, R. E.; Kochi, J. K. *Organometallics* **1991**, *10*, 190.
(84) Beck, W.; Niemer, B.; Breimair, J.; Heidrich, J. *J. Organomet. Chem.* **1989**, *372*, 79.
(85) Niemer, B.; Breimair, J.; Wagner, B.; Polborn, K.; Beck, W. *Chem. Ber.* **1991**, *124*, 2227.
(86) Hüffer, S.; Wieser, M.; Polborn, K.; Sünkel, K.; Beck, W. *Chem. Ber.* **1994**, *127*, 1369.
(87) Parlier, A.; Rose, F.; Rudler, M.; Rudler, H. *J. Organomet. Chem.* **1982**, *235*, C13.
(88) Macomber, D. W.; Liang, M.; Roger, R. D. *Organometallics* **1988**, *7*, 416.
(89) Stone, F. G. A. *Angew. Chem., Int. Ed. Engl.* **1984**, *23*, 89.
(90) Howard, J. A. K.; Mead, K. A.; Moss, J. R.; Navarro, R.; Stone, F. G. A.; Woodward, P. *J. Chem. Soc., Dalton Trans.* **1981**, 743.
(91) Kreiter, C. G.; Wendt, G.; Kaub, J. *Chem. Ber.* **1989**, *122*, 215.
(92) Kühn, A.; Werner, H. *J. Organomet. Chem.* **1979**, *179*, 421.
(93) Kühn, A.; Burschka, Ch.; Werner, H. *Organometallics* **1982**, *1*, 496.
(94) Gregg, M. R.; Powell, J.; Sawyer, J. F. *J. Organomet. Chem.* **1988**, *352*, 357.
(95) Dyke, A. F.; Knox, S. A. R.; Naish, P. J. *J. Organomet. Chem.* **1980**, *199*, C47.
(96) Beck, W. *Polyhedron*, **1988**, *7*, 2255.
(97) Beck, W.; Raab, K.; Nagel, U.; Sacher, W. *Angew. Chem., Int. Ed. Engl.* **1985**, *24*, 505.
(98) Müller, H.-J.; Nagel, U.; Steiman, M.; Polborn, K.; Beck, W. *Chem. Ber.* **1989**, *122*, 1387.
(99) Müller, H.-J.; Beck, W. *J. Organomet. Chem.* **1987**, *330*, C13.
(100) Ehrlich, K.; Emerson, G. F. *J. Am. Chem. Soc.* **1972**, *94*, 2464.
(101) Kemmitt, R. D. W.; Platt, A. W. G. *J. Chem. Soc., Dalton Trans.* **1986**, 1603.
(102) Kreiter, C. G. *Adv. Organomet. Chem.* **1986**, *26*, 297.

(103) Kreiter, C. G.; Lipps, W. *Angew. Chem., Int. Ed. Engl.* **1981**, *20*, 201.
(104) Kreiter, C. G.; Lipps, W. *Chem. Ber.* **1982**, *115*, 973.
(105) Leyendecker, M.; Kreiter, C. G. *J. Organomet. Chem.* **1983**, *249*, C31.
(106) Baralt, E.; Lukehart, C. M.; McPhail, A. T.; McPhail, D. R. *Organometallics* **1991**, *10*, 516.
(107) Lukehart, C. M.; Owen, M. D. *J. Cluster Sci.* **1991**, *2*, 71.
(108) Campora, J.; Gutierrez, E.; Monge, A.; Poveda, M. L.; Carmona, E. *Organometallics* **1992**, *11*, 2644.
(109) Wu, I.-Y.; Tseng, T.-W.; Lin, Y.-C.; Cheng, M.-C.; Wang, Y. *Organometallics* **1993**, *12*, 478.
(110) Wojcicki, A.; Shuchart, C. E. *Coord. Chem. Rev.* **1990**, *105*, 35.
(111) Schuchart, C. E.; Young, G. H.; Wojcicki, A.; Calligaris, M.; Nardin, G. *Organometallics* **1990**, *9*, 2417.
(112) Behr, A.; Ilsemann, G.; Keim, W.; Krüger, C.; Tsay, Y.-H. *Organometallics* **1986**, *5*, 514.
(113) Niemer, B.; Steimann, M.; Beck, W. *Chem. Ber.* **1988**, *121*, 1767.
(114) Niemer, B.; Weidmann, T.; Beck, W. *Z. Naturforsch. B: Chem. Sci.* **1992**, *47B*, 509.
(115) Beck, W.; Niemer, B.; Wagner, B. *Angew. Chem., Int. Ed. Engl.* **1989**, *28*, 1705.
(116) Carty, A. J.; Cherkas, A. A.; Randall, L. H. *Polyhedron* **1988**, *7*, 1045.
(117) Nast, R. *Coord. Chem. Rev.* **1982**, *47*, 89.
(118) Seyferth, D.; Hoke, J. B.; Wheeler, D. R. *J. Organomet. Chem.* **1988**, *341*, 421.
(119) Sappa, E.; Tiripicchio, A.; Braunstein, P. *Chem. Rev.* **1983**, *83*, 203.
(120) Ogawa, H.; Joh, T.; Takahashi, S.; Sonogashira, K. *J. Chem. Soc., Chem. Commun.* **1985**, 1220.
(121) Heidrich, J.; Steimann, M.; Appel, M.; Beck, W.; Phillips, J. R.; Trogler, W. C. *Organometallics* **1990**, *9*, 1296.
(122) Weidmann, T.; Weinrich, V.; Wagner, B.; Robl, C.; Beck, W. *Chem. Ber.* **1991**, *124*, 1363.
(123) Beck, W. personal communication, 1993.
(124) Appel, M.; Heidrich, J.; Beck, W. *Chem. Ber.* **1987**, *120*, 1087.
(125) Cross, R. J.; Davidson, M. F. *J. Chem. Soc., Dalton Trans.* **1986**, 411.
(126) Bruce, M. I.; Swincer, A. G. *Adv. Organomet. Chem.* **1983**, *22*, 59.
(127) Fritz, P. M.; Polborn, K.; Steimann, M.; Beck, W. *Chem. Ber.* **1989**, *122*, 889.
(128) Schubert, U.; Ackermann, K.; Rustemeyer, P. *J. Organomet. Chem.* **1982**, *231*, 323.
(129) Kolobova, N. E.; Antonova, A. B.; Khitrova, O. M.; Antipin, M. Y.; Struchkov, Y. T. *J. Organomet. Chem.* **1977**, *137*, 69.
(130) Kolobova, N. E.; Skripkin, V. V.; Rozantseva, T. V.; Struchkov, Y. T.; Aleksandrov, G. G.; Andrianov, V. G. *J. Organomet. Chem.* **1981**, *218*, 351.
(131) Akita, M.; Terada, M.; Oyama, S.; Moro-oka, Y. *Organometallics* **1990**, *9*, 816.
(132) Akita, M.; Terada, M.; Oyama, S.; Sugimoto, S.; Moro-oka, Y. *Organometallics* **1991**, *10*, 1561.
(133) Akita, M.; Ishii, N.; Takabuchi, A.; Tanaka, M.; Moro-oka, Y. *Organometallics* **1994**, *13*, 258.
(134) Frank, K. G.; Selegue, J. P. *J. Am. Chem. Soc.* **1990**, *112*, 6414.
(135) Carriedo, G. A.; Miguel, D.; Riera, V.; Solans, X.; Font-Altaba, M.; Coll, M. *J. Organomet. Chem.* **1986**, *299*, C43.
(136) Carriedo, G. A.; Miguel, D.; Riera, V. *J. Chem. Soc., Dalton Trans.* **1987**, 2867.
(137) Solans, J.; Solans, X.; Miravitlles, C.; Miguel, D.; Riera, V.; Rubio-Gonzalez, J. M. *Acta Crystallogr. Sect. C: Cryst. Struct. Commun.* **1986**, *42*, 975.
(138) Bruce, M. I.; Abu Salah, O. M.; Davis, R. E.; Raghavan, N. V. *J. Organomet. Chem.* **1974**, *64*, C48.

(*139*) Carriedo, G. A.; Howard, J. A. K.; Stone, F. G. A. *J. Chem. Soc., Dalton Trans.* **1984**, 1555.
(*140*) Abu Salah, O. M.; Bruce, M. I. *J. Chem. Soc., Dalton Trans.* **1974**, 2302.
(*141*) Bruce, M. I.; Clark, R.; Howard, J.; Woodward, P. *J. Organomet. Chem.* **1972**, *42*, C107.
(*142*) Chaudret, B.; Delavaux, B.; Poilblanc, R. *Coord. Chem. Rev.* **1988**, *86*, 191.
(*143*) Erker, G.; Frömberg, W.; Benn, R.; Mynott, R.; Angermund, K.; Krüger, C. *Organometallics* **1989**, *8*, 911.
(*144*) McDonald, W. S.; Pringle, P. G.; Shaw, B. L. *J. Chem. Soc., Chem. Commun.* **1982**, 861.
(*145*) Langrick, C. R.; McEwan, D. M.; Pringle, P. G.; Shaw, B. L. *J. Chem. Soc., Dalton Trans.* **1983**, 2487.
(*146*) Hutton, A. T.; Pringle, P. G.; Shaw, B. L. *Organometallics* **1983**, *2*, 1889.
(*147*) Cooper, G. R.; Hutton, A. T.; Langrick, C. R.; McEwan, D. M.; Pringle, P. G.; Shaw, B. L. *J. Chem. Soc., Dalton Trans.* **1984**, 855.
(*148*) Blagg, A.; Hutton, A. T.; Pringle, P. G.; Shaw, B. L. *J. Chem. Soc., Dalton Trans.* **1984**, 1815.
(*149*) Blagg, A.; Hutton, A. T.; Pringle, P. G.; Shaw, B. L. *Inorg. Chim. Acta* **1983**, *76*, L265.
(*150*) Cooper, G. R.; Hutton, A. T.; McEwan, D. M.; Pringle, P. G.; Shaw, B. L. *Inorg. Chim. Acta* **1983**, *76*, L267.
(*151*) Hutton, A. T.; Langrick, C. R.; McEwan, D. M.; Pringle, P. G.; Shaw, B. L. *J. Chem. Soc., Dalton Trans.* **1985**, 2121.
(*152*) Blagg, A.; Robson, R.; Shaw, B. L.; Thornton-Pett, M. *J. Chem. Soc., Dalton Trans.* **1987**, 2171.
(*153*) Deranlyagala, S. P.; Grundy, K. R. *Organometallics* **1985**, *4*, 424.
(*154*) Cowie, M.; Loeb, S. J. *Organometallics* **1985**, *4*, 852.
(*155*) Carr, S. W.; Pringle, P. G.; Shaw, B. L. *J. Organomet. Chem.* **1988**, *341*, 543.
(*156*) McEwan, D. M.; Markham, D. P.; Pringle, P. G.; Shaw, B. L. *J. Chem. Soc., Dalton Trans.* **1986**, 1809.
(*157*) Yam, V. W.-W.; Chan, L.-P.; Lai, T.-F. *J. Chem. Soc., Dalton Trans.* **1993**, 2075.
(*158*) Abu Salah, O. M.; Bruce, M. I.; Churchill, M. R.; DeBoer, B. G.; *J. Chem. Soc., Chem. Commun.* **1974**, 688.
(*159*) Churchill, M. R.; DeBoer, B. G. *Inorg. Chem.* **1975**, *14*, 2630.
(*160*) Lang, H.; Herres, M.; Zsolnai, L.; Imhof, W. *J. Organomet. Chem.* **1991**, *409*, C7.
(*161*) Lang, H.; Zsolnai, L. *J. Organomet. Chem.* **1991**, *406*, C5.
(*162*) Lang, H.; Imhof, W. *Chem. Ber.* **1992**, *125*, 1307.
(*163*) Lang, H.; Herres, M.; Zsolnai, L. *Organometallics* **1993**, *12*, 5008.
(*164*) Ciriano, M.; Howard, J. A. K.; Spencer, J. L.; Stone, F. G. A.; Wadepohl, H. *J. Chem. Soc., Dalton Trans.* **1979**, 1749.
(*165*) Lang, H.; Herres, M.; Zsolnai, L. *Bull. Chem. Soc. Jpn.* **1993**, *66*, 1.
(*166*) Lang, H.; Herres, M.; Imhof, W. *J. Organomet. Chem.* **1994**, *465*, 283.
(*167*) Berenguer, J. R.; Falvello, L. R.; Forniés, J.; Lalinde, E.; Tomas, M. *Organometallics* **1993**, *12*, 6.
(*168*) Yamazaki, S.; Deeming, A. J. *J. Chem. Soc., Dalton Trans.* **1993**, 3051.
(*169*) Espinet, P.; Forniés, J.; Martinez, F.; Sotes, M.; Lalinde, E.; Moreno, M. T.; Ruiz, A.; Welch, A. J. *J. Organomet. Chem.* **1991**, *403*, 253.
(*170*) Forniés, J.; Lalinde, E.; Martinez, F.; Moreno, M. T.; Welch, A. J. *J. Organomet. Chem.* **1993**, *455*, 271.
(*171*) Berenguer, J. R.; Forniés, J.; Lalinde, E.; Martinez, F.; Urriolabeitia, E.; Welch, A. J. *J. Chem. Soc., Dalton Trans.* **1994**, 1291.

(172) Berenguer, J. R.; Forniés, J.; Lalinde, E.; Martinez, F. *J. Organomet. Chem.* **1994,** *470,* C15.
(173) Berenguer, J. R.; Forniés, J.; Martinez, F.; Cubero, J. C.; Lalinde, E.; Moreno, M. T.; Welch, A. J. *Polyhedron* **1993,** *12,* 1797.
(174) Müller, J.; Tschampel, M.; Pickardt, J. *J. Organomet. Chem.* **1988,** *355,* 513.
(175) Metler, N.; Nöth, H. *J. Organomet. Chem.* **1993,** *454,*C5.
(176) Wood, G. L.; Knobler, C. B.; Hawthorne, M. F. *Inorg. Chem.* **1989,** *28,* 382.
(177) Erker, G.; Frömberg, W.; Mynott, R.; Gabor, B.; Krüger, C. *Angew. Chem., Int. Ed. Engl.* **1986,** *25,* 463.
(178) Kumar, P. N. V. P.; Jemmis, E. D. *J. Am. Chem. Soc.* **1988,** *110,* 125.
(179) Teuben, J. H.; de Liefde Meijer, H. J.; *J. Organomet. Chem.* **1969,** *17,* 87.
(180) Sekutowski, D. G.; Stucky, G. D. *J. Am. Chem. Soc.* **1976,** *98,* 1376.
(181) Cuenca, T.; Gomez, R.; Gomez-Sal, P.; Rodriguez, G. M.; Royo, P. *Organometallics* **1992,** *11,* 1229.
(182) Rosenthal, U.; Görls, H. *J. Organomet. Chem.* **1992,** *439,* C36.
(183) Evans, W. J.; Keyer, R. A.; Ziller, J. W. *Organometallics* **1990,** *9,* 2628
(184) Forniés, J.; Lalinde, E.; Martin, A.; Moreno, M. T. *J. Chem. Soc., Dalton Trans.* **1994,** 135.
(185) Berry, M.; Howard, J. A. K.; Stone, F. G. A.; *J. Chem. Soc., Dalton Trans.* **1980,** 1601.
(186) Ozawa, F.; Park, J. W.; Mackenzie, P. B.; Schaefer, W. P.; Henling, L. M.; Grubbs, R. H. *J. Am. Chem. Soc.* **1989,** *111,* 1319.
(187) Mackenzie, P. B.; Coots, R. J.; Grubbs, R. H. *Organometallics* **1989,** *8,* 8.
(188) Casey, C. P. *CHEMTECH* **1979,** 378.
(189) Casey, C. P.; Brunsvold, W. R. *J. Organomet. Chem.* **1974,** *77,* 345.
(190) Xu, Y. C.; Wulff, W. D. *J. Org. Chem.* **1987,** *52,* 3263.
(191) Macomber, D. W.; Hung, M. H.; Verma, A. G.; Rogers, R. D. *Organometallics* **1988,** *7,* 2072.
(192) Macomber, D. W.; Hung, M. H.; Madhukar, P.; Liang, M.; Rogers, R. D. *Organometallics* **1991,** *10,* 737.
(193) Macomber, D. W.; Madhukar, P.; Rogers, R. D. *Organometallics* **1991,** *10,* 2121.
(194) Casey, C. P.; Shusterman, A. J. *Organometallics* **1985,** *4,* 736.
(195) Beck, W.; Breimair, J.; Fritz, P.; Knauer, W.; Weidmann, T.; In "Transition Metal Carbyne Complexes"; Kreissl, F. R., Ed.; Kluwer: Dordrecht, The Netherlands, 1993; p. 189.
(196) Breimair, J.; Weidmann, T.; Wagner, B.; Beck, W. *Chem. Ber.* **1991,** *124,* 2431.
(197) Kelley, C.; Terry, M. R.; Kaplan, A. W.; Geoffroy, G. L.; Lugan, N.; Mathieu, R.; Haggerty, B. S.; Rheingold, A. L. *Inorg. Chim. Acta* **1992,** *198–200,* 601.
(198) Aumann, R.; Runge, M. *Chem. Ber.* **1992,** *125,* 259.
(199) Schubert, U. "Transition Metal Carbene Complexes"; Verlag Chemie: Weinheim, 1983; p. 74.
(200) Weidmann, T.; Sünkel, K.; Beck, W. *J. Organomet. Chem.* **1993,** *459,* 219.
(201) Wadepohl, H.; Pritzkow, H. *Angew. Chem., Int. Ed. Engl.* **1987,** *26,* 127.
(202) Wadepohl, H.; Galm, W.; Pritzkow, H.; Wolf, A. *Angew. Chem., Int. Ed. Engl.* **1992,** *31,* 1058.
(203) Weng, W.; Ramsden, J. A.; Arif, A. M.; Gladysz, J. A. *J. Am. Chem. Soc.* **1993,** *115,* 3824.
(204) Roger, C.; Peng, T.-S.; Gladysz, J. A. *J. Organomet. Chem.* **1992,** *439,* 163.
(205) Weng, W.; Arif, A. M.; Gladysz, J. A. *Angew. Chem., Int. Ed. Engl.* **1993,** *32,* 891.
(206) Erker, G.; Krüger, C.; Müller, G. *Adv. Org. Met. Chem.* **1985,** *24,* 1.

(207) Erker, G. In "Organometallics in Organic Synthesis"; de Meijere A.; tom Dieck, H. Eds., Springer-Verlag: Heidelberg; 1987.

(208) Erker, G. *Polyhedron* **1988**, *7*, 2451.

(209) Erker, G.; Sosna, F.; Hoffmann, U. *J. Organomet. Chem.* **1989**, *372*, 41.

(210) Erker, G.; Lecht, R.; Schlund, R.; Angermund, K.; Krüger, C. *Angew. Chem., Int. Ed. Engl.* **1987**, *26*, 666.

(211) Erker, G.; Pfaff, R.; Krüger, C.; Werner, S. *Organometallics* **1991**, *10*, 3559.

(212) Erker, G.; Dorf, U.; Benn. R.; Reinhardt, R.-D.; Petersen, J. L. *J. Am. Chem. Soc.* **1984**, *106*, 7649.

(213) Erker, G.; Lecht, R.; Sosna, F.; Uhl, S.; Tsay, Y.-H.; Krüger, C.; Grondey, H.; Benn, R. *Chem. Ber.* **1988**, *121*, 1069.

(214) Erker, G.; Sosna, F.; Pfaff, R.; Noe, R.; Sarter, C.; Kraft, A.; Krüger, C.; Zwettler, R. *J. Organomet. Chem.* **1990**, *394*, 99.

(215) Erker, G.; Sosna, F.; Zwettler, R.; Krüger, C. *Organometallics* **1989**, *8*, 450.

(216) Erker, G.; Lecht, R. *J. Organomet. Chem.* **1986**, 311, 45.

(217) Erker, G.; Lecht, R.; Petersen, J. L.; Bönnemann, H. *Organometallics* **1987**, *6*, 1962.

(218) Erker, G.; Lecht, R.; Tsay, Y.-H.; Krüger, C. *Chem. Ber.* **1987**, *120*, 1763.

(219) Erker, G.; Menjón, B. *Chem. Ber.* **1990**, *123*, 1327.

(220) Berlekamp, M.; Erker, G.; Petersen, J. L. *J. Organomet. Chem.* **1993**, *458*, 97.

(221) Erker, G.; Mühlenbernd, T.; Benn, R.; Rufińska, A. *Organometallics* **1986**, *5*, 402.

(222) Noe, R.; Wingbermühle, D.; Erker, G.; Krüger, C.; Bruckmann, J. *Organometallics* **1993**, *12*, 4993.

(223) Erker, G.; Sosna, F.; Noe, R. *Chem. Ber.* **1990**, *123*, 821.

(224) Erker, G.; Sosna, F.; Petersen, J. L.; Benn, R.; Grondey, H. *Organometallics* **1990**, *9*, 2462.

(225) Dötz, K. H.; Schäfer, T.; Kroll, F.; Harms, K. *Angew. Chem., Int. Ed. Engl.* **1992**, *31*, 1236.

(226) Spaniol, T. P.; Stone, F. G. A. *Polyhedron* **1989**, *8*, 2271.

(227) Stone, F. G. A. *Adv. Orgmet. Chem.* **1990**, *31*, 53.

(228) Mansuy, D.; Lecomte, J.-P.; Chottard, J.-C.; Bartoli, J.-F. *Inorg. Chem.* **1981**, *20*, 3119.

(229) Ercolani, C.; Gardini, M.; Goedken, V. L.; Pennesi, G.; Rossi, G.; Russo, U.; Zanonato, P. *Inorg. Chem.* **1989**, *28*, 3097.

(230) Latesky, S. L.; Selegue, J. P. *J. Am. Chem. Soc.* **1987**, *109*, 4731.

(231) Beck, W.; Knauer, W.; Robl, C. *Angew. Chem., Int. Ed. Engl.* **1990**, *29*, 318.

(232) Desmond, T.; Parvez, M.; Lalor, F. J.; Ferguson, G. *J. Chem. Soc., Chem. Commun.* **1983**, 457.

(233) Etienne, M.; White, P. S.; Templeton, J. L. *J. Am. Chem. Soc.* **1991**, *113*, 2324.

(234) Desmond, T.; Parvez, M.; Lalor, F. J.; Ferguson, G. *J. Chem. Soc., Chem. Commun.* **1984**, 75.

(235) Coville, N. J.; du Plooy, K. E.; Pickl, W. *Coord. Chem. Rev.* **1992**, *116*, 1 and references therein.

(236) Batsanov, A. S.; Struchkov, Yu. T. *J. Organomet. Chem.* **1984**, *266*, 295.

(237) Orlova, T. Yu.; Setkina, V. N.; Andrianov, V. G.; Struchkov, Yu. T. *Izv. Akad. Nauk SSSR Ser. Khim.* **1986**, *437*, 87.

(238) Orlova, T. Yu.; Setkina, V. N.; Petrovskii, P. V.; Yanovskii, A. I.; Batsanov, A. S.; Struchkov, Yu. T. *Organomet. Chem. USSR* **1988**, *1*, 725.

(239) Orlova, T. Yu.; Setkina, V. N.; Petrovskii, P. V.; Zagorevskii, D. V.; *Organomet. Chem. USSR* **1992**, *5*, 535.

(240) Tsyryapkin, V. A.; Antonova, T. V.; Timofeeva, T. V.; Shil'nikov V. I.; Orlova,

T. Yu.; Solodova, M. Ya.; Rozantseva, T. V.; Nedospasova, L. V.; Shirokov, V. A.; Kalinin, V. N.; Struchkov, Yu. T. *Organomet. Chem. USSR* **1992**, *5*, 433.

(241) Sünkel, K.; Birk, U. *J. Organomet. Chem.* **1993**, *458*, 181.

(242) Carneiro, T. M. G.; Matt, D.; Braunstein, P. *Coord. Chem. Rev.* **1989**, *96*, 49.

(243) Pregosin, P. S.; Togni, A.; Venanzi, L. M. *Angew. Chem., Int. Ed. Engl.* **1981**, *20*, 668.

(244) Albinati, A.; Togni, A.; Venanzi, L. M. *Organometallics* **1986**, *5*, 1785.

(245) Howarth, O. W.; McAteer, C. H.; Moore, P.; Morris, G. E.; Alcock, N. W. *J. Chem. Soc., Dalton Trans.* **1982**, 541.

(246) Fryzuk, M. D.; Haddad, T. S.; Berg, D. J. *Coord. Chem. Rev.* **1990**, *99*, 137, and references therein.

(247) Blandy, C.; Locke, S. A.; Young, S. J.; Schore, N. E. *J. Am. Chem. Soc.* **1988**, *110*, 7540.

(248) Kool, L. B.; Rausch, M. D.; Alt, H. G.; Herberhold, M.; Thewalt, U.; Honold, B. *J. Organomet. Chem.* **1986**, *310*, 27.

(249) Gell, K. I.; Harris, T. V.; Schwartz, J. *Inorg. Chem.* **1981**, *20*, 481.

(250) Gambarotta, S.; Chiang, M. Y. *Organometallics* **1987**, *6*, 897.

(251) Wielstra, Y.; Gambarotta, S.; Meetsma, A.; de Boer, J. L. *Organometallics* **1989**, *8*, 250.

(252) Wielstra, Y.; Meetsma, A.; Gambarotta, S.; Khan, S. *Organometallics* **1990**, *9*, 876.

(253) Wielstra, Y.; Gambarotta, S.; Spek, A. L.; Smeets, W. J. J. *Organometallics* **1990**, *9*, 2142.

(254) Rosenthal, U.; Ohff, A.; Michalik, M.; Görls, H.; Burlakov, V. V.; Shur, V. B. *Angew. Chem., Int. Ed. Engl.* **1993**, *32*, 1193.

(255) Rosenthal, U.; Ohff, A.; Michalik, M.; Görls, H.; Burlakov, V. V.; Shur, V. B. *Organometallics* **1993**, *12*, 5016.

(256) Berry, M.; Cooper, N. J.; Green, M. L. H.; Simpson, S. J. *J. Chem. Soc., Dalton Trans.* **1980**, 29.

(257) Barral, M.; Green, M. L. H.; Jimenez, R. *J. Chem. Soc., Dalton Trans.* **1982**, 2495.

(258) Bashkin, J.; Green, M. L. H.; Poveda, M. L.; Prout, K. *J. Chem. Soc., Dalton Trans.* **1982**, 2485.

(259) Hoxmeier, R. J.; Blickensderfer, J. R.; Kaez, H. D. *Inorg. Chem.* **1979**, *18*, 3453.

(260) Vollhardt, K. P. C.; Weidman, T. W. *J. Am. Chem. Soc.* **1983**, *105*, 1676.

(261) Casey, C. P.; Palermo, R. E.; Jordan, R. F.; Rheingold, A. L. *J. Am. Chem. Soc.* **1985**, *107*, 4597.

(262) Daroda, R. J.; Wilkinson, G.; Hursthouse, M. B.; Abdul Malik, K. M.; Thornon-Pett, M. *J. Chem. Soc., Dalton Trans.* **1980**, 2315.

(263) Cullen, W. R.; Woollins, J. D. *Coord. Chem. Rev.* **1981**, *39*, 1.

(264) Bürger, H.; Kluess, C. *J. Organomet. Chem.* **1973**, *56*, 269.

(265) Wedler, M.; Roesky, H. W.; Edelmann, F. T.; Behrens, U. *Z. Naturforsch. B: Chem. Sci.* **1988**, *43B*, 1461.

(266) Herberhold, M.; Kniesel, H.; Haumaier, L.; Gieren, A.; Ruiz-Pérez, C. *Z. Naturforsch. B: Chem. Sci.* **1986**, *41B*, 1431.

(267) Herberhold, M.; Kniesel, H. *J. Organomet. Chem.* **1987**, *334*, 347.

(268) Pannell, K. H.; Cassias, J. B.; Crawford, G. M.; Flores, A. *Inorg. Chem.* **1976**, *15*, 2671.

(269) Herberhold, M.; Feger, W.; Kölle, U. *J. Organomet. Chem.* **1992**, *436*, 333.

(270) Osborne, A. G.; Whiteley, R. H. *J. Organomet. Chem.* **1979**, *181*, 425.

(271) Lehmkuhl, H.; Schwickardi, R.; Krüger, C.; Raabe, G. *Z. Anorg. Allg. Chem.* **1990**, *581*, 41.

(272) Broussier, R.; Da Rold, A.; Gautheron, R.; Dromzee, Y.; Jeannin, Y. *Inorg. Chem.* **1990,** *29,* 1817.

(273) Seyferth, D.; Withers, H. P. *J. Organomet. Chem.* **1980,** *185,* C1.

(274) Seyferth, D.; Withers, H. P. *Organometallics* **1982,** *1,* 1275.

(275) Butler, I. R.; Cullen, W. R.; Elnstein, F. W. B.; Willis, A. C. *Organometallics* **1985,** *4,* 603.

(276) Butler, I. R.; Cullen, W. R. *Organometallics* **1984,** *3,* 1846.

(277) Sokolov, V. I.; Troitskaya, L. L.; Reutov, O. A. *J. Organomet. Chem.* **1979,** *182,* 537.

(278) Sokolov, V. I.; Nechaeva, K. S.; Reutov, O. A. *Russ. J. Org. Chem.* **1983,** *19,* 1003.

(279) Sokolov, V. I.; Nechaeva, K. S.; Reutov, O. A. *J. Organomet. Chem.* **1983,** *253,* C55.

(280) Kotz, J. C.; Getty, E. E.; Lin, L. *Organometallics* **1985,** *4,* 610.

(281) Nonoyama, M.; Hamamura, K. *J. Organomet. Chem.* **1991,** *407,* 271.

(282) Knox, G. R.; Pauson, P. L.; Willison, D. *J. Organomet. Chem.* **1993,** *450,* 177.

(283) Crawford, S. S.; Kaesz, H. D. *Inorg. Chem.* **1977,** *16,* 3193.

(284) Butler, I. R. *Organometallics* **1992,** *11,* 74.

(285) Butler, I. R.; Cullen, W. R.; Rettig, S. J. *Organometallics* **1987,** *6,* 872.

(286) Razuvaev, G. A.; Domrachev, G. A.; Sharutin, V. V.; Suvorova, O. N. *J. Organomet. Chem.* **1977,** *141,* 313.

(287) Krüger, C.; Thiele, K.-H.; Dargatz, M.; Bartik, T. *J. Organomet. Chem.* **1989,** *362,* 147.

(288) Köhler, F. H.; Geike, W. A.; Hofmann, P.; Schubert, U. *Chem. Ber.* **1984,** *117,* 904.

(289) Izumi, T.; Maemura, M.; Endoh, K.; Oikawa, T.; Zakozi, S.; Kasahara, A. *Bull. Chem. Soc. Jpn.* **1981,** *54,* 836.

(290) Fischer, E. O.; Postnov, V. N.; Kreissl, F. R. *J. Organomet. Chem.* **1977,** *127,* C19.

(291) Fischer, E. O.; Postnov, V. N.; Kreissl, F. R. *J. Organomet. Chem.* **1982,** *231,* C73.

(292) Connor, J. A.; Lloyd, J. P. *J. Chem. Soc., Dalton Trans.* **1972,** 1470.

(293) Fischer, E. O.; Gammel, F. J.; Besenhard, J. O.; Frank, A.; Neugebauer, D. *J. Organomet. Chem.* **1980,** *191,* 261.

(294) Fischer, E. O.; Wanner, J. K. R.; *Chem. Ber.* **1985,** *118,* 2489.

(295) Fischer, E. O.; Wanner, J. K. R.; Müller, G.; Riede, J. *Chem. Ber.* **1985,** *118,* 3311.

(296) Hermann, R.; Ugi, I. *Angew. Chem., Int. Ed. Engl.* **1982,** *21,* 788.

(297) Pilette, D.; Ouzzine, K.; Le Bozec, H.; Dixneuf, P. H.; Rickard, C. E. F.; Roper, W. R. *Organometallics* **1992,** *11,* 809.

(298) Macomber, D. W.; Madhukar, P.; Rogers, R. D. *Organometallics* **1989,** *8,* 1275.

(299) Fischer, E. O.; Schluge, M.; Besenhard, J. O.; Friedrich, P.; Huttner, G.; Kreissl, F. R. *Chem. Ber.* **1978,** *111,* 3530.

(300) Fischer, E. O.; Schluge, M.; Besenhard, J. O. *Angew. Chem., Int. Ed. Engl.* **1976,** *15,* 683.

(301) Schubert, U. "Carbyne Complexes", VCH Verlagsgesellschaft, Weinheim, 1988; p. 40.

(302) Anderson, S.; Hill, A. F. *J. Chem. Soc., Dalton Trans.* **1993,** 587.

(303) Fischer, E. O.; Gammel, F. J.; Neugebauer, D. *Chem. Ber.* **1980,** *113,* 1010.

(304) Heppert, J. A.; Thomas-Miller, M. E.; Swepston, P. N.; Extine, M. W. *J. Chem. Soc., Chem. Commun.* **1988,** 280.

(305) Lotz, S.; Schindehutte, M.; van Rooyen, P. H. *Organometallcis* **1992,** *11,* 629.

(306) Heppert, J. A.; Morgenstern, M. A.; Scherubel, D. M.; Takusagawa, F.; Shaker, M. R.; *Organometallics* **1988,** *7,* 1715.

(307) Dessey, R. E.; Pohl, R. C.; King, R. B. *J. Am. Chem. Soc.* **1966,** *88,* 5121.

(308) Mahaffy, C. A. L.; Pauson, P. L. *Inorg. Synth.* **1979,** *19,* 154.

(309) Rausch, M. D.; Moser, G. A.; Zaiko, E. J.; Lipman, A. L. *J. Organomet. Chem.* **1970,** *23,* 185.

(310) Smirnov, A. S.; Kasatkina, T. G.; Artemov, A. N. *Organomet. Chem. USSR* **1988**, *1*, 642.

(311) Heppert, J. A.; Thomas-Miller, M. E.; Scherubel, D. M.; Takusagawa, F.; Morgenstern, M. A.; Shaker, M. R. *Organometallics* **1989**, *8*, 1199.

(312) Richter-Addo, G. B.; Hunter, A. D. *Inorg. Chem.* **1989**, *28*, 4063.

(313) Richter-Addo, G. B.; Hunter, A. D.; Wichrowska, N. *Can. J. Chem.* **1990**, *68*, 41.

(314) Hunter, A. D.; Ristic-Petrovic, D.; McLernon, J. L. *Organometallics* **1992**, *11*, 864.

(315) van Rooyen, P. H.; Schindehutte, M.; Lotz, S. *Inorg. Chim. Acta.* **1993**, *208*, 207.

(316) van Rooyen, P. H.; Schindehutte, M.; Lotz, S. *Organometallics* **1992**, *11*, 1104.

(317) Meyer, R.; Lotz, S. unpublished results, 1993.

(318) Hunter, A. D.; Shilliday, L.; Furey, W. S.; Zaworotko, M. J. *Organometallics* **1992**, *11*, 1550.

(319) Lotz, S.; van Rooyen, P. H.; Waldbach, T. A.; Meyer, R.; Schindehutte, M. In "Contributions to the Development of Coordination Chemistry"; Proceedings of the 14th Conference on Coordination Chemistry; Odrejovic, G., Ed.; Slovak Technical University Press: Smolenice, 1993; p. 167.

(320) Dufaud, V.; Thivolle-Cazat, J.; Basset, J.-M.; Mathieu, R.; Jaud, J.; Waissermann, J. *Organometallics* **1991**, *10*, 4005.

(321) Butler, I. R.; Cullen, W. R.; Einstein, F. W. B.; Jones, R. H. *J. Organomet. Chem.* **1993**, *463*, C6.

(322) Butler, I. R.; Gill, U.; Lepage, Y.; Lindsell, W. E.; Preston, P. N. *Organometallics* **1990**, *9*, 1964.

(323) Hunter, A. D.; McLernon, J. L. *Organometallics* **1989**, *8*, 2679.

(324) Hunter, A. D. *Organometallics* **1989**, *8*, 1118.

(325) Galamb, V.; Palyi, G.; Ungvary, F.; Marko, L.; Boese, R.; Schmid, G. *J. Am. Chem. Soc.* **1986**, *108*, 3344.

(326) Galamb, V.; Palyi, G. *J. Chem. Soc., Chem. Commun.* **1982**, 487.

(327) Tasi, M.; Palyi, G. *Organometallics* **1985**, *4*, 1523.

(328) Galamb, V.; Palyi, G. *Coord. Chem. Rev.* **1984**, *59*, 203.

(329) Domingo, M. R.; Irving, A.; Liao, Y.-H.; Moss, J. R.; Nash, A. *J. Organomet. Chem.* **1993**, *443*, 233.

(330) Kaganovich, V. S.; Kudinov, A. R.; Rybinskaya, M. I. *Organomet. Chem. USSR* **1990**, *3*, 465.

(331) Wieser, M.; Karaghiosoff, K.; Beck. W. *Chem. Ber.* **1993**, *126*, 1081.

(332) Guerchais, V. *J. Chem. Soc., Chem. Commun.* **1990**, 534.

(333) Fischer, E. O.; Gammel, F. J.; Besenhard, J. O.; Frank, A.; Neugenbauer, D. *J. Organomet. Chem.* **1980**, *191*, 261.

(334) Fernandez, J. R.; Stone, F. G. A. *J. Chem. Soc., Dalton Trans.* **1988**, 3035.

(335) Zheng, P. Y.; Nadasdi, T. T.; Stephan, D. W. *Organometallics* **1989**, *8*, 1393.

(336) Dick, D. G.; Hou, Z.; Stephen, D. W. *Organometallics* **1992**, *11*, 2378.

(337) Dick, D. G.; Stephan, D. W. *Organometallics* **1900**, *9*, 1910.

(338) Longato, B.; Norton, J. R.; Huffman, J. C.; Marsella, J. A.; Caulton, K. G. *J. Am. Chem. Soc.* **1981**, *103*, 209.

(339) Marsella, J. A.; Huffman, J. C.; Caulton, K. G.; Longato, B.; Norton, J. R. *J. Am. Chem. Soc.* **1982**, *104*, 6360.

(340) Barger, P. T.; Bercaw, J. E. *J. Organomet. Chem.* **1980**, *201*, C39.

(341) Barr, R. D.; Green, M.; Howard, J. A. K.; Marder, T. B.; Moore, I.; Stone, F. G. A. *J. Chem. Soc., Chem. Commun.* **1983**, 746.

(342) Dawkins, G. M.; Green, M.; Mead, K. A.; Salaün, J.-Y.; Stone, F. G. A.; Woodward, P. *J. Chem. Soc., Dalton Trans.* **1983**, 527.

(*343*) Awang, M. R.; Barr, R. D.; Green, M.; Howard, J. A. K.; Marder, T. B.; Stone, F. G. A. *J. Chem. Soc., Dalton Trans.* **1985**, 2009.
(*344*) Herrmann, W. A.; Menjon, B.; Herdtweck, E. *Organometallics* **1991**, *10*, 2134.
(*345*) Herrmann, W. A.; Cuenca, T.; Menjon, B.; Herdtweck, E. *Angew. Chem., Int. Ed. Engl.* **1987**, *26*, 697.
(*346*) Lemke, F. R.; Szalda, D. J.; Bullock, R. M. *Organometallics* **1992**, *11*, 876.
(*347*) Kemke, F. R.; Szalda, D. J.; Bullock, R. M. *J. Am. Chem. Soc.* **1991**, *113*, 8466.
(*348*) Mashima, K.; Jyodoi, K.; Ohyoshi, A.; Takaya, H. *Organometallics* **1987**, *6*, 885.
(*349*) Nixon, J. F. *Chem. Rev.* **1988**, *88*, 1327.
(*350*) Mathey, F. *Chem. Rev.* **1988**, *88*, 429.
(*351*) Albert, A. "Heterocyclic Chemistry"; Athlone Press: London, 1968.
(*352*) Pearson, R. G. *J. Am. Chem. Soc.* **1963**, *85*, 3533.
(*353*) Pannell, K. H.; Kalsotra, B. L.; Parkanyi, C. *J. Heterocycl. Chem.* **1978**, *15*, 1057.
(*354*) Sadimenko, A. P.; Garnovskii, A. D.; Retta, N. *Coord. Chem. Rev.* **1993**, *126*, 237.
(*355*) Geilich, K.; Stumpf, K.; Pritzkow, H.; Siebert, W. *Chem. Ber.* **1987**, *120*, 911.
(*356*) Rauchfuss, T. B. *Prog. Inorg. Chem.* **1991**, *39*, 259.
(*357*) Angelici, R. J. *Coord. Chem. Rev.* **1990**, *105*, 61.
(*358*) Waldbach, T. A.; Lotz, S. unpublished results, 1993.
(*359*) Waldbach, T. A.; van Rooyen, P. H.; Lotz, S. *Organometallics* **1993**, *12*, 4250.
(*360*) Kolobova, N. E.; Goncharenko, L. V. *Izv. Akad. Nauk SSSR, Ser. Kim* **1979**, 900.
(*361*) Johnson, C. K. ORTEP, Report ORNL-3794, Oak Ridge National Laboratory, Oak Ridge, TN, 1965.
(*362*) Allen, F. H.; Kennard, O.; Taylor, R. *Acc. Chem. Res.* **1983**, *16*, 146.

Index

A

Cumulative List of Contributors for Volumes 1-36

Abel, E. W., **5**, 1; **8**, 117
Aguiló, A., **5**, 321
Akkerman, O. S., **32**, 147
Albano, V. G., **14**, 285
Alper, H., **19**, 183
Anderson, G. K., **20**, 39; **35**, 1
Angelici, R. J., **27**, 51
Aradi, A. A., **30**, 189
Armitage, D. A., **5**, 1
Armor, J. N., **19**, 1
Ash, C. E., **27**, 1
Ashe III, A. J., **30**, 77
Atwell, W. H., **4**, 1
Baines, K. M., **25**, 1
Barone, R., **26**, 165
Bassner, S. L., **28**, 1
Behrens, H., **18**, 1
Bennett, M. A., **4**, 353
Bickelhaupt, F., **32**, 147
Birmingham, J., **2**, 365
Blinka, T. A., **23**, 193
Bockman, T. M., **33**, 51
Bogdanović, B., **17**, 105
Bottomley, F., **28**, 339
Bradley, J. S., **22**, 1
Brew, S. A., **35**, 135
Brinckman, F. E., **20**, 313
Brook, A. G:, **7**, 95; **25**, 1
Bowser, J. R., **36**, 57
Brown, H. C., **11**, 1
Brown, T. L., **3**, 365
Bruce, M. I., **6**, 273, **10**, 273; **11**, 447;
    12, 379; **22**, 59
Brunner, H., **18**, 151
Buhro, W. E., **27**, 311
Byers, P. K., **34**, 1
Cais, M., **8**, 211
Calderon, N., **17**, 449
Callahan, K. P., **14**, 145
Canty, A. J., **34**, 1
Cartledge, F. K., **4**, 1
Chalk, A. J., **6**, 119
Chanon, M., **26**, 165
Chatt, J., **12**, 1
Chini, P., **14**, 285

Chisholm, M. H., **26**, 97; **27**, 311
Chiusoli, G. P., **17**, 195
Chojinowski, J., **30**, 243
Churchill, M. R., **5**, 93
Coates, G. E., **9**, 195
Collman, J. P., **7**, 53
Compton, N. A., **31**, 91
Connelly, N. G., **23**, 1; **24**, 87
Connolly, J. W., **19**, 123
Corey, J. Y., **13**, 139
Corriu, R. J. P., **20**, 265
Courtney, A., **16**, 241
Coutts, R. S. P., **9**, 135
Coville, N. J., **36**, 95
Coyle, T. D., **10**, 237
Crabtree, R. H., **28**, 299
Craig, P. J., **11**, 331
Csuk, R., **28**, 85
Cullen, W. R., **4**, 145
Cundy, C. S., **11**, 253
Curtis, M. D., **19**, 213
Darensbourg, D. J., **21**, 113; **22**, 129
Darensbourg, M. Y., **27**, 1
Davies, S. G., **30**, 1
Deacon, G. B., **25**, 237
de Boer, E., **2**, 115
Deeming, A. J., **26**, 1
Dessy, R. E., **4**, 267
Dickson, R. S., **12**, 323
Dixneuf, P. H., **29**, 163
Eisch, J. J., **16**, 67
Ellis, J. E., **31**, 1
Emerson, G. F., **1**, 1
Epstein, P. S., **19**, 213
Erker, G., **24**, 1
Ernst, C. R., **10**, 79
Errington, R. J., **31**, 91
Evans, J., **16**, 319
Evans, W. J., **24**, 131
Faller, J. W., **16**, 211
Farrugia, L. J., **31**, 301
Faulks, S. J., **25**, 237
Fehlner, T. P., **21**, 57; **30**, 189
Fessenden, J. S., **18**, 275
Fessenden, R. J., **18**, 275

Fischer, E. O., **14,** 1
Ford, P. C., **28,** 139
Forniés, J., **17,** 219
Forster, D., **17,** 255
Fraser, P. J., **12,** 323
Friedrich, H., **36,** 229
Friedrich, H. B., **33,** 235
Fritz, H. P., **1,** 239
Fürstner, A., **28,** 85
Furukawa, J., **12,** 83
Fuson, R. C., **1,** 221
Gallop, M. A., **25,** 121
Garrou, P. E., **23,** 95
Geiger, W. E., **23,** 1; **24,** 87
Geoffroy, G. L., **18,** 207; **24,** 249; **28,** 1
Gilman, H., **1,** 89; **4,** 1; **7,** 1
Gladfelter, W. L., **18,** 207; **24,** 41
Gladysz, J. A., **20,** 1
Glänzer, B. I., **28,** 85
Green, M. L. H., **2,** 325
Grev, R. S., **33,** 125
Griffith, W. P., **7,** 211
Grovenstein, Jr., E., **16,** 167
Gubin, S. P., **10,** 347
Guerin, C., **20,** 265
Gysling, H., **9,** 361
Haiduc, I., **15,** 113
Halasa, A. F., **18,** 55
Hamilton, D. G., **28,** 299
Handwerker, H., **36,** 229
Harrod, J. F., **6,** 119
Hart, W. P., **21,** 1
Hartley, F. H., **15,** 189
Hawthorne, M. F., **14,** 145
Heck, R. F., **4,** 243
Heimbach, P., **8,** 29
Helmer, B. J., **23,** 193
Henry, P. M., **13,** 363
Heppert, J. A., **26,** 97
Herberich, G. E., **25,** 199
Herrmann, W. A., **20,** 159
Hieber, W., **8,** 1
Hill, A. F., **36,** 131
Hill, E. A., **16,** 131
Hoff, C., **19,** 123
Hoffmeister, H., **32,** 227
Holzmeier, P., **34,** 67
Honeyman, R. T., **34,** 1
Horwitz, C. P., **23,** 219
Hosmane, N. S., **30,** 99

Housecroft, C. E., **21,** 57; **33,** 1
Huang, Y. Z., **20,** 115
Hughes, R. P., **31,** 183
Ibers, J. A., **14,** 33
Ishikawa, M., **19,** 51
Ittel, S. D., **14,** 33
Jain, L., **27,** 113
Jain, V. K., **27,** 113
James, B. R., **17,** 319
Janiak, C., **33,** 291
Jastrzebski, J. T. B. H., **35,** 241
Jenck, J., **32,** 121
Jolly, P. W., **8,** 29; **19,** 257
Jonas, K., **19,** 97
Jones, M. D., **27,** 279
Jones, P. R., **15,** 273
Jordan, R. F., **32,** 325
Jukes, A. E., **12,** 215
Jutzi, P., **26,** 217
Kaesz, H. D., **3,** 1
Kalck, P., **32,** 121; **34,** 219
Kaminsky, W., **18,** 99
Katz, T. J., **16,** 283
Kawabata, N., **12,** 83
Kemmitt, R. D. W., **27,** 279
Kettle, S. F. A., **10,** 199
Kilner, M., **10,** 115
Kim, H. P., **27,** 51
King, R. B., **2,** 157
Kingston, B. M., **11,** 253
Kisch, H., **34,** 67
Kitching, W., **4,** 267
Kochi, J. K., **33,** 51
Köster, R., **2,** 257
Kreiter, C. G., **26,** 297
Krüger, G., **24,** 1
Kudaroski, R. A., **22,** 129
Kühlein, K., **7,** 241
Kuivila, H. G., **1,** 47
Kumada, M., **6,** 19; **19,** 51
Lappert, M. F., **5,** 225; **9,** 397; **11,** 253;
    14, 345
Lawrence, J. P., **17,** 449
Le Bozec, H., **29,** 163
Lednor, P. W., **14,** 345
Linford, L., **32,** 1
Longoni, G., **14,** 285
Luijten, J. G. A., **3,** 397
Lukehart, C. M., **25,** 45
Lupin, M. S., **8,** 211

McGlinchey, M. J., **34**, 285
McKillop, A., **11**, 147
McNally, J. P., **30**, 1
Macomber, D. W., **21**, 1; **25**, 317
Maddox, M. L., **3**, 1
Maguire, J. A., **30**, 99
Maitlis, P. M., **4**, 95
Mann, B. E., **12**, 135; **28**, 397
Manuel, T. A., **3**, 181
Markies, P. R., **32**, 147
Mason, R., **5**, 93
Masters, C., **17**, 61
Matsumura, Y., **14**, 187
Mayr, A., **32**, 227
Meister, G., **35**, 41
Mingos, D. M. P., **15**, 1
Mochel, V. D., **18**, 55
Moedritzer, K., **6**, 171
Molloy, K. C., **33**, 171
Monteil, F., **34**, 219
Morgan, G. L., **9**, 195
Morrison, J. A., **35**, 211
Moss, J. R., **33**, 235
Mrowca, J. J., **7**, 157
Müller, G., **24**, 1
Mynott, R., **19**, 257
Nagy, P. L. I., **2**, 325
Nakamura, A., **14**, 245
Nesmeyanov, A. N., **10**, 1
Neumann, W. P., **7**, 241
Norman, N. C., **31**, 91
Ofstead, E. A., **17**, 449
Ohst, H., **25**, 199
Okawara, R., **5**, 137; **14**, 187
Oliver, J. P., **8**, 167; **15**, 235; **16**, 111
Onak, T., **3**, 263
Oosthuizen, H. E., **22**, 209
Otsuka, S., **14**, 245
Pain, G. N., **25**, 237
Parshall, G. W., **7**, 157
Paul, I., **10**, 199
Peres, Y., **32**, 121
Petrosyan, W. S., **14**, 63
Pettit, R., **1**, 1
Pez, G. P., **19**, 1
Poland, J. S., **9**, 397
Poliakoff, M., **25**, 277
Popa, V., **15**, 113
Pourreau, D. B., **24**, 249
Powell, P., **26**, 125

Pratt, J. M., **11**, 331
Prokai, B., **5**, 225
Pruett, R. L., **17**, 1
Rao, G. S., **27**, 113
Raubenheimer, H. G., **32**, 1
Rausch, M. D., **21**, 1; **25**, 317
Reetz, M. T., **16**, 33
Reutov, O. A., **14**, 63
Rijkens, F., **3**, 397
Ritter, J. J., **10**, 237
Rochow, E. G., **9**, 1
Rokicki, A., **28**, 139
Roper, W. R., **7**, 53; **25**, 121
Roundhill, D. M., **13**, 273
Rubezhov, A. Z., **10**, 347
Salerno, G., **17**, 195
Salter, I. D., **29**, 249
Satgé, J., **21**, 241
Schade, C., **27**, 169
Schaverien, C. J., **36**, 283
Schmidbaur, H., **9**, 259; **14**, 205
Schrauzer, G. N., **2**, 1
Schubert, U., **30**, 151
Schulz, D. N., **18**, 55
Schumann, H., **33**, 291
Schwebke, G. L., **1**, 89
Seppelt, K., **34**, 207
Setzer, W. N., **24**, 353
Seyferth, D., **14**, 97
Shapakin, S. Yu., **34**, 149
Shen, Y. C., **20**, 115
Shriver, D. F., **23**, 219
Siebert, W., **18**, 301; **35**, 187
Sikora, D. J., **25**, 317
Silverthorn, W. E., **13**, 47
Singleton, E., **22**, 209
Sinn, H., **18**, 99
Skinner, H. A., **2**, 49
Slocum, D. W., **10**, 79
Smallridge, A. J., **30**, 1
Smeets, W. J. J., **32**, 147
Smith, J. D., **13**, 453
Speier, J. L., **17**, 407
Spek, A. L., **32**, 147
Stafford, S. L., **3**, 1
Stańczyk, W., **30**, 243
Stone, F. G. A., **1**, 143; **31**, 53; **35**, 135
Su, A. C. L., **17**, 269
Suslick, K. M., **25**, 73
Süss-Fink, G., **35**, 41

ISBN 0-12-031137-2